Micro and Nano Mechanical Testing of Materials and Devices

Fuqian Yang · James C.M. Li

Editors

Micro and Nano Mechanical Testing of Materials and Devices

 Springer

Editors
Fuqian Yang
Department of Chemical and Materials
Engineering
University of Kentucky
Lexington, KY 40506
fyang0@engr.uky.edu

James C.M. Li
Department of Mechanical Engineering
University of Rochester
Rochester, NY 14627
li@me.rochester.edu

ISBN: 978-0-387-78700-8   e-ISBN: 978-0-387-78701-5

Library of Congress Control Number: 2008923469

Printed on acid-free paper

9 8 7 6 5 4 3 2 1

springer.com

Foreword

Micro- and nano- mechanical devices are certain to play an ever more important role in future technologies. Already, sensors and actuators based on MEMS technologies are common, and new devices based on NEMS are just around the corner. These developments are part of a decade-long trend to build useful engineering devices and structures on a smaller and smaller scale. Since the invention of the integrated circuit 50 years ago we have been building microelectronic devices on a smaller and smaller scale, with some of the critical dimensions now reaching just a few nanometers. Similar developments have occurred in thin film magnetic storage technologies and related optical devices. The creation of such small structures and devices calls for an understanding of the mechanical properties of materials at these small length scales, not only because these functional devices are primarily load bearing structures, but also because mechanical durability and reliability are required for their successful operation. In the macroscale (bulk), the mechanical properties of materials are commonly described by single-valued parameters (e.g., yield stress, hardness, etc.), which are largely independent of the size of the specimen. However, the dimensions of micro- and nanoscale devices are comparable to the microstructural features that control mechanical properties, and so the mechanical behavior of these devices cannot be predicted using the known properties of bulk materials. Thus, micro- and nanoscale mechanical testing techniques, as described in this book, are required to characterize the mechanical properties of the materials used for these functional devices. In addition, the materials that comprise these devices and structures may not even exist in bulk form, thus making small-scale mechanical testing the only way to determine their mechanical properties. Beyond functional devices, MEMS and NEMS structures comprise micro- and nanoscale devices that have important mechanical functions well, with mechanical actuation and mechanical resonance being critical features of these devices.

The study of mechanical properties of materials in small dimensions is important for another, very different, reason. The ability to study experimentally the basic mechanical properties of materials with sample dimensions in the range of tens to hundreds of nanometers opens up the possibility of conducting mechanical deformation experiments on length scales that can be accessed by atomistic and multiscale modeling, even if there still exists a huge discrepancy in the timescale of the experiments compared to the atomistic modeling. Thus, the mechanical testing techniques described in this book are expected to provide not only new knowledge on the mechanical properties of materials for micro- and nanoscale devices but also experimental validation for computational modeling of length-scale-dependent mechanical properties.

In the last two decades we have seen an explosion of research on the mechanical properties of thin films and materials in small dimensions. Much of this work was necessarily devoted to the development of new techniques for studying the mechanical properties at the micro- and nanoscales, as testing instruments and methods were not previously available. The present book brings many of these new developments together for the first time and allows new workers to enter this field quickly and extend it to new, emerging, nanoscale materials and devices. With the very rapid development of completely new materials in the form of nanowires, nanotubes, nanobelts, and other nanoscale objects, it is likely that the readers of this book will be the ones to bring about the next transformation of this exciting field.

It is fitting that this book be organized and edited by professors J.C.M. Li and F. Yang, as they were among the first to recognize the importance of studying mechanical properties of materials in small dimensions. Their seminal work on impression creep inspired a new generation to focus on the mechanical properties of materials at smallscales.

Many important new developments in the mechanical testing of materials for micro- and nanoscale devices are described in this book. The development of commercial instruments for nanoindentation was accompanied by seminal papers describing the techniques and methodologies through which these new instruments could be used to determine the fundamental mechanical properties in small volumes. Therefore, it is fitting that the basic principles and applications of indentation are described in the first chapter by M. Sakai. Other chapters dealing with the properties and effects revealed by nanoindentation include Chap. 2 on size effects in nanoindentation by X. Feng et al., and Chap. 7 on the determination of residual stresses by nanoindentation by Z.-H. Xu and X. Li. In Chap. 3, Y.-T. Cheng and D.S. Grummon make good use of indentation in their study of shape memory and superelastic effects.

Chapter 10 by T.-Y. Zhang shows how fundamental mechanical properties of thin film materials can be found by deflecting microbridge structures, geometries that arise naturally in MEMS and NEMS devices. Other MEMS and NEMS devices frequently involve contacting surfaces wherein adhesion becomes very important. This topic is covered by F.Q. Yang in Chap. 4. These micro- and nanomechanical devices are often electrostatically actuated, and this too has led to new techniques for mechanical testing of materials at the chip level, as described by A. Corigliano et al. in Chap. 13. Still other kinds of loading can occur in thin film devices; they produce important mechanical deformation effects that play a central role in understanding device reliability. Some of these developments are described by R.R. Keller et al. in Chap. 12.

Size effects in plasticity are important for low-dimensional materials such as nanoparticles, nanowires, and thin films, and can also control the mechanical properties of macroscopic 3D materials with unusual microstructures such as nanoporous metals. The mechanical behavior of these nanoscale materials can now be studied using the new testing methodologies described in this book. In particular, size effects in plasticity of nanoporous metals are described in Chap. 6 by J. Biener et al. Also, micro- and nanoscale contacts can produce important nonmechanical effects such as the piezoelectric effects described by F.Q. Yang in Chap. 8.

As noted above, the development of nanowires and related nanoscale objects represents one of the most exciting developments in this field. This has naturally led to the exploration of the mechanical properties of these new nanomaterials. Chapters 5 and 11 by Y.S. Zhang et al. and B. Peng et al., respectively, focus on the mechanical testing and characterization of these new materials. L.C. Zhang also discusses the mechanics of carbon nanotubes in Chap. 9 and considers their use in composite materials.

Readers of this book will greatly benefit from the experience of these authors and will be well prepared to read the original literature and make their own contributions to this important field.

Stanford University   *William D. Nix*
January 17, 2008

Contents

Contributors

Juergen Biener
Nanoscale Synthesis and Characterization Laboratory, Lawrence Livermore
National Laboratory, Livermore, CA 94550, USA, biener2@llnl.gov

Fabrizio Cacchione
Department of Structural Engineering, Politecnico di Milano, piazza Leonardo da
Vinci 32, 20133 Milano, Italy

Yang-Tse Cheng
Materials and Processes Laboratory, General Motors R&D Center, Warren, MI
48090, USA, Yangtcheng@aol.com

Alberto Corigliano
Department of Structural Engineering, Politecnico di Milano, piazza Leonardo da
Vinci 32, 20133 Milano, Italy, alberto.corigliano@polimi.it

Horacio D. Espinosa
Department of Mechanical Engineering, Northwestern University, 2145 Sheridan
Rd., Evanston, IL 60208-3111, USA, espinosa@northwestern.edu

Xue Feng
Department of Engineering Mechanics, Tsinghua University, Beijing 100084,
China, fengxue@mail.tsinghua.edu.cn

David S. Grummon
Department of Chemical Engineering and Materials Science, Michigan State
University, East Lansing, Michigan 48823, USA, grummon@egr.msu.edu

Alex V. Hamza
Aerospace and Mechanical Engineering Department, University of Southern
California, Los Angeles, CA 90089, USA

Andrea M. Hodge
Nanoscale Synthesis and Characterization Laboratory, Lawrence Livermore
National Laboratory, Livermore, CA 94550, USA

Yonggang Huang
Department of Civil and Environmental Engineering, Department of
Mechanical Engineering, Northwestern University, Evanston, IL 60208,
USA, y-huang@northwestern.edu

Donna C. Hurley
Materials Reliability Division, National Institute of Standards and Technology,
Boulder, CO 80305, USA

Keh-chih Hwang
Department of Engineering Mechanics, Tsinghua University, Beijing 100084,
China

Robert R. Keller
Materials Reliability Division, National Institute of Standards and Technology,
Boulder, CO 80305, USA, keller@boulder.nist.gov

Xiaodong Li
Department of Mechanical Engineering, University of South Carolina, 300 Main
Street, Columbia, SC 29208, USA, LIXIAO@engr.sc.edu

Chwee Teck Lim
Nanoscience and Nanotechnology Initiative, Division of Bioengineering,
Department of Mechanical Engineering, National University of Singapore, 9
Engineering Dr. 1, Singapore 117576, ctlim@nus.edu.sg

Bei Peng
Department of Mechanical Engineering, Northwestern University, 2145 Sheridan
Rd., Evanston, IL 60208-3111, USA, bpeng@northwestern.edu

David T. Read
Materials Reliability Division, National Institute of Standards and Technology,
Boulder, CO 80305, USA

Paul Rice
Materials Reliability Division, National Institute of Standards and Technology,
Boulder, CO 80305, USA

Mototsugu Sakai
Department of Materials Science, Toyohashi University of Technology,
Tempakucho, Toyohashi 441-8580, Japan, msakai@tutms.tut.ac.jp

Chorng Haur Sow
Nanoscience and Nanotechnology Initiative, Department of Physics, Blk S12,
Faculty of Science, National University of Singapore, 2 Science Drive 3, Singapore
117542, physowch@nus.edu.sg

Yugang Sun
Argonne National Laboratory, Argonne, IL 60439, USA

Eunice Phay Shing Tan
Division of Bioengineering, National University of Singapore, 9 Engineering Dr. 1,
Singapore 117576, tanphayshing@yahoo.com

Hsien-Hau Wang
Argonne National Laboratory, Argonne, IL 60439, USA

Zhi-Hui Xu
Department of Mechanical Engineering, University of South Carolina, 300 Main
Street, Columbia, SC 29208, USA

Fuqian Yang
Department of Chemical and Materials Engineering, University of Kentucky,
Lexington, KY 40506, USA, fyang0@engr.uky.edu

Sarah Zerbini
MEMS Product Division, STMicroelectronics, via Tolomeo 1. 20010 Cornaredo,
Milano, Italy

Liangchi Zhang
School of Aerospace, Mechanical and Mechatronic Engineering, The University of
Sydney, Sydney, NSW 2006, Australia, zhang@aeromech.usyd.edu.au

Tong-Yi Zhang
Department of Mechanical Engineering, Hong Kong University of Science and
Technology, Clear Water Bay, Kowloon, Hong Kong, China, mezhangt@ust.hk

Yousheng Zhang
Nanoscience and Nanotechnology Initiative, National University of Singapore,
2 Science Dr. 3, Singapore 117542, nnizys@nus.edu.sg

Yong Zhu
Department of Mechanical Engineering, Northwestern University, 2145 Sheridan
Rd., Evanston, IL 60208-3111, USA

Chapter 1
Principles and Applications of Indentation

Mototsugu Sakai

1.1 Introduction

The microscopic characterization of mechanical properties through indentation contact hardness has a long history exceeding more than one century. After intensive as well as extensive studies on the "plasticity" based on the contact hardness of ductile metals in the mid of twentieth century, the mechanical/physical understanding of contact hardness has well been advanced, since 1980s, along with the development of sophisticated instrumented indentation test systems that are capable of measuring the indentation load P and the penetration depth h in various mechanical modes. This has been enhanced by the upsurge of the electronics engineering and technology that made possible to provide several types of highly sophisticated electronic sensors for detecting load/displacement in accurate manners combined with the developments of microcomputing systems.

In ordinary mechanical testing for characterizing material properties, the test conditions, and the geometries and dimensions of test specimens are carefully designed that realize the "simplest" mechanical boundary conditions, resulting in the straightforward measurement of the material properties having their sound scientific significances. As an example, a well-designed uniaxial tension has been utilized to measure the Young's modulus, for the test result gives the "well-defined physical basis" of the Young's modulus. In fracture toughness testing, as another example for mechanical testing, the ASTM-prescribed dimensions, specimen's geometries, and the test conditions are conventionally recommended to give the *well-defined fracture mode*, leading straightforward to the fracture toughness with its sound physical basis.

M. Sakai
Department of Materials Science, Toyohashi University of Technology, Tempakucho, Toyohashi 441-8580, Japan
msakai@tutms.tut.ac.jp

F. Yang and J.C.M. Li (eds.), *Micro and Nano Mechanical Testing of Materials and Devices*, doi: 10.1007/978-0-387-78701-5, © Springer Science+Business Media, LLC, 2008

However, in contrast to these conventional mechanical testing, as has been well recognized, the stress/strain field beneath the contact point of indentation impression is crucially complicated even in an elastic contact. Plastic and/or viscous flows add further mechanical complications to the contact filed. Such an extremely complex mechanical environment associated with indentation contact processes inevitably yields undesirable complexity and ambiguity to the "*physical meanings*" of the measured mechanical properties, such as contact hardness. Accordingly, in generally speaking, an indentation contact test should not be recommended for characterizing the mechanical properties of engineering materials as a standard test technique having a sound basis of physics. Gilman addressed in his review article [1] that "Hardness measurements are at once among the most maligned and the most magnificent of physical measurements. Maligned because they are often misinterpreted by *the uninitiated*, and magnificent because they are so efficient in generating information for *the skilled practitioner*." In this context, therefore, thoroughly understanding the physical insight and the meaning of indentation-derived mechanical properties must be the most essential prior to utilizing indentation contact test techniques as efficient micro/nanoprobes. In this chapter, the *contact physics* will be intensively emphasized of the various mechanical parameters obtained in instrumented indentation tests.

1.2 History of Indentation Contact Mechanics

In 1881, Hertz, 24 years old at that time, published a historical paper, *On the contact of elastic solids*, in which he made theoretical considerations for the contact pressure distributions of spherical solids on the basis of the analogy to electrostatic potential theory to understand the optical interference fringe patterns at contact [2, 3]. This paper founded the present contact mechanics as Hertzian contact theory. Four years after his publication, in 1885, Boussinesq made a classical approach to finding the elastic stresses and their distributions induced by arbitrary surface tractions by the use of the theory of potential [4]. Though the Boussinesq's theoretical framework included the Hertzian contact problem as a specific case, its analytical solution was first derived by Love in 1930s [5], and then by Sneddon in 1960s [6, 7] by applying the Hankel transforms to potential functions, having led to the basis of the present elastic theory for axisymmetric indentation contact problems.

Due to the large elastic moduli, the contact deformation of ductile metals, even for a spherical contact, is predominantly out of the elastic regime, yielding significant plastic flows and leaving a finite residual impression after unload. Furthermore, the tip-acuity-induced stress concentration in cone/pyramid contact inevitably yields a finite plastic deformation beneath the contact even for very elastic solids. On the basis of the experimental results for spherical indentation on various metallic alloys, along with the considerations on the geometrical similarity of contact, in 1908, Meyer proposed the concept of "contact hardness, H_M (Meyer hardness)" as a mean contact pressure defined by the applied indentation load P divided by the *projected* area A_r of residual impression, $H_M = P/A_r$ [8]. The Meyer hardness for *ductile* materials is closely related to their yield strength Y as $H_M = C \cdot Y$ with the constraint factor of C. Prandtl in 1920s examined the constraint factor related to the plastic flow problem beneath a flat punch, and then Hill extended it into what is now well known as slip line field theory. They predicted the C-value to be 2.57; in fact it varies from about 2.5 to 3.2, depending on the tip-geometry of indenter and the frictional condition at the interface between the indenter and the material contacted [9, 10].

Among numbers of outstanding scientists and researches, Flory in 1940s [11] and Ferry in 1950s [12] made great contributions to polymer physics/rheology, having led to an upsurge of the modeling and the characterization of time-dependent viscoelastic behaviors of engineering materials. This historical upsurge progressively expanded to other fields of mechanical science and engineering, including indentation contact mechanics. Radock and Lee published pioneering papers, in which they examined the time-dependent contact mechanics for spherical indentation [13, 14]. By the use of the well-known fact that the Laplace transform of time-dependent viscoelastic constitutive equation results in the purely elastic equation, they derived the viscoelastic solution in spherical indentation contact problem through utilizing the Laplace-transform inversion of the elastic Hertzian solution. Their theories were then more generalized to other contact geometries by several researchers [15–18]. However, due to the experimental difficulties associated with time-dependent deformations and flows of viscoelastic contacts, the experimental studies have not been conducted until the end of 1990s [19–24].

The hysteresis relationship of indentation load P and the associated penetration depth h during a loading/unloading cycle, i.e., the $P - h$ hysteresis curve, includes all of the information on the microscopic processes and mechanisms of the contact surface deformation, flow, and failure. Tabor in 1951 conducted the earliest experiments for determining the $P - h$ hysteresis curves during a spherical indenter pressed into contact with metals [9]. Stilwell and Tabor in 1961 emphasized the importance of the $P - h$ curve in terms of the elastic recovery of the hardness impression during unloading, and related it to the elastic modulus of the material indented [25]. In 1981, Shorshorov et al. made an experimental determination of the Young's modulus of the material by the use of the unloading stiffness S (the initial slope of unloading $P - h$ curve at the maximum penetration), and suggested the importance of constructing an instrumented indentation apparatus to measure $P - h$ hysteresis curve [26]. An experimental work was then reported by Newey et al. in 1982 for constructing an instrumented micro/nanoindenter to obtain the continuous records of P and h in time, and applied it to the study of the contact behavior of an ion-implanted layer [27]. The dramatic progresses in electronic sensor devices with high precision started in the middle of 1980s for measuring displacement in nm and load in μN, enabling to develop several types of instrumented micro/nanoindentation test systems, and to apply them to the studies on the characteristic material parameters in micro/nanoscales [28–62].

Instrumented indentation test systems have been widely utilized in materials science and engineering fields as an efficient micro/nanoprobe. In particular, an instrumented indentation apparatus plays the most efficient as well as powerful testing roles in characterizing the mechanical properties of electro-optical thin films coated on substrate. In characterizing the film-only properties of coating in indentation test, the key is the quantitative separation of the substrate effect from the observed $P - h$ hysteresis curve of the coating/substrate system to single out the film-only properties, for the substrate beneath a *thin* film significantly affects the observed $P - h$ curve. Theoretical studies on the *elastic* contact mechanics of coatings started at mid-1970s, and was successfully accomplished with the paper published by Yu et al. in 1990 [63–67]. They reduced the elastic Boussinesq problem for an axisymmetric indentation on coating/substrate systems to a simple Fredholm integral equation of the second kind, and then solved the equation in a numerical manner. In *elastoplastic* contact regimes, however, none of analytical approaches to the theoretical works on the coating/substrate systems have been successfully achieved due to the very complicated behaviors of plastic flows within the coating film. The substrate plays an important role in controlling the degree of constraint of the plastic flow of the coating; a very compliant substrate having a considerable amount of the capacity for elastic strain accommodation, in one hand, makes the film more elastic and brittle, resulting in a sinking-in profile of impression. A very stiff substrate, in other hand, significantly constrains the elastic deformation, yielding a significant enhancement of plastic flow within the coating film, and resulting in the film more plastic and ductile. This enhanced plastic flow within the coating film creeps up along the faces of the indenter, yielding a piling-up profile of impression. Numerical studies by the use of finite element analysis (FEA) along with experimental as well as analytical examinations are now extensively conducted to establish the scientific frameworks of time-independent elastoplastic contact mechanics of coating/substrate composites [34, 68–80].

1.3 Theory

1.3.1 Elastic Contact Mechanics [81, 82]

On the basis of potential theory, Boussinesq examined the contact problems for the penetration of a rigid solid of revolution with its axis normal to the surface of an elastic half-space [4]. The *practical* computations were successfully made by Sneddon as the solutions of axisymmetric Boussinesq problems using the theory of the Hankel transforms of an arbitrary potential function ψ [6, 7]. The equations of equilibrium for the stresses within an elastic body combined with the compatibility equations for the elastic displacements lead to an arbitrary function ψ satisfying the biharmonic equation of $\nabla^4 \psi = 0$. This biharmonic equation can be reduced to a fourth-order ordinary differential equation in cylindrical coordinate (r, θ, z) through the Hankel transform $\psi_0^H(\xi, z)$ of this biharmonic function defined by $\psi_0^H(\xi, z) = \int_0^\infty r\psi(r, z)J_0(\xi r)\mathrm{d}r$ in terms of the zero-order Bessel function $J_0(\xi r)$ of the first kind. The transformed function $\psi_0^H(\xi, z)$ can be analytically expressed by

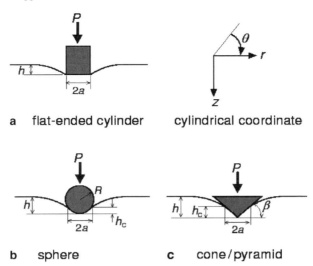

Fig. 1.1 Coordinate system and the parameters for the contact profiles of axisymmetric indentations

$\psi_0^H(\xi,z) = (A + B\xi z)e^{-\xi z}$, where the unknown parameters A and B are determined through the fact that the stresses approach zero at infinity. Since all of the components of stresses and strains within the elastic body contacted with an axisymmetric rigid indenter punch can be given in terms of $\psi_0^H(\xi,z)$, we can finally determine the indentation load P and the penetration depth h as the experimentally observable indentation parameters, after we conduct the spatial integrations of these stresses and/or strains.

Figure 1.1 illustrates three of the representative geometries of axisymmetric indentations with their coordinate system and the contact parameters of P, h, contact radius a, and the contact depth h_c. The origin of the (r, θ, z)-coordinate is located at the first contact point of indenter-tip on the original flat surface (the center of contact circle for the flat-ended cylinder).

We define the function $f(x)$ that quantitatively represents the shape of an axisymmetric indenter as a function of the normalized radius $x(\equiv r/a)$. The vertical displacement $u(x,0)$ of the contacted surface beneath the indenter is, therefore, expressed by $u(x,0) = h - f(x)$ in $0 \leqslant x \leqslant 1$ with $f(0) = 0$. The detailed functional forms of $f(x)$ for the representative tip-geometries (flat-ended cylinder, sphere and cone) are given in Table 1.1. The function $\chi(t)$ is then defined by the use of the indenter's shape-function $f(x)$ as follows [6, 82]

$$\chi(t) = \frac{2}{\pi}\left[h - t\int_0^t \frac{f'(x)}{\sqrt{t^2 - x^2}}dx\right]. \qquad (1.1)$$

Using the functions $f(x)$ and $\chi(t)$, the indentation load P, penetration depth h, and the pressure distribution $p(x)$ beneath the indenter in the contact area $(x \leqslant 1)$ are given in the following formulas [6, 82]

Table 1.1 Elastic indentation contact parameters ($a \ll R$ is assumed for spherical indentation)

	Flat-ended cylinder	Sphere	Cone
$f(x)$	0	$(a^2/2R)x^2$	$a\tan\beta \cdot x$
$p_N(x)$	$1/2\sqrt{1-x^2}$	$3\sqrt{1-x^2}/2$	$\cosh^{-1}(1/x)$
n	1	3/2	2
A	$2a$	$4\sqrt{R}/3$	$2\cot\beta/\pi$
B	a	$\sqrt{R}$	$2\cot\beta/\pi$
nA/B	2	2	2
$\gamma_e(=h/h_c)$	–	2	$\pi/2$

$$P = \pi a E' \int_0^1 \chi(t)\mathrm{d}t, \tag{1.2}$$

$$h = \int_0^1 \frac{f'(x)}{\sqrt{1-x^2}}\mathrm{d}x + \frac{\pi}{2}\chi(1), \tag{1.3}$$

and

$$p(x) = \frac{E'}{2a}\left[\frac{\chi(1)}{\sqrt{1-x^2}} - \int_x^1 \frac{\chi'(t)}{\sqrt{t^2-x^2}}\mathrm{d}t\right]. \tag{1.4}$$

In (1.2)–(1.4), E' is the elastic modulus ($E' = E/(1-v^2)$) in terms of the Young's modulus E and the Poisson's ratio v; $f'(x)$ and $\chi'(t)$ are, respectively, defined by the first-order derivative of the functions $f(x)$ and $\chi(x)$ as $f'(x) = \mathrm{d}f(x)/\mathrm{d}x$ and $\chi'(t) = \mathrm{d}\chi(t)/\mathrm{d}t$. It should be noted that we must have $\chi(1) = 0$ if the function $\chi(t)$ is differentiable in the neighborhood of $t = 1$, i.e., $p(x)$ tends to a finite value along the periphery of the contact at $x = 1.0$. These are the cases of spherical and conical indentations, whereas we have $\chi(t) \equiv \chi(1) = 2h/\pi$ in a flat-ended cylindrical punch, where the contact stresses at the contact periphery become infinite [82].

Substituting the respective shape functions $f(x)$ (see Table 1.1) into (1.2)–(1.4), the relations of P to h and a to h are, respectively, given in the following general expressions *for a given penetration depth of h*

$$P = AE' \cdot h^n \tag{1.5}$$

and

$$a = Bh^{n-1}. \tag{1.6}$$

The parameters of n, A, and B are listed in Table 1.1 for the respective shapes of indenters. The profiles of contact pressure distribution (1.4) for the respective contact geometries are also listed in Table 1.1 as their normalized expressions of $p_N(x)$ defined by $p_N(x) = p(x)/p_m$ in which the mean contact pressure p_m is defined by $P/\pi a^2$.

Using (1.5) and the parameter A listed in Table 1.1, the $P-h$ relation, that is the key in determining the elastic modulus E' in indentation contact test, is given by

$$P = (2a)E'h \tag{1.7}$$

for flat-ended cylindrical punch indentation, yielding the linear relationship between P and h. For spherical indentation contact (Hertzian contact), the $P - h$ relation is

$$P = \frac{4\sqrt{R}}{3} E' h^{3/2} \tag{1.8}$$

in which P is proportional to $h^{3/2}$. As shown in the following $P - h$ relationship for conical indentation, P is related to h in a quadratic manner

$$P = \frac{2\cot\beta}{\pi} E' h^2, \tag{1.9}$$

where β is the inclined faced-angle of cone indenter (see Fig. 1.1c).

As clearly illustrated in Fig. 1.1, the contact surface always sinks in for *elastic* indentation contact. The amount of sinking-in of the contact surface can be quantitatively described by the contact impression profile parameter γ_e defined by the ratio of the total depth h to the contact depth h_c, as follows

$$\gamma_e = h/h_c. \tag{1.10}$$

The subscript "e" indicates "elastic." As the elastic contact surface always sinks-in, the value of γ_e is always larger than 1.0. Substituting the relationships of $a \approx \sqrt{2Rh_c}$ ($h_c \ll R$ or $a \ll R$) for spherical indentation and $a = h_c\cot\beta$ for conical indentation into (1.6), we find that the γ_e-values are the constant of 2.0 and $\pi/2$, respectively, as listed in Table 1.1.

The contact impression profile changes from "sinking-in" to "piling-up," when the indentation behavior of solids becomes progressively plastic with the increase in the ratio of E'/Y, i.e., the decrease in the capacity of elastic strain accommodation beneath the indenter, as well as the increase in the externally applied strains of a/R for spherical and $\tan\beta$ for conical indentations [10, 52, 55, 60, 61]. The $\gamma(= h/h_c)$-value becomes less than 1.0 for the piling-up profiles of impression, the details of which will be given in Sect. 1.3.2.

1.3.2 Elastoplastic Contact Mechanics

1.3.2.1 Geometrical Similarity and the Concept of Meyer Hardness

Two solids are geometrically similar when one of two can be transformed to the other in a linear way without any distortions [59]. Cone and polyhedral pyramids (such as the conventional trihedral Berkovich and tetrahedral Vickers indenters) are geometrically similar if their apex angles are identical. Spherical contacts are also geometrically similar, provided that the ratio of the contact radius a (see Fig. 1.1b) to the radius of sphere R, a/R, is the same.

Meyer found in the experiments of spherical indentation for various metallic alloys that the mean contact pressure defined by the indentation load P divided by the projected contact area of impression $A_c(= \pi a^2)$, P/A_c (referred to as the Meyer hardness, H_M) is given by the following expression [8,9],

$$H_M(\equiv P/\pi a^2) = f(a/R). \tag{1.11}$$

As the ratio a/R is constant for geometrically similar impressions, (1.11) indicates an important experimental fact that the mean pressure (Meyer hardness) is constant for geometrically similar impressions. The ratio of a/R increases during penetration with the increase in the contact depth h_c through the relationship of $a/R = \sqrt{1 - (1 - h_c/R)^2}$. Accordingly, in spherical indentation for hardness measurement, the penetration depth must be controlled to yield the recommended a/R-value; in the Brinell test, the penetration depth is adjusted so that the a/R-value is between 0.3 and 0.5, the most recommended value being 0.4 [83].

Due to the geometrical self-similarity in conical/pyramidal indentations, we have a similar relation of (1.11), as follows

$$H_M(\equiv P/\pi a^2) = g(\beta), \tag{1.12}$$

where a is the radius in conical/pyramidal indentation impression and β represents the inclined face angle of indenter (see Fig. 1.1c). As readily recognized in conical/pyramidal indentation, the Meyer hardness H_M (mean contact pressure; $P/\pi a^2$) is simply described by the inclined face angle only. Accordingly, the Meyer hardness is constant *independent of the penetration depth h or the indentation load P*. This fact is practically important for determining the indentation hardness value in experiments, whereas it is highly affected by the depth of penetration h in spherical indentation.

1.3.2.2 Conical/Pyramidal Indentation

A conical/pyramidal indentation contact accommodates elastic as well as plastic deformations over the contact region to avert the stress singularity induced by the acute tip which highly activates plastic flows even at the onset of indentation contact. The contact stresses are highly concentrated beneath the tip and decrease significantly in magnitude with distance away from the contact region. The Meyer's principle of geometrical similarity represented in (1.12) for conical/pyramidal indentation contact suggests the following quadratic relation between the indentation load P and the penetration depth h [29,31,32,41,43,49,52,55,60],

$$P = k_1 h^2, \tag{1.13}$$

where the indentation loading parameter k_1 is related to the Meyer hardness H_M in the following formula [41,49,52],

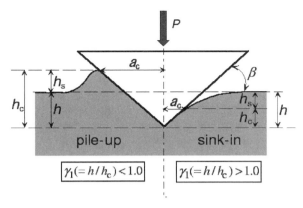

Fig. 1.2 Contact profiles (pileup and sink-in) of the conical/pyramidal indentation. h, h_s, and h_c are, respectively, the total depth, surface depth, and the contact depth of penetration. The contact radius is denoted by a_c

$$k_1 = \left(\frac{g}{\gamma_1^2} \right) H_M. \qquad (1.14)$$

In (1.14), g is the area function of the indenter defined by $A_c = gh_c^2$ (g is expressed by $\pi \cot^2 \beta$, $3\sqrt{3} \cot^2 \beta$, and $4\cot^2 \beta$, respectively, for cone, trihedral Berkovich, and tetrahedral Vickers indenters). The g-value was historically designed to be 24.5 both for the Berkovich ($\beta = 24.7°$) and Vickers ($\beta = 22.0°$) indenters. The γ_1 in (1.14) represents the parameter of contact impression profile that is defined in the *loading process* of elastoplastic indentation contact as $\gamma_1 = h/h_c$ (refer to γ_e in (1.10) for the sinking-in profile of *elastic* impression) [29, 41, 49]. As readily seen in Fig. 1.2, the γ_1-value is always larger than 1.0 for the impression profile of sinking-in and less than 1.0 for the impressions having piling-up profile. When a solid indented is more elastic, due to its large capacity for elastic accommodation, the surface of contact always tends to sink in. In contrast to elastically compliant solids, if the elastic accommodation is limited, that are the cases of ductile metals, the major of contact-induced deformations is accommodated through plastic flows along the faces of the indenter, overcoming the mechanical constraints of the free surface of the solid, and resulting in the contact profile of piling-up around the impression, as shown in Fig. 1.2.

Upon comparing the quadratic $P - h$ relations of elastic (1.9) and elastoplastic (1.13) indentations, as an extreme of elastoplastic contacts, the *purely elastic* contact results in the following expression for the Meyer hardness,

$$H_M = \frac{\tan\beta}{2} E', \qquad (1.15)$$

where use has been made of the relations of $g = \pi \cot^2\beta$ and $\gamma_1 (\equiv \gamma_e) = \pi/2$ for conical indentation. Equation (1.15) was first derived by Love [5] and then by Sneddon [6, 7]. Equation (1.15) indicates that *the Meyer hardness is an elastic measure in purely elastic indentation contact*.

For *purely plastic* indentation contact, the Meyer's principle of geometrical similarity in conical/pyramidal indentation suggests that H_M is simply related to the yield strength Y as

$$H_M = C \cdot Y \tag{1.16}$$

in terms of the constraint factor $C(\approx 3.0)$. The C-value ranges from about 2.5 to 3.2, depending on the inclined face angle β (C increases with the decrease in β) and the frictional coefficient μ at the interface between the indenter and the material indented (C increases with the increase in μ) [10,83,84].

Modeling of elastoplastic indentation contact is addressed in the following to gain further physical insights into the Meyer hardness.

1.3.2.2 (a) Cavity Model [10,84]

The model is based on the assumption that the subsurface displacements produced by a blunt indenter are approximately radial from the point of first contact with hemispherical contours of equal strain. The contact surface of the indenter is encased in a hemispherical cavity of radius a, within which there is a hydrostatic pressure of $\bar{p}$ assumed. The stresses and strains outside the cavity are assumed to be the same as those in an infinite elastic-perfectly plastic body having a spherical cavity under a pressure of $\bar{p}$. In the plastic core of radius c that encloses the cavity, the radial stresses are proportional to $\ln(1/r)$ where r is the radial distance from the point of contact. In the elastic zone extending outside the plastic core, the radial stresses are proportional to $1/r^3$. The conservation rule of the volume displaced by indentation contact locates the elastic/plastic boundary as a function of the plastic index of $E' \tan \beta / Y$, and then the Meyer hardness $H_M (\equiv \bar{p} + 2Y/3)$ of the present model for an elastoplastic body is given by

$$\frac{H_M}{Y} = \frac{2}{3} \left[2 + \ln \left[\left(\frac{(1+\nu)E'\tan\beta}{6Y} + \frac{2(1-2\nu)}{3(1-\nu)} \right) \right] \right]. \tag{1.17}$$

Equation (1.17) indicates that the normalized Meyer hardness H_M/Y is not a measure of plasticity, but an elastoplastic parameter dependent not only on the elastic strain capacity Y/E', but also on the external strain $\tan \beta$ imposed by the cone/pyramid indenter. Accordingly, the Meyer hardness is not a characteristic material parameter; it increases with the increase in the inclined face angle β of the indenter utilized (see Fig. 1.3).

1.3.2.2 (b) Elastoplastic Maxwell Model [41,49,52]

Consider a simple model comprising a purely elastic component having the elastic modulus E' and a purely plastic component having the yield strength Y, being connected in series, like as the Maxwell's linear viscoelastic model. The respective

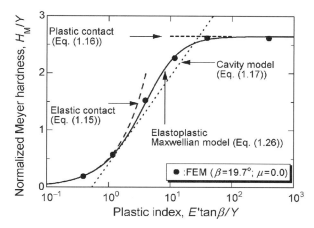

Fig. 1.3 Normalized Meyer hardness vs. plastic index for the cavity model (the *dotted line*) and the elastoplastic Maxwell model (the *solid line*). The *closed circles* are the FEM-results for the Vickers-equivalent cone indentation (frictionless contact)

elastic and plastic constitutive equations in indentation contact (i.e., $P - h$ relationships) can be derived from their expressions for the Meyer hardness given in (1.15) and (1.16) combined with (1.13) and (1.14), as follows,

$$P_e = k_e h_e^2 \tag{1.18}$$

for purely elastic indentation and

$$P_p = k_p h_p^2 \tag{1.19}$$

for purely plastic indentation, respectively. The subscripts "e" and "p" indicate elastic and plastic, respectively. In (1.18) and (1.19), the frontal parameters of k_e and k_p are, respectively, given by the following formulas in terms of E', Y, and the respective contact profile parameters of γ_e and γ_p

$$k_e = \frac{E'\tan\beta}{2} \frac{g}{\gamma_e^2} \tag{1.20}$$

and

$$k_p = CY \frac{g}{\gamma_p^2}. \tag{1.21}$$

As addressed in the previous sections, for conical indentation, γ_e-value is $\pi/2$ (see (1.10) and Table 1.1). However, the analytically closed-form solution for the contact profile parameter γ_p has not yet been provided, whereas it may be less than 1.0, due to its impression profile of piling-up (see Fig. 1.2).

The nominal indentation load P and the total penetration depth h of this elastoplastic model are then expressed in terms of the loads (P_e, P_p) and the depths (h_e, h_p) of the respective elastic (e) and plastic (p) component of materials through the following relations for the Maxwellian serial combination

$$P = P_e = P_p \tag{1.22}$$

and

$$h = h_e + h_p. \tag{1.23}$$

Substituting (1.18)–(1.21) into (1.22) and (1.23), we have finally the following *quadratic* $P - h$ expression for the present elastoplastic model

$$P = k_1 h^2, \tag{1.24}$$

where the frontal coefficient k_1 is expressed by

$$k_1 = \frac{C \cdot Y}{\left(1 + \sqrt{2C}\sqrt{Y/E'\tan\beta}\right)^2} \left(\frac{g}{\gamma_1^2}\right) \tag{1.25}$$

and then, we have the Meyer hardness as a function of the plastic index of $E' \tan \beta/Y$:

$$H_M\left(\equiv k_1 \frac{\gamma_1^2}{g}\right) = \frac{C \cdot Y}{\left(1 + \sqrt{2C}\sqrt{Y/E'\tan\beta}\right)^2}. \tag{1.26}$$

As readily seen in both of the elastoplastic contact models ((1.17) and (1.26)), the Meyer hardness is expressed by the plastic index of $E' \tan \beta/Y$; suggesting that *the Meyer hardness is by no means an indenter-geometry-invariant characteristic material parameter*. It should be noted that (1.26) is naturally reduced to (1.15) in one extreme of purely elastic contact ($E' \tan \beta/Y \leqslant 1$), and to (1.16) in another extreme of purely plastic contact ($E' \tan \beta/Y \gg 1$).

The analytical predictions of the models are shown in Fig. 1.3 along with the numerical results of elastoplastic FEA. In Fig. 1.3, the two extremes of purely elastic and purely plastic indentation contacts ((1.15) and (1.16)) are also indicated. The cavity model ((1.17); the dotted line) well predicts the Meyer hardness in the elastoplastic regime of $1 \leqslant E' \tan \beta/Y \leqslant 20$, while the Maxwellian serial model ((1.26); the solid line) satisfactorily realizes the FEA-results (the closed circles) from purely elastic to purely plastic regimes. It should be repeatedly emphasized that the Meyer hardness is not a characteristic material parameter for expressing plasticity, but is an elastoplastic parameter, increasing with the increase in the plastic index of $E' \tan \beta/Y$ (i.e., with the decrease in the elastic strain capacity of Y/E', and with the increase in the external strain of $\tan \beta$ imposed by the cone/pyramid indenter).

1.3.2.3 Energy-Based Considerations on the Elastoplastic $P - h$ Hysteresis Curve [26, 41]

The indentation contact behavior is well characterized in the load P vs. penetration depth h hysteresis in the loading and the subsequent unloading processes. The schematic diagram of an elastoplastic $P - h$ hysteresis for conical/pyramidal

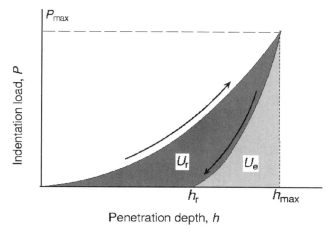

Fig. 1.4 Indentation load vs. penetration depth hysteresis during a loading/unloading cycle for an elastoplastic indentation

indentation is depicted in Fig. 1.4. The penetration depth at the maximum load is subjected to a finite amount of elastic recovery, yet leaving a finite residual depth h_r after unloading. The hysteresis loop energy U_r enclosed with the loading and unloading paths is the indentation work consumed in the loading/unloading cycle. This irreversible energy is consumed to create the elastoplastic hardness impression, if the contact-induced cracking is insignificant.

The integral of the indentation load P with respect to the penetration depth h along the loading path yields the external work U_t (total indentation energy) applied to the material indented to the maximum penetration depth h_{max}, i.e., $U_t = \int_0^{h_{max}} P \, dh$. Applying the quadratic $P - h$ relation (1.24) to this integral affords to

$$U_t = \frac{1}{3} k_1 (h_{max})^3. \tag{1.27}$$

The elastic energy U_e released during unloading that is associated with the elastic recovery of penetration from h_{max} to h_r is expressed by the integral of the indentation load P with respect to the penetration depth h along the unloading path, i.e., $U_e = \int_{h_r}^{h_{max}} P \, dh$, resulting in

$$U_e = \frac{1}{3} k_2 (h_{max})^3 (1 - \xi_r)^3, \tag{1.28}$$

where use has been made of the *quadratic approximation* for the unloading $P - h$ path of $P = k_2(h - h_r)^2$, and the definition for the relative residual depth of $\xi_r = h_r/h_{max}$. The hysteresis loop energy U_r consumed for creating the residual impression is, therefore, given by

$$U_r (\equiv U_t - U_e) = \frac{1}{3} k_1 (h_{max})^3 \xi_r. \tag{1.29}$$

The mechanical compatibility of $k_1 h_{\max}^2 = k_2 (h_{\max} - h_r)^2$ (or its alternative expression of $\xi_r = 1 - \sqrt{k_1/k_2}$) at the maximum penetration was utilized in deriving (1.29).

The most essential elastoplastic parameter among the energy-based indentation characteristics will be the work-of-indentation, W_I; that is defined as the indentation work required to create a unit volume of residual impression [41]

$$W_I = U_r/V_r, \tag{1.30}$$

where V_r is the volume of residual impression approximated by

$$V_r \approx \frac{1}{3} \frac{g}{\gamma_I^2} (h_{\max})^3 \xi_r \tag{1.31}$$

under the assumptions that (1) the side surfaces and edges of pyramid impression are straight, (2) piling-up is not taken into account in calculating V_r, and (3) the projected contact area A_c at $P_{\max}$ coincides with that of residual impression. Substituting (1.29) and (1.31) into (1.30) and comparing the result with (1.14), we can find the equivalence between the work-of-indentation W_I and the Meyer hardness H_M

$$W_I = H_M. \tag{1.32}$$

The numerical results of FEA of the Meyer hardness and the work-of-indentation for a conical indentation (having the Vickers-equivalent inclined face angle β of 19.7°) are plotted in Fig. 1.5 against the plastic index of $E' \tan \beta / Y$ to gain a good physical insight into W_I as well as to examine the significance of the approximations included in (1.31). In Fig. 1.5, for calculating W_I from (1.30), the volume V_r of the respective residual impression was correctly determined *without any*

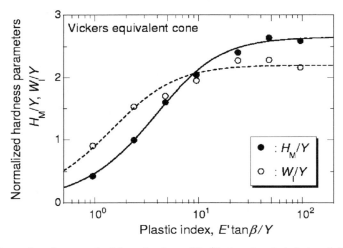

Fig. 1.5 Comparison between the Meyer hardness (H_M/Y; the *closed circles*) and the work-of-indentation (W_I/Y; the *open circles*)

geometrical approximations included in (1.31) by the use of the three-dimensional spatial coordinates of the resultant FEA-nodes of the residual impression. As readily recognized in Fig. 1.5, although both of the hardness parameters are closely correlated, W_I is always larger than the corresponding values of H_M in rather elastic region of $E' \tan\beta/Y (< 2)$, and both coincides in the elastoplastic region of $5 < E' \tan\beta/Y < 20$, while H_M exceeds W_I in ductile contacts of $E' \tan\beta/Y > 30$. The FEA results indicate that, in rather *elastic* region, the side-surfaces of the residual impression tend to be convex due to significant elastic recovery along the penetration axis, suggesting that the straight-side assumption utilized in (1.31) always overestimates the residual impression volume of V_r. However, in ductile contacts, the side-surfaces of the residual impressions are rather straight, but exhibit a significant pileup of impression, implying that the assumption of (1.31) always underestimates the value of V_r. These FEA findings for the profile of residual impression combined with the assumptions and approximations included in (1.31) suggest that the W_I-value via the V_r-value *approximated* in (1.31) will be considerably close to or agree with the Meyer hardness H_M, confirming the equivalence given in (1.32). Accordingly, in conclusion, the present energy-based considerations suggest that *the Meyer hardness approximately represents the elastoplastic work for creating a unit volume of hardness impression.*

The Meyer hardness H_M is reduced to the yield strength Y via the expression of $H_M = C \cdot Y$ in purely plastic/ductile extreme of $(E' \tan\beta)/Y \geqslant 50$ (see Figs. 1.3 and 1.5). In the extreme of ductile flow, therefore, the work-of-indentation is reduced to the yield strength Y through the approximated expression of $Y \approx W_I/C$.

"Ductility" is simply defined as the degree of plastic flow. It plays an important role in tribological science and engineering. A measure for ductility can be defined and determined in an elastoplastic indentation test by the use of the energy-derived observable parameters. Define the ductility, D, with the following nondimensional expression,

$$D = U_r/U_t \tag{1.33}$$

as the ratio of the irreversible hysteresis energy U_r to the total energy applied to the body indented. D is 0.0 for a purely elastic body with a complete elastic recovery during unloading ($U_r = 0.0$), and 1.0 for a purely ductile material ($U_r \equiv U_t$). Equations (1.27) and (1.29) lead to a simple relation of $D = \xi_r$. This simple relation is acceptable only when the exponent n in the indentation loading path of $P = k_1 h^n$ coincides with the exponent m in the subsequent unloading path of $P = k_2(h - h_r)^m$, not only for conical/pyramidal indentation ($n = 2.0$) but also for spherical indentation ($n = 3/2$). Accordingly, instead of conducting the numerical integrations for U_r and $U_t (= U_r + U_e)$ (see Fig. 1.4) to determine the ductility D, we can simply determine the dimensionless residual depth $\xi_r (= h_r/h_{max})$ (see Fig. 1.4) as the measure for ductility. It should be noted that D is independent of the depth of penetration in conical/pyramidal indentation, and be described uniquely as a function of the plastic index of $E' \tan\beta/Y$ [41, 52]. One of the practical applications of the ductility index D was made through the combination with the energy-derived fracture toughness $(K_{Ic})^2/E'$ for examining the machinability and the machining-induced contact damages of engineering ceramics [85].

1.3.3 Viscoelastic Contact Deformation and Flow

A body combining liquid-like and solid-like characteristics does not maintain a time-independent elastic/plastic deformation under constant stress, but goes on slowly creeping with time (creep deformation). When the body is constrained at a constant deformation, the induced stress gradually relaxes with time (stress relaxation). In a body being not quite liquid, some of the energy applied to the system is partially stored during flowing under constant stress, resulting in elastic recovery of a part of deformation when the applied stress is removed. Materials exhibiting such time-dependent characteristics are called viscoelastic. Noncrystalline amorphous materials including organic polymers, inorganic glasses, glass-forming simple liquids, and metallic glasses are classified as viscoelastic materials [12].

It is referred to as *viscoelastic linearity*, provided that an increase in stress σ by a constant factor n, i.e., $n\sigma$, results in an increase in the strain response ε by the same factor, i.e., $n\varepsilon$, and vice versa. Furthermore, for viscoelastic linearity, the resultant strain for different stress histories *simultaneously* applied to the system is identical to the sum of the stress histories acting *separately*, i.e., $\varepsilon(\sigma_1 + \sigma_2) = \varepsilon(\sigma_1) + \varepsilon(\sigma_2)$. These facts of linear characteristics naturally lead to the Boltzmann's hereditary integral on the basis of the principle of superposition of *hereditary memories*. As a memory function of this hereditary integral, the most common is the use of the stress relaxation function $Y(t)$ for given strains and the creep compliance function $D(t)$ for given stresses [86]. If their viscoelastic stress–strain relationships are *linear*, these stresses and strains can be straightforward incorporated into a theoretical framework of contact mechanics [13–18].

A linear viscoelastic theory for spherical indentation is credited to the pioneering work of Radok and Lee [13, 14]. They suggested a simple approach to finding time-dependent viscoelastic solution of indentation contact in cases where the corresponding solution for a purely elastic body is known. Consider an indenter pressed into contact with a linear viscoelastic body under a constant load. The penetration of the indenter and its contact area both grow with time during the indentation loading. It has been well known that the constitutive equations for linear viscoelastic materials can always be reduced to the corresponding elastic constitutive equation via the Laplace transform. The viscoelastic solutions are, therefore, obtained in terms of the associated elastic solutions through their Laplace-transform inversion. Accordingly, in viscoelastic contact problems, the relaxation modulus $Y(t)$ and the creep compliance function $D(t)$ can be well utilized as the substitutes for the elastic modulus E' and the elastic compliance $C(=1/E')$, respectively, in elastic contact problems. Upon applying the principle of the Laplace transform and its inversion to the relationship between elastic and linear viscoelastic indentations on the basis of elastic $P-h$ relation of (1.5), the following integral-type constitutive equations are derived in viscoelastic contact problems by the use of the Boltzmann hereditary integral with respect to the past times t' from $t' = 0$ to the present time $t' = t$. The time-dependent indentation load $P(t)$ resulting from a *monotonic* penetration depth history $h(t'; 0 \leqslant t' \leqslant t)$ is given by [19, 21–23]

$$P(t) = A \int_0^t Y(t - t') \left[\frac{d\{h(t')\}^n}{dt'} \right] dt'. \tag{1.34}$$

In a similar way, the time-dependent penetration depth $h(t)$ resulting from a *monotonic* indentation load history $P(t'; 0 \leqslant t' \leqslant t)$ is expressed as follows:

$$[h(t)]^n = \frac{1}{A} \int_0^t D(t - t') \left[\frac{dP(t')}{dt'} \right] dt', \tag{1.35}$$

where A- and n-values for the respective indenter's geometries are the same as those given for the elastic solutions (1.5), and listed in Table 1.1.

In a specific case of indentation prescribed by a constant rate of penetration V_0, i.e., $h(t') = V_0 t'$, (1.34) is recast into the following formula

$$P(t) = A \cdot n V_0^n \int_0^t Y(t - t')(t')^{n-1} dt'. \tag{1.36}$$

In indentation creep for stepwise loading with $P(t') = P_0 H(t')$ [$H(t')$: Heaviside step function], (1.35) leads to the following creep equation

$$h(t) = [P_0 D(t)/A]^{1/n}. \tag{1.37}$$

In contrast to constant indentation load testing, the indentation load relaxation under a constant penetration depth of h_0 is described in terms of the depth history of $h(t') = h_0 H(t')$ substituted into (1.34) resulting in the following load relaxation

$$P(t) = AY(t)h_0^n. \tag{1.38}$$

Comparing (1.38) for the load relaxation with the corresponding purely elastic relation of (1.5), it is readily seen that a simple replacement of the elastic modulus E' in (1.5) with the stress relaxation function $Y(t)$ yields the constitutive expression for time-dependent indentation load relaxation. This fact that the simple replacement of E' to $Y(t)$ yields an elastic-to-viscoelastic conversion plays an important role in the later section for the indentation contact rheology of *coating/substrate system*.

The indentation contact behavior of viscoelastic *liquid* subjected to a stepwise deformation/load is always elastic at short times, and then viscoelastic at intermediate times, whereas it becomes viscous at times long enough for attaining a steady-state flow. The time-dependent indentation load $P(t)$ of a viscoelastic *liquid* under a constant penetration rate of V_0 is expressed by

$$P(t) = An V_0^n [2(1 + \nu)\eta] \cdot t^{n-1} \tag{1.39}$$

in its steady-state flow, where η is the steady-state shear viscosity and ν is the Poisson's ratio. For a constant indentation load of P_0, on the other hand, the time-dependent penetration $h(t)$ in its steady state is related to the shear viscosity in the following expression

$$h(t) = \left[\frac{P_0}{A} \frac{t}{2(1+\nu)\eta} \right]^{1/n}. \tag{1.40}$$

Accordingly, using the viscoelastic constitutive relations given in this section, we can determine in a quantitative manner the rheological parameters and functions in indentation contact tests. Further details of the theoretical considerations as well as the experimental procedures and test analyses of viscoelastic indentation contact mechanics are reported in the literature [13–24, 87].

1.3.4 Coating/Substrate System

Micro/nanoindentation techniques are the most efficient in mechanically characterizing film/substrate systems, in which the key issue is the establishment of a quantitative procedure for correctly evaluating the substrate effect, and then singling the film-only mechanical properties out of the indentation contact data obtained for coating/substrate systems. Numbers of theoretical models have been proposed and experimental studies have been conducted to examine the substrate effect on the elastic modulus, contact hardness, and viscoelastic functions, etc. obtained in indentation tests. However, in general, any rigorous test techniques and procedures have not yet been established for determining the film-only characteristic mechanical properties. In this section, elastic contact mechanics is first addressed, and then followed by linear viscoelastic contact mechanics of coating/substrate systems. Due to crucial difficulties associated with plastic flows, the understanding on the contact mechanics including the physics of contact hardness of elastoplastic coating/substrate systems is still immature, and its scientific basis has not yet been established. In this section, therefore, no arguments will be made for the elastoplastic contact mechanics, whereas some of the potentially important studies are listed in the references [75–80].

1.3.4.1 Elastic Contact [63–67, 69]

Dhaliwal solved the Boussinesq contact problem for a flat-ended cylinder contacted with an elastic layer resting in a frictionless manner on a rigid foundation [63]. Dhaliwal and Rau subsequently extended this analysis to axisymmetric indenters pressed into an elastic coating perfectly bonded to an elastic substrate [64]. The contact parameters are shown in Fig. 1.6a, b for a cone/pyramid (inclined face angle β) as the representative geometry of axisymmetric indentations. Figure 1.6a illustrates the indentation contact of a *homogeneous film* at a *given penetration depth of h*. The indentation contact for a film having the thickness of t_f coated on a substrate is shown in Fig. 1.6b with *the same penetration depth of h* as that applied in Fig. 1.6a for *homogeneous* film. The respective indentation loads induced are denoted by P_f and P. The resultant contact radii at the penetration of h are, respectively, denoted by a_H (the subscript H indicates "homogeneous") and a.

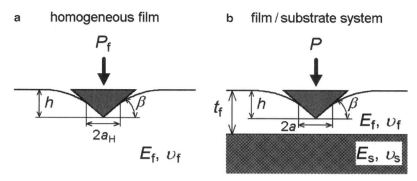

Fig. 1.6 Indentation contact profile and the contact parameters of (**a**) a homogeneous film and (**b**) a film/coating system

The elastic Boussinesq problems for an axisymmetric indenter pressed into a coating/substrate system can be solved through a similar mathematical algorithm for a homogeneous elastic half-space via the Hankel transforms of harmonic functions (see Sect. 1.3.1). Utilizing a pair of the Pabkovich–Neuber functions of $\varphi(r,z)$ and $\psi(r,z)$ applied to the equations of equilibrium for stresses combined with the compatibility equations for displacements, and expressing these harmonic functions in terms of the Hankel integrals, and then applying them to the boundary conditions both at the indenter/film contact surface and at the interface of film/substrate, Yu, et al. derived the following Fredholm integral equation of the second kind [67]

$$H(x) - \frac{1}{\pi} \int_0^1 [K(y+x) + K(y-x)] H(y) \mathrm{d}y = F_0(x), \qquad (1.41)$$

where x represents the normalized radius r/a using the contact radius a. $F_0(x)$ is the dimensionless function related to the tip-geometry of indenter; $F_0(x) = 1 - (a/a_H)x$ for a conical indenter as an example, where a_H is the contact radius at the penetration depth h for the homogeneous half-space (see Fig. 1.6a). Since a_H for a *homogeneous* film is expressed by the relation of $a_H = Bh^{n-1}$ with the elastic modulus-independent parameter B (see (1.6) and Table 1.1), *it must be noted that the contact radius a_H is independent of the elastic modulus E_f of the coating film for a given penetration depth of h.*

The continuous symmetrical kernel $K(y)$ in (1.41) is dependent on the contact conditions (perfectly bonded or frictionless contact) at the film/substrate interface, and given as a function of the elastic moduli of E_f' and E_s', and the Poisson's ratios of v_f and v_s of the coating film and substrate, respectively. By varying the a/a_H-value included in the function $F_0(x)$ in an iterative manner until the boundary condition of $H(1) = 0$ is satisfied, the numerical solution to the Fredholm integral equation yields $H(x)$ in the form of a Chebyshev series [67].

The normalized indentation load P/P_f (the load P for the coating/substrate system normalized with the load P_f for the homogeneous film both *at a given penetration depth of h*; see Fig. 1.6) is given by integrating the contact stress distributions over the contact area, and described by the following formula for axisymmetric indentation

$$\frac{P}{P_{\mathrm{f}}} = n \left(\frac{a}{a_{\mathrm{H}}} \right) \int_0^1 H(x)\mathrm{d}x, \tag{1.42}$$

where n is 1.0, 3/2, and 2.0 for a flat-ended cylinder, sphere, and a cone, respectively (see Table 1.1) (note that a/a_{H} is always 1.0 for flat-ended cylinder). Since the $P-h$ relation of a coating/substrate system can be written by simply replacing E' with its effective value E'_{eff} in (1.5)

$$P = AE'_{\mathrm{eff}}h^n \tag{1.5'}$$

and the relation for a homogeneous film by $P_{\mathrm{f}} = AE'_{\mathrm{f}}h^n$, the alternative expression of (1.42) is given by [88],

$$\frac{E'_{\mathrm{eff}}}{E'_{\mathrm{f}}} = n \left(\frac{a}{a_{\mathrm{H}}} \right) \int_0^1 H(x)\mathrm{d}x. \tag{1.42'}$$

The numerical results of $E'_{\mathrm{eff}}/E'_{\mathrm{f}}$ and a/a_{H} in the conical indentation, as a representative for the conventional axisymmetric indentation contacts, are plotted in Figs. 1.7 and 1.8 (marked by the symbols) against the normalized film thickness t_{f}/a for various values of the modulus mismatch $E'_{\mathrm{f}}/E'_{\mathrm{s}}$ from 0.0 to 10.0. Figures 1.7 and 1.8 readily show that both of $E'_{\mathrm{eff}}/E'_{\mathrm{f}}$ and a/a_{H} are reduced to 1.0, namely E'_{eff} and a approach E'_{f} and a_{H}, respectively, in the extreme of large t_{f}/a-values (i.e., the film

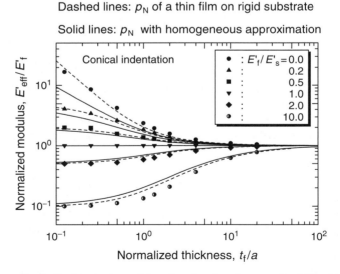

Fig. 1.7 Normalized effective elastic modulus plotted against the normalized film thickness. The *symbols* are the numerical solutions of the Fredholm integral equation (1.41) and (1.42')

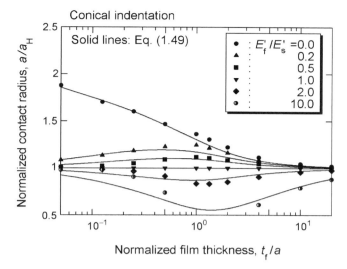

Fig. 1.8 Normalized plots of the contact radius vs. the film thickness for various values of the modulus mismatch E'_f/E'_s. The *solid lines* indicate the numerical predictions of (1.49)

thickness is sufficiently larger than the contact radius, implying the contact on a homogeneous film). In another extreme of small t_f/a-values, where the film thickness becomes negligibly small or the penetration is significant, the substrate-effect progressively becomes dominant, and the substrate exclusively represents the indentation contact behavior; the E'_{eff}/E'_f-value, therefore, approaches E'_s/E'_f-value.

As readily recognized in Fig. 1.8, in the extreme of $t_f/a \downarrow 0$, the contact radius a again reaches its *homogeneous* value of a_H, and so the ratio of a/a_H is again reduced to 1.0 (note that a_H is independent of the material characteristics through the relation of $a_H = Bh^{n-1}$ for a *given depth of penetration* h (see (1.6) and B listed in Table 1.1).

Due to a significant substrate effect, Fig. 1.8 shows that the contact radius a is considerably different from the corresponding homogeneous value of a_H, if the modulus mismatch E'_f/E'_s is significant. For example, an indentation on a soft-coating/hard-substrate system (see Fig. 1.8 for $E'_f/E'_s < 1.0$) results in the enhancement of pileup even in elastic contact; the coating film elastically deformed beneath the indenter is constrained by the rigid substrate, enhancing the elastic deformations along the indenter's faces upward to the regions near the contact periphery, subsequently enhancing the *elastic pileup*. In contrast, for a coating film overlying a compliant substrate ($E'_f/E'_s > 1.0$), a sink-in contact profile is more enhanced than that for the homogeneous semi-infinite half-space. Accordingly, a is significantly larger than a_H (i.e., $a/a_H > 1.0$) in the former and smaller (i.e., $a/a_H < 1.0$) in the latter for a given penetration depth in the region of about $0.1 \leqslant t_f/a \leqslant 10$.

Although the Fredholm integral equation (1.41) yields the highly reliable numerical predictions for the elastic contact deformations of coating/substrate systems, the extremely tedious and complicated mathematical algorithm that is required for solving the integral equation reduces its practical importance. To circumvent the mathematical as well as numerical difficulties included in (1.41), an approximated approach was made by Hsueh and Miranda to solve the elastic contact problems of coating/substrate systems [89], where they reckoned on the use of Boussinesq's Green function $G(r,z; E, v)$ as the elastic solution of the contact surface deformation $w(r,z)$ induced by *point loading* of P on a *homogeneous* semi-infinite half-space (elastic modulus E and Poisson's ratio v) [91]

$$G(r,z; E, v)(\equiv w(r,z)) = \frac{P}{2\pi} \frac{1+v}{E} \left[\frac{r^2}{\sqrt{(r^2+z^2)^3}} + \frac{3-2v}{\sqrt{r^2+z^2}} \right]. \qquad (1.43)$$

This Green function $G(r,z; E, v)$ applied to a coating/substrate layered composite yields the analytical expression for the surface displacements for point loading, and then the penetration depth h is derived by applying the superposition principle to the distributed contact stresses $p(r;a)$ beneath a solid indenter, as follows

$$h = \int_0^a p(r,a) \left[\int_0^{t_f} G(r,z; E_f, v_f)dz + \int_{t_f}^\infty G(r,z; E_s, v_s)dz \right] 2\pi r dr. \qquad (1.44)$$

The integration of (1.44) results in the following general expression between h and P for axisymmetric indentation geometries [88]

$$h = \frac{n/2}{E'_{eff}/E'_f} \frac{P}{aE'_f} \qquad (1.45)$$

in terms of the exponent n ($n = 1.0$, $3/2$, and 2.0 for flat-ended cylinder, sphere, and cone indentations (see Table 1.1)). The normalized elastic modulus E'_{eff}/E'_f included in (1.45) can then be expressed by [88]

$$\frac{E'_{eff}}{E'_f} = \frac{1}{1 + \frac{1}{2I_0} \frac{1}{1-v}[(3-2v)I_1(t_f/a) - I_2(t_f/a)] \left(\frac{E'_f}{E'_s} - 1\right)} \qquad (1.46)$$

with the nondimensional functions of $I_1(t_f/a)$ and $I_2(t_f/a)$ defined by

$$I_1(t_f/a) = \int_0^1 p_N(x) \frac{x}{\sqrt{x^2 + (t_f/a)^2}} dx \qquad (1.47)$$

and

$$I_2(t_f/a) = \int_0^1 p_N(x) \frac{x^3}{(\sqrt{x^2 + (t_f/a)^2})^3} dx. \qquad (1.48)$$

In the derivation of (1.46), for simplicity, the Poisson's ratios of v_f and v_s are supposed to be similar in their magnitude, being assumed to be a single value of v. In (1.47) and (1.48), x represents the normalized radius r/a, and $p_N(x)$ is the normalized contact pressure distribution defined by $p_N(x) = p(r,a)/p_m$ in terms of the mean contact pressure of $p_m = P/(\pi a^2)$.

Both of the functions of $I_1(t_f/a)$ and $I_2(t_f/a)$ coincide with each other in the extreme of thin coating of $t_f/a \approx 0.0$, giving the expression of $I_0\left(= I_1(0) = I_2(0) = \int_0^1 p_N(x)dx\right)$. Accordingly, utilizing (1.46), we can *analytically* express the composite modulus E'_{eff} as a function of the normalized film thickness t_f/a, when we know the modulus ratio of E'_f/E'_s, or analytically determine the film modulus E'_f once we measure the composite modulus E'_{eff} in indentation test, if the substrate modulus E'_s is known.

The analytical expression for the normalized contact pressure distribution $p_N(x)$ is required in calculating the functions of $I_1(t_f/a)$, $I_2(t_f/a)$, and also I_0-value (see (1.47) and (1.48)). Hsueh and Miranda, and Sakai made a use of the *homogeneous approximation* of the normalized contact pressure distribution $p_N(x)$ listed in Tables 1.1 and 1.2 [88–90]. This approximated distribution is only applicable in the extreme of thick film of $t_f/a \gg 1$, or for the coating with $E'_f \simeq E'_s$.

However, in general, $p_N(x)$ is significantly affected by the presence of substrate if the modulus mismatch of E'_f/E'_s becomes significant. Yang examined the analytical expression of $p_N(x)$ for the thin extreme of $t_f/a \ll 1$ (the extreme of a thin film overlying a *rigid* substrate) [92]. These formulas of $p_N(x)$ for three different axisymmetric indentation contacts are listed in Table 1.2 along with their homogeneous approximations. Figure 1.9 compares the profiles of $p_N(x)$ for the thick

Table 1.2 Contact pressure distribution $p_N(x)$ and the related integrals, $I_1(\xi)$ and $I_2(\xi)$

	Homogeneous half-space	Thin film coated on a rigid substrate
	(a) Flat-ended cylinder	
$P_N(x)$	$\frac{1}{2\sqrt{1-x^2}}$	1
$I_1(\xi)$	$\frac{1}{2}\left[\frac{\pi}{2} - \sin^{-1}\left(\frac{\xi}{\sqrt{1+\xi^2}}\right)\right]$	$\sqrt{1+\xi^2} - \xi$
$I_2(\xi)$	$\frac{1}{2}\left[\frac{\pi}{2} - \sin^{-1}\left(\frac{\xi}{\sqrt{1+\xi^2}}\right) - \frac{\xi}{1+\xi^2}\right]$	$2\left(\sqrt{1+\xi^2} - \xi\right) - \frac{1}{\sqrt{1+\xi^2}}$
	(b) Sphere	
$P_N(x)$	$\frac{3}{2}\sqrt{1-x^2}$	$2(1-x^2)$
$I_1(\xi)$	$\frac{3}{2}\left[(1+\xi^2)\left\{\frac{\pi}{2} - \sin^{-1}\left(\frac{\xi}{\sqrt{1+\xi^2}}\right)\right\} - \xi\right]$	$2\left[\frac{2}{3}(1+\xi^2)\sqrt{1+\xi^2} - \xi - \frac{2}{3}\xi^3\right]$
$I_2(\xi)$	$\frac{3}{2}\left[(1+3\xi^2)\left\{\frac{\pi}{2} - \sin^{-1}\left(\frac{\xi}{\sqrt{1+\xi^2}}\right)\right\} - 3\xi\right]$	$4\left[\frac{1}{3}(1+4\xi^2)\sqrt{1+\xi^2} - \xi - \frac{4}{3}\xi^3\right]$
	(c) Cone	
$P_N(x)$	$\cosh^{-1}\left(\frac{1}{x}\right)$	$3(1-x)$
$I_1(\xi)$	$\frac{\pi}{2} - \tan^{-1}\xi + \frac{\xi}{2}\ln\frac{\xi^2}{1+\xi^2}$	$3\left[\frac{1}{2}\sqrt{1+\xi^2} - \xi - \frac{1}{2}\xi^2\ln\frac{\xi}{1+\sqrt{1+\xi^2}}\right]$
$I_2(\xi)$	$\frac{\pi}{2} - \tan^{-1}\xi + \xi\ln\frac{\xi^2}{1+\xi^2}$	$3\left[\frac{1}{2}\sqrt{1+\xi^2} - 2\xi - \frac{3}{2}\xi^2\ln\frac{\xi}{1+\sqrt{1+\xi^2}}\right]$

x and ξ are defined by r/a and t_f/a, respectively

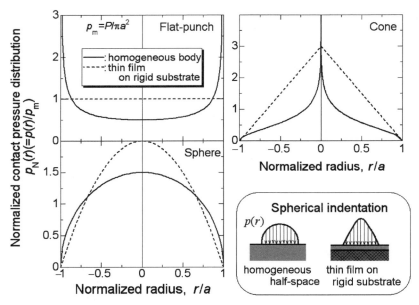

Fig. 1.9 Distributions of normalized contact pressure of a cylindrical flat-punch, sphere, and cone indentations. The *solid lines* and the *dashed lines* represent, respectively, the distributions for the homogeneous half-space and the thin film overlying a rigid substrate

extreme (homogeneous half-space) with those for the thin extreme, indicating that the singularities, i.e., the infinite stresses at the tip $(r/a = 0)$ of cone and at the periphery $(r/a = 1)$ of flat-ended cylinder observed in the thick extreme (the solid lines) are extinct in the thin extreme (the dashed lines). The local strain singularities (i.e., the infinite elastic deformations) at the tip of cone and at the periphery of flat-punch give rise to the stress singularities in the thick extreme. The *rigid* substrate, however, always suppresses these elastic deformations to finite values in thin films, resulting in the nonsingular profiles of $p_N(x)$ over the contact area [93].

The normalized modulus of E'_{eff}/E'_f for conical indentation as an example thus analytically calculated using (1.46) is plotted against the normalized film thickness of t_f/a in Fig. 1.7, where the dashed lines indicate the results for the normalized pressure $p_N(x)$ of a thin film on a rigid substrate, and the solid lines are calculated using $p_N(x)$ for the homogeneous approximation [88,93]. As clearly seen in Fig. 1.7, the analytical results (1.46) with $p_N(x)$ for a thin film on a rigid substrate (the solid lines) well reproduce the rigorous solutions of the Fredholm integral equation ((1.41); the symbols in Fig. 1.7). In particular, the successful fitting is made of the analytical prediction to the rigorous solution in the region of $E'_f/E'_s < 2.0$, where the *rigid* substrate well approximates the mechanical environment of the layered composites. On the other hand, the homogeneous approximation for $p_N(x)$ leads to finite discrepancies (the solid lines in Fig. 1.7) when the modulus mismatch becomes significant ($E'_f/E'_s > 2$ and $E'_f/E'_s < 0.2$). Accordingly, instead of solving the

Fredholm integral equation (1.41) in a numerical manner, adopting an appropriate approximation of $p_N(x)$, we can analytically estimate the elastic modulus of coating film E_f' from the composite modulus E_{eff}' determined in a micro/nanoindentation test as a function of the dimensionless film thickness t_f/a, when the substrate modulus E_s' is known.

In this experimental procedure for estimating E_f' from the observed E_{eff}'-value, there still exists a crucial problem for analytically or experimentally estimating the contact radius a of coating/substrate composite system, if there are no experimental or analytical ways to estimate the contact radius a from the experimentally observable penetration depth h. As mentioned in the preceding considerations, the substrate significantly affects on the contact radius a, resulting in a sizable discrepancy from the contact radius $a_H = Bh^{n-1}$ of homogeneous half-space, as clearly demonstrated in Fig. 1.8 for conical indentation, as an example.

Instead of numerically solving the Fredholm integral equation (1.41) *in an iterating manner* to obtain E_{eff}'/E_f' and a/a_H as functions of t_f/a (see the symbols in Figs. 1.7 and 1.8), Hsueh and Miranda examined an "empirical" formula of a/a_H by best-fitting this radius-ratio to the results of FEA for *spherical* indentation [91]. Sakai extended their formula to include other axisymmetric indentation contact problems [88]

$$\frac{a}{a_H} = \left(\frac{E_{eff}'}{E_f'}\right)^{(n-1)/n} \left(1 - \frac{1 - (E_f/E_s)^{(n-1)/n} + \chi(t_f/a; E_f/E_s)}{1 + \Sigma_j C_j (t_f/a)^j}\right), \qquad (1.49)$$

where $\chi(t_f/a; E_f/E_s)$ and C_j are the fitting function and parameters, respectively, the details of which have been given in the literature [91].

The solid lines in Fig. 1.8 illustrate the results of (1.49) with $n = 2$ for conical indentation, where the rigorous solutions of the Fredholm integral equation (1.41) are indicated by the symbols. The empirical expression of (1.49) well represents the rigorous solution of the Fredholm integral equation, except for the coatings overlaying a very compliant substrate of $E_f'/E_s' \geqslant 10$ in the regions of extremely thin film or very deep penetration of $t_f/a \leqslant 0.5$. Using (1.49) in combination with the relation of $a_H = Bh^{n-1}$, we can estimate the contact radius a of a layered composite for a given penetration depth h that is an observable parameter in indentation test.

The elastic contact behaviors of conical indentation were demonstrated, as an example, in the preceding considerations. However, due to the tip-acuity of conical/pyramidal indentation, even in a shallower penetration, a finite plastic deformation is inevitably induced beneath the indenter. Accordingly, the considerations on spherical indentation problems will be more practical, when the analytical prediction is compared with the experimental results in *elastic* regime. In this context, the a vs. h relationship thus analytically derived using (1.49) in combination with the relation of $a_H = Bh^{n-1}$ is plotted in Fig. 1.10 for *spherical* indentation ($n = 3/2$) in its dimensionless relation of a/t_f vs. $\sqrt{h/t_f}$. The numerical results shown in Fig. 1.10 are very essential from the practical point of view, for we can estimate the *unobservable* contact radius of a from the experimentally *observable* penetration depth h, and then using the normalized film thickness of t_f/a, the

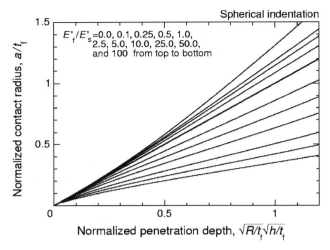

Fig. 1.10 The normalized relations between the resultant contact radius a and the applied penetration depth h for spherical indentations. The *thick solid line* with $E_f'/E_s' = 1.0$ represents the a vs. h relation for the homogeneous half-space, i.e., $a_H = Bh^{n-1}$ ($n = 3/2$). The normalized radius of spherical indenter is defined as R/t_f

composite modulus E_{eff}' as a function of h is analytically predicted in (1.46), and so the indentation load P vs. penetration depth h relation can be analytically described by the use of the relationship of $P = (4/3)\sqrt{R}E_{eff}'h^{3/2}$.

Figure 1.11 illustrates the elastic $P - h$ loading curves of methylsilsesquioxane films having various thicknesses of t_f ranging from 0.5 μm to 20 μm coated on a soda-lime silicate glass plate that are pressed into contact with spherical indenters (the tip radius R of 4.5 μm and 50.0 μm) [88]. The solid lines in Fig. 1.11 are the analytical predictions, showing an excellent agreement with the observed results (the symbols). This excellent coincidence between the observation and the analytical prediction indicates that one can estimate the elastic modulus of coating film E_f' from the observed $P - h$ loading curve of the composite system through inversely solving the analytical problem demonstrated in this section.

1.3.4.2 Linear Viscoelastic Contact [93]

As emphasized in Sect. 1.3.3 for the indentation load relaxation test with a stepwise penetration of h_0, a simple replacement of the elastic modulus E' with the stress relaxation function $Y(t)$ in the elastic $P - h$ relationship yields the constitutive relation for the time-dependent load relaxation, given in (1.38). Accordingly, a similar replacement of E_{eff}' in the elastic relation of $P = AE_{eff}'h^n$ ((1.5') in Sect. 1.3.4) with the effective relaxation modulus of $Y_{eff}(t)$ leads to the viscoelastic expression for the stress relaxation of a coating/substrate composite [94]

$$P(t) = AY_{eff}(t)h_0^n. \qquad (1.50)$$

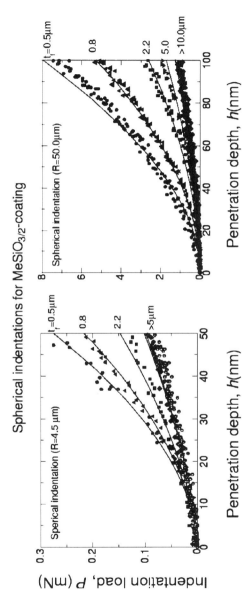

Fig. 1.11 The load vs. penetration depth relationships for the spherical indentation of methylsilsesquioxane films coated on a soda-lime glass plate. The indentation was conducted using two different spherical indenters with their radii of 4.5 and 50µm. The *solid lines* are the theoretical predictions for the respective values of the film thickness t_f [88]

Applying this elastic-to-viscoelastic conversion rule to (1.46) and simply replacing the respective elastic moduli of E_f' and E_s' with the corresponding relaxation functions of $Y_f(t)$ and $Y_s(t)$, the following general expression is obtained for the stress relaxation function $Y_{eff}(t)$ of *linear viscoelastic coating/substrate composites*:

$$\frac{Y_{eff}(t)}{Y_f(t)} = \frac{1}{1 + \frac{1}{2I_0}\frac{1}{1-v}\left[(3-2v)I_1(t_f/a) - I_2(t_f/a)\right]\left(\frac{Y_f(t)}{Y_s(t)} - 1\right)}, \qquad (1.51)$$

where both of the coating and the substrate are supposed to be viscoelastic. In a specific case where the film is viscoelastic while the substrate is elastic, $Y_s(t)$ in (1.51) must be E_s', by way of example.

An organic polymer film overlaying a metal or a ceramic is an example for a viscoelastic film on an elastic substrate. A sol–gel derived brittle hybrid film coated on an organic polymer substrate will be an example for an elastic coating on a viscoelastic substrate. Several interesting numerical simulations using (1.51) were reported in the literature to gain a deep physical insight into the stress relaxation behaviors of various types of viscoelastic coating/substrate composites [93].

As the numerical examples of (1.51), the effects of elastic substrate on the contact behavior of a viscoelastic film are illustrated in Figs. 1.12 and 1.13, where the normalized film thickness t_f/a in the former, and the elastic modulus E_s' of the substrate in the latter are changed in a systematic manner, while the viscoelastic nature of the film is fixed with its relaxation modulus $Y_f(t)$ of

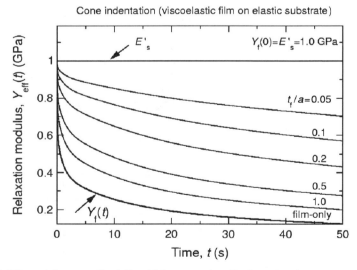

Fig. 1.12 Effect of the normalized film thickness on the effective relaxation modulus in conical indentation. A viscoelastic film having the instantaneous modulus of $Y_f(0) = 1.0$ GPa is coated on an elastic substrate of $E_s' = 1.0$ GPa. The observed stress relaxation behavior becomes progressively sluggish when the elastic substrate-effect becomes significant, where the film is thinner (smaller t_f) or the penetration is deeper (larger a)

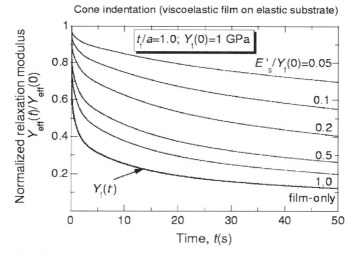

Fig. 1.13 Effect of the elastic modulus of the substrate on the observed effective relaxation modulus. A viscoelastic film having the instantaneous modulus of $Y_f(0) = 1.0$ GPa is coated on elastic substrates having various values of the modulus E_s'. The stress relaxation is examined at a fixed value of $t_f/a = 1.0$. The more compliant the elastic substrate becomes, the more sluggish the apparent relaxation behavior is

$$Y_f(t) = \sum_{i=1}^{n} E_i' \exp(-t/\tau_i) + E_e', \qquad (1.52)$$

where E_e' is the elastic modulus in the equilibrium state. This model of viscoelastic film comprises numbers of the elastic elements (springs) with the modulus of E_i' $(i = 1, 2, 3, \ldots, m)$ and the viscous elements (dashpot) with the viscosity η_i $(i = 1, 2, 3, \ldots, m)$. Accordingly, the relaxation time of the ith component is represented by τ_i with the definition of $\tau_i = \eta_i/E_i'$. At the onset of relaxation $(t = 0)$, the value of the relaxation function of $Y_f(0)$ $(\equiv E_e' + \sum_{i=1}^{n} E_i')$ represents the glassy modulus, being equivalent to the elastic modulus (instantaneous modulus) E_f' of this viscoelastic film. The model material is a viscoelastic *liquid*, provided that E_e' is zero, leading to a complete stress relaxation.

A specific model viscoelastic film having five viscoelastic components $(m = 5)$ is supposed in the present simulation. Assumed are the relaxation times of $\tau_1 = 0.01$ s, $\tau_2 = 0.1$ s, $\tau_3 = 1$ s, $\tau_4 = 10$ s, and $\tau_5 = 100$ s. Furthermore, a box-type spectrum for the internal elastic moduli E_i' is assumed for simplicity; all of E_i'-values $(i = 1, 2, \ldots, 5)$, therefore, have a single value of E_o'. This model film has, therefore, the glassy modulus of $Y_f(0)(\equiv E_f') = 5E_o' + E_e'$.

Figure 1.12 shows the effect of the elastic substrate (the elastic modulus of $E_s' = 1.0$ GPa) on the stress relaxation behavior for various values of the normalized film thickness of t_f/a. The coating film is supposed to be a viscoelastic *liquid*, i.e., $E_e' = 0.0$ GPa. It is clearly seen in Fig. 1.12 that, with the decrease in the normalized film thickness t_f/a, the rheological transition progressively proceeds from the purely viscoelastic relaxation behavior $Y_f(t)$ of the film to the time-independent elastic behavior (E_s') of the substrate, resulting in the progressive increases in the *apparent* relaxation times and the progressively sluggish relaxation behaviors with the decrease in t_f/a, i.e., the decrease in the film thickness and/or the increase in the penetration.

Figure 1.13 shows the relaxation behaviors of the same film examined in Fig. 1.12, but the normalized film thickness t_f/a is fixed to be 1.0, whereas the elastic modulus E_s' of the substrate is systematically changed in its normalized values from $E_s'/Y_f(0) = 0.05$ $(Y_f(0) \equiv 1\,\text{GPa})$ to 1.0. As readily seen in Fig. 1.13, the viscoelastic behavior of the system is significantly affected by the modulus value of the substrate; the more compliant the substrate becomes, the more sluggish the stress relaxation is. When the elastic modulus of the substrate exceeds about ten times larger than the modulus of the film, i.e., $E_s'/Y_f(0) > 10$ (i.e., when the substrate is about ten times stiffer than the film), even at $t_f/a = 1.0$, the relaxation behavior of the composite is reduced to that of the homogeneous film. In other words, *the relaxation time of the film is significantly and always overestimated* whenever the substrate is more compliant than the film $(Y_f(0) \geqslant E_s')$. This apparently sluggish relaxation behavior is essentially the same rheological response as that we observe for bulk viscoelastic materials which are tested on an instrument having *elastically compliant* test frames [95].

1.4 Instrumented Indentation Apparatus and the Analysis of Experimental Data

1.4.1 Geometry of the Conventional Pyramid Indenters

The geometries of the most widely utilized pyramidal indenters, Vickers tetrahedral pyramid and Berkovich trihedral pyramid, are depicted in Fig. 1.14. As pointed out in Sect. 1.3.2 for geometrical similarity, the inclined face angle β of the Vickers indenter is 22.0°, and thus the apex angle (the diagonal face-to-face angle 2θ) is designed to be 136°, since this is the angle subtended by the tangent of a Brinell sphere when the ratio of the contact diameter $2a$ (see Fig. 1.1b) to the diameter of Brinell sphere $2R$ is 0.375, that is the most recommended ratio in the Brinell hardness tests [83]. The diagonal edge-to edge angle 2ψ is 148.1°. Accordingly, the area function g defined by $A_c = gh_c^2$ (see Sect. 1.3.2) is 24.5. As the projected edge length $a(= h_c \tan \psi)$ is expressed by $3.50h_c$, the resultant edge length of impression $2a$ is seven times of the contact penetration depth.

The trihedral Berkovich indenter has been much more preferably used in *nano*indentation tests than the Vickers, because of much more easily machining a sharp tip of indenter due to its specific geometry of trihedron rather than tetrahedron. The g-value is prescribed to give the same g-value of 24.5 as that of Vickers, resulting in the inclined face-angle β of 24.7°, $\psi = 77.1°$, and then $a(= h_c \tan \psi) = 4.37h_c$; the equivalent g-value of Vickers and Berkovich indenters indicates that both of Vickers and Berkovich indenters displace the same volume of the material beneath the indenters during their penetrations.

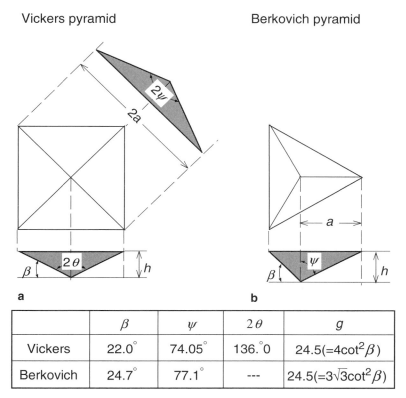

Vickers pyramid   **Berkovich pyramid**

	β	ψ	2θ	g
Vickers	22.0°	74.05°	$136.^\circ0$	$24.5(=4\cot^2\beta)$
Berkovich	24.7°	77.1°	---	$24.5(=3\sqrt{3}\cot^2\beta)$

Fig. 1.14 The tip-geometry of the conventional pyramid indenters; (**a**) Vickers and (**b**) Berkovich

Upon machining a pyramidal tip on a tiny diamond crystal, it is impossible to make an ideally sharp tip (i.e., atomistic tip), always resulting in a rounded or a truncated tip. Commercially available Berkovich indenter, in general, has the rounded tip radius of about 100 nm. This rounded tip may yield crucial difficulties in *nano*indentation testing, when the penetration depths are several tens of nanometers or less [40, 53, 73, 96–100]. A schematic of a conical indenter with a truncated (rounded) tip is shown in Fig. 1.15, where the tip truncation Δh_t and the depth of spherical part h_{sphere} are expressed in terms of the tip-radius r_0 and the inclined face-angle β of the cone. Suppose a rounded Vickers-equivalent cone (the face-angle β of 19.7°, the tip-radius r_0 of 100 nm, and thus $\Delta h_t = 6.22$ nm and $h_{\mathrm{sphere}} = 5.85$ nm) is pressed into contact with an elastic half-space. At the onset of penetration, the indentation is purely spherical up to the total penetration h of $\gamma_e h_{\mathrm{sphere}}$ ($\gamma_e = 2.0$ for elastic Hertzian contact; see Table 1.1), and then the surface of the material starts contacting with the face of cone. The results of FEA are shown in Fig. 1.16a, b (Fig. 1.16b is the magnified plot of the hatched area indicated in Fig. 1.16a) for this rounded cone contacted with an elastic half-space having the Young's modulus E of 75 GPa and the Poisson's ratio ν of 0.3 (a soda-lime silicate glass is supposed). As clearly seen in Fig. 1.16, in the initial stage of penetration, the $P - h$ relation (the

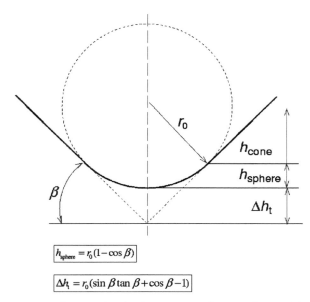

Truncated cone with a round tip

$$h_{\text{sphere}} = r_0(1-\cos\beta)$$

$$\Delta h_t = r_0(\sin\beta\tan\beta+\cos\beta-1)$$

Fig. 1.15 Tip-geometry and penetration depth parameters of truncated cone with a round tip

open circles) faithfully goes along the path of spherical contact (the solid line) to the penetration h of about 12 nm $(= \gamma_e h_{\text{sphere}})$, and then gradually deviates from the path of spherical contact, approaching the path of conical contact (the dashed line). The present FEA study demonstrated in Fig. 1.16 suggests that the effect of rounded tip (tip-radius of about 100 nm) will be insignificant when the penetration depth h exceeds about 200 nm $(h > 40\Delta h_t)$.

As suggested in the preceding considerations on the significance of a truncated tip of cone indentation, a precise calibration for the tip-geometry must be made, prior to quantitative nanoindentation testing. This calibration can be made through empirically modifying the relation of $A_c = gh_c^2$ (see Sect. 1.3.2); g is 24.5 for Vickers and Berkovich indentations that express the projected contact area A_c vs. contact depth h_c relation for an *ideally sharp* pyramid/cone indentation. Two of the representative formulas have been proposed for the calibration [40, 97, 99, 100]

$$A_c = g \cdot h_c(h_c + 2\Delta h_t) \tag{1.53}$$

and

$$A_c = g \cdot h_c^2 + a_1 h_c + a_2 h_c^{1/2} + a_3 h_c^{1/4} + \cdots + a_8 h_c^{1/128}. \tag{1.54}$$

The area function given in (1.53) is equivalent to that of (1.54) with $a_1 = 2g\Delta h_t$ and $a_2 = a_3 = \cdots = a_8 = 0$. The experimental details for determining the area function will be given in Sect. 1.4.2 along with the calibration for the frame compliance of instrumented indentation apparatus.

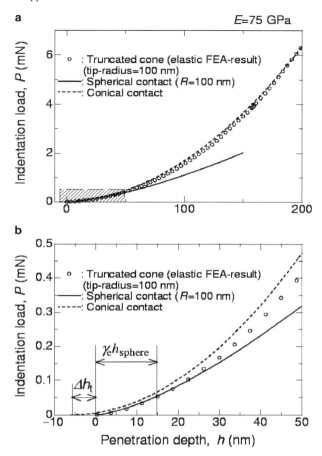

Fig. 1.16 Analytical results of the elastic $P - h$ relationships of a sphere (the *solid line*) and cone (the *dashed line*) indentations, and the FEA-result (the *circles*) of a truncated cone. The magnified plot of the hatched region in (**a**) is illustrated in (**b**)

1.4.2 Instrumented Indentation Apparatus

The schematic of an instrumented micro/nanoindentation test system is depicted in Fig. 1.17. A piezo or an electromagnetic actuator is widely utilized as the load/displacement actuator. Load sensors with precision of about $\pm 0.5\,\mu$N and displacement sensors with precision of about ± 0.1 nm are commercially available; a strain-gage-mounted load cell for the former and a linear transducer or a capacitive displacement gage for the latter are, in general, mounted in the conventional micro/nanoindentation test systems.

The calibration for the load frame compliance of the instrumented indentation apparatus is essential because the observed total displacement h_{obs} is the sum of

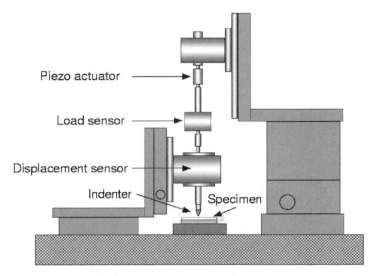

Fig. 1.17 Basic setup of an instrumented indentation test system

the displacements of the load frame and the test specimen, h_f and h, respectively; $h_{obs} = h_f + h$. In purely elastic indentation, these displacements can then be related to the applied indentation load P as $h_{obs} = C_{obs} \cdot P$, $h_f = C_f \cdot P$, and $h = C \cdot P$, where C_{obs}, C_f, and C are the elastic compliances of the test system, load frame, and the test specimen, respectively. Accordingly, we have the following serial expression for the respective elastic compliances [40,49]

$$C_{obs} = C_f + C. \tag{1.55}$$

An elastoplastic indentation loading/unloading $P - h$ hysteresis diagram is depicted in Fig. 1.18. At the onset of unloading, the deformation is purely elastic due to the exclusively elastic recovery processes at the surface of indented body. The observed unloading stiffness S_{obs} (the initial slope dP/dh of the unloading path at h_{max}) shown in Fig 1.18, therefore, dictates the *elastic* displacement of the test system, being inversely related to C_{obs} as $C_{obs} = 1/S_{obs}$.

Upon combining the elastic constitutive relations of (1.5) and (1.6), the stiffness S defined by the slope dP/dh of the elastic $P - h$ path is given by

$$S(= dP/dh) = \frac{nA}{B} E' \frac{\sqrt{A_c}}{\sqrt{\pi}}, \tag{1.56}$$

where the contact area is described by $A_c = \pi a^2$. It should be noted in (1.56) that the frontal factor nA/B is always 2.0 (see Table 1.1) independent of the indenter shape having a solid of revolution [39]. These considerations on the elastic indentation contact stiffness lead to the following relationship for an *elastic indentation with axisymmetric geometry*

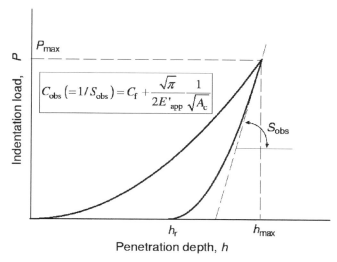

Fig. 1.18 The loading/unloading hysteresis of an elastoplastic indentation defines the unloading stiffness S_{obs} and the residual depth h_{r}. The compliance C_{obs} of the test system is given by the inverse of the unloading stiffness S_{obs}

$$C_{\text{obs}} = C_{\text{f}} + \frac{\sqrt{\pi}}{2E'_{\text{app}}} \frac{1}{\sqrt{A_{\text{c}}}}. \tag{1.56'}$$

In (1.56'), A_{c} is the contact area of the indenter at P_{max}. E'_{app} is the apparent elastic modulus (the composite modulus of indenter and test specimen) defined by $(E'_{\text{app}})^{-1} = (E')^{-1} + (E'_i)^{-1}$ in terms of the elastic modulus E' of the test specimen, and the elastic modulus of the indenter E'_i (the value is 1,147 GPa for a diamond indenter). When the observed modulus C_{obs} plotted against the various values of $1/\sqrt{A_{\text{c}}}$ is linear for a given material, the intercept of this linear plot yields the load frame compliance C_{f}.

In the first step of the simultaneous calibration for the frame compliance and the area function defined in (1.53) or (1.54), relatively large indentations at two of different indentation loads are made in a ductile metal, such as aluminum, copper, etc. as the standard material for calibration, since the area function for a perfect pyramid indenter of $A_{\text{c}} = g \cdot h_{\text{c}}^2$ (see the first term in the right-hand side of (1.53), and (1.54)) can be used to calculate A_{c} for *large* indentations, where the experimental estimate of the contact depth h_{c} can be made by the use of the Oliver–Pharr approximation using P_{max} and the unloading stiffness $S(= 1/(C_{\text{obs}} - C_{\text{f}}))$, the details of the approximation is given in the following section (see (1.59) and (1.62)). The linear plot of C_{obs} vs. $1/\sqrt{A_{\text{c}}}$ for these two large indentations thus provides the initial estimate of C_{f}-value from the intercept and E'_{app}-value from the slope. These two values combined with the C_{obs}-values obtained in additional several indentation sizes can then be utilized to calculate the contact area A_{c} by rewriting (1.56') as

$$A_{\text{c}} = \frac{\pi}{4} \frac{1}{E'^2_{\text{app}}} \frac{1}{(C_{\text{obs}} - C_{\text{f}})^2} \tag{1.57}$$

from which we can make an initial estimate for the area function by fitting the A_c vs. h_c data to the relations of (1.53) or (1.54).

In the second step of calibration, as the initially assumed area function for a perfect pyramid of $A_c = g \cdot h_c^2$ has significant influences on the initially estimated values of C_f and E'_{app}, the new area function thus determined in the first step of calibration is applied again to (1.56') to estimate C_f and E'_{app}, and iterated several times until convergence is achieved. Similar procedures must be also subsequently applied to several standard materials including brittle materials such as fused silica ($E' = 74.1\,\text{GPa}$), etc. to extend the area function to smaller depths. Some more details of the calibration procedures are reported in the literature [40, 97, 99, 100].

1.4.3 Oliver–Pharr [40] and Field–Swain [42] Approximations for Estimating the Contact Depth h_c

Elastic approximations/assumptions have been widely used for estimating h_c and then A_c, where we always assume that the contact periphery sinks-in in a similar manner that can be given by an axisymmetric rigid indenter pressed into contact with a flat elastic half-space. This elastic assumption includes a critical limit, for it does not account for pileup of material at the contact periphery occurring in ductile materials. An elastoplastic unloading process from the maximum penetration at P_{max} is schematically shown in Fig. 1.19a for a hardness impression with a sinking-in

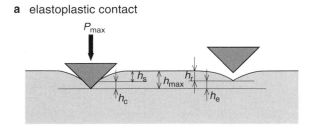

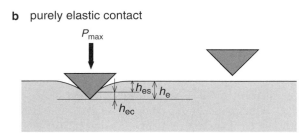

Fig. 1.19 Contact profiles and the penetration depth parameters of (**a**) an elastoplastic and (**b**) a purely elastic indentations at their maximum penetration depths and after complete unloading

profile, where a cone indentation is demonstrated as the representative for indenters of solid of revolution (cone/pyramid, sphere, paraboloid of revolution, etc.), where the sum of the contact depth h_c and the sink-in depth h_s gives the total penetration depth h_{max} at the maximum load P_{max}. After complete unloading, the residual impression has the residual depth h_r resulted from the elastically recovered penetration of h_e. There exist the following relations between these penetration depths $h_{max} = h_s + h_c = h_r + h_e$.

Consider a purely elastic indentation shown in Fig. 1.19b, where the total penetration depth at P_{max} is supposed to be the recovered elastic depth h_e of the elastoplastic indentation given in Fig. 1.19a. Furthermore, the P_{max}-value in this purely elastic indentation contact is supposed to be the P_{max}-value applied to the elastoplastic contact shown in Fig. 1.19a. The sink-in depth and the contact depth of this elastic indentation are expressed by h_{es} and h_{ec}, respectively, with the relation of $h_e = h_{es} + h_{ec}$. The following elastic assumption has been made by Oliver–Pharr for pyramidal indentation and by Field–Swain for spherical indentation

$$h_s = h_{es}. \tag{1.58}$$

Equation (1.58) means that the sink-in depth of elastoplastic impression is assumed to be equal to that of the elastic impression when the depth of *elastic recovery in elastoplastic indentation is the same as that of the total penetration depth in purely elastic indentation*. Combining (1.58) with the preceding relations ($h_{max} = h_s + h_c = h_r + h_e$ and $h_e = h_{es} + h_{ec}$) for the respective penetration depths in elastoplastic and purely elastic indentations, the elastoplastic contact depth h_c can be described by the experimentally observable depths of h_r and h_{max}, as follows

$$h_c = \left(1 - \frac{1}{\gamma_e}\right) h_r + \frac{1}{\gamma_e} h_{max}, \tag{1.59}$$

where γ_e represents the surface profile of elastic impression, being defined by $\gamma_e = h_e/h_{ec}$ (the ratio of the total penetration depth to the contact depth in elastic contact). As shown in Table 1.1, the γ_e-value is 2.0 for spherical contact (Hertzian contact), and then (1.59) coincides with the original formula of the Field–Swain approximation [42]. Equation (1.59) with $\gamma_e = \pi/2$ for conical/pyramidal indentation leads to an alternative expression for the Oliver–Pharr approximation (the original formula is given in (1.62)) [40].

Equation (1.59) is practically important, for the unobservable contact depth h_c can be given in terms of the experimentally observable depths of h_r and h_{max}, and then this value of h_c thus estimated affords to the contact area A_c by substituting it to the area function (1.53) or (1.54). Under the assumption of (1.58) combined with the relations of $h_{max} = h_s + h_c$ and (1.59), the sink-in depth h_s of elastoplastic indentation can be described in the following formula

$$h_s = \left(1 - \frac{1}{\gamma_e}\right) (h_{max} - h_r) \tag{1.60}$$

by the use of the experimentally observable depths of h_r and h_{max}. If the elastoplastic $P - h$ unloading expression is simply described by $P = \alpha(h - h_r)^m$, (1.60) is recast into the following relation,

$$h_s = m \left(1 - \frac{1}{\gamma_e}\right) \frac{P_{max}}{S}. \tag{1.61}$$

In the original Oliver–Pharr expression [40], the frontal factor was expressed with ε as $\varepsilon = m(1 - 1/\gamma_e)$, and then $\varepsilon = 0.726$ (that is, $m = 2.0$ and $\gamma_e = \pi/2$) for cone/pyramid indentation, and $\varepsilon = 0.750$ (that is, $m = 3/2$ and $\gamma_e = 2.0$) for spherical indentation. Accordingly, the contact depth h_c, that is required for calculating the contact area A_c via the area function of (1.53) or (1.54), is given by the following formula

$$h_c = h_{max} - m \left(1 - \frac{1}{\gamma_e}\right) \frac{P_{max}}{S} \tag{1.62}$$

using only the observable parameters of P_{max}, h_{max}, and the unloading stiffness S. Both (1.59) and (1.62) are derived on the basis of the critical assumption of the sinking-in profile of elastoplastic hardness impression with the relation of (1.58). Furthermore, in the derivation of the Oliver–Pharr approximation, it is assumed that the unloading path is subjected to the relation of $P = \alpha(h - h_r)^m$ with $m = 2.0$ for conical/pyramidal indentation and $m = 3/2$ for spherical indentation.

Equation (1.59) (the Oliver–Pharr/Field–Swain approximation) can be recast into the following alternative expression

$$\frac{1}{\gamma_1} = \frac{1}{\gamma_e} + \left(1 - \frac{1}{\gamma_e}\right) \xi_r \tag{1.59'}$$

using the elastoplastic impression profile γ_1 defined by $\gamma_1 = h_{max}/h_c$, and the relative residual depth of penetration, $\xi_r(= h_r/h_{max})$. Equation (1.59') indicates that γ_1-value approaches its elastic value of γ_e, when ξ_r meets to the elastic extreme of 0.0, whereas γ_1-value decreases monotonically with the increase in the relative residual depth ξ_r, reducing to 1.0 in the plastic extreme of $\xi_r = 1.0$, *suggesting none of piling-up profile of impression even in perfectly plastic extreme.*

Equation (1.59') (Oliver–Pharr approximation with $\gamma_e = \pi/2$ for conical/pyramidal indentation and Field–Swain approximation with $\gamma_e = 2$ for spherical indentation) is, respectively, plotted in Fig. 1.20 as the dashed lines along with the elastoplastic FEA results (the open circles) and the experimental observations obtained by the indentation microscope (the filled circles; the details of the apparatus are given in the following section [101, 102]). As clearly seen in Fig. 1.20, these approximations (dashed lines), *in generally speaking*, well represents the experimental observations in qualitative manners. However, these elasticity-based approximations (the dashed lines) always overestimates the γ_1-value (i.e., underestimates h_c, and so A_c), when the indentation behavior becomes progressively ductile in the regions of $\xi_r \geqslant 0.6$. This overestimate in γ_1 results from the purely elastic assumption (i.e., the assumption of always sinking-in profile of impression even for very ductile indentations; refer to (1.58)).

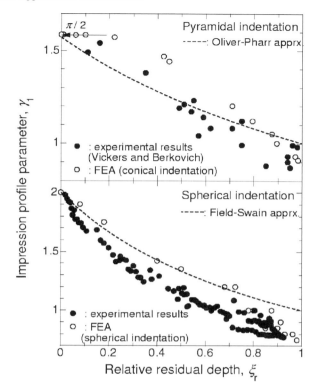

Fig. 1.20 The relations between the impression profile parameter $\gamma_1 \, (= h_{\max}/h_c)$ and the relative residual depth $\xi_r (= h_r/h_{\max})$ for pyramidal and spherical indentations. The *dashed lines* are the elastic approximations (1.59')

The contact area A_c is related to the impression profile parameter γ_1 via the following relationships

$$A_c = \frac{g}{\gamma_1^2}(h_{\max})^2 \tag{1.63}$$

for cone/pyramid indentations ($g = 24.5$ in Vickers/Berkovich indentation), and

$$A_c = \frac{2\pi R}{\gamma_1} h_{\max} \tag{1.64}$$

for spherical indentation. Accordingly, we can readily estimate A_c from the experimentally observable parameter of $h_{\max}$, once we know the reliable value of the impression profile parameter γ_1. As suggested in Fig. 1.20, γ_1 is uniquely related to the experimentally observable parameter $\xi_r (= h_r/h_{\max})$. This fact indicates that we can easily estimate A_c by the use of the experimentally observable indentation parameters of h_r and $h_{\max}$, and then successfully determine the Meyer hardness H_M

through the relation of $H_M = P/A_c$, and the elastic modulus E' by rewriting (1.56) with $nA/B = 2.0$ into the following formula

$$E' = \frac{\sqrt{\pi}}{2} \frac{1}{\sqrt{A_c}} S. \tag{1.65}$$

1.4.4 Indentation Microscope [101, 102]

In the preceding discussions, emphasized are the reliability and the limit of the elastic approximations (Oliver–Pharr and Field–Swain approximations), and recognized is that these approximations yield rather good estimates for the A_c-values only for considerably elastic indentations, whereas for rather ductile materials with $\xi_r \geqslant 0.6$, these approximations always result in an underestimate of A_c, and then overestimate of H_M and E' (see Figs. 1.20 and 1.26). Furthermore, it must be noted that these elastic approximations are only applicable to elastoplastic *homogeneous half-space*; they cannot be extended to coating/substrate systems, where the estimated values of A_c may include crucial errors as well as physical insignificances. As an example, even for a rather elastic film, the induced plastic flow beneath the indenter highly enhances the pileup of impression, if the film is coated on a stiff substrate. This is resulted from the spatially constrained displacements of the coating material against the stiff substrate to flow upward along the faces of the indenter to the free surface. To overcome the difficulties associated with the A_c-estimate not only for ductile homogeneous bulks but also for coating/substrate composites, Miyajima et al. designed and constructed a novel instrumented indentation apparatus, *an indentation microscope*, to directly observe the contact regions and quantitatively determine the contact area A_c [101]. In other words, the conventional instrumented indentation apparatus equips a "*haptic* sensor," while the indentation microscope adds an "*optic* function" to it.

The schematic of an instrumented indentation microscope is depicted in Fig. 1.21 [102]. An optical beam reflected at the indentation contact region passes through the transparent indenter (made of diamond, sapphire, etc.), and the optical image is magnified/recorded by a microscope combined with a charge coupled device (CCD) camera. The details for the instrumented apparatus and the bright- (dark-) filed viewing techniques combined with the appropriate geometries and configurations of pyramidal (spherical) indenter are reported in the literature [101, 102]. Examples of the bright-field images (tetragonal pyramid indentation with the inclined face angle $\beta = 45°$) are shown in Fig. 1.22 for the materials ranging from a brittle glassy carbon to a ductile stainless steel (SUS). As well demonstrated in Fig. 1.22, we can directly determine the contact area A_c, and then the characteristic material parameters of H_M and E' in a precise manner without utilizing any approximation and assumption such as the Oliver–Pharr and Field–Swain approximations.

The evolution of the contact area is shown in Fig. 1.23 in a sequential manner during the Vickers-indentation loading on a ZrO_2 (3Y-TZP; 3 mol% yttria partially

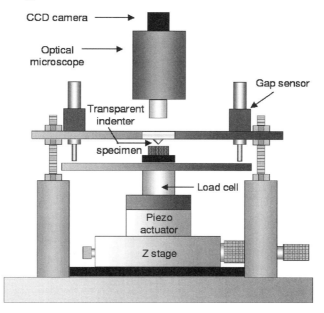

Fig. 1.21 Schematic of the instrumented indentation microscope

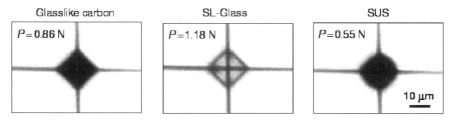

Fig. 1.22 Examples of the in situ bright-field images of the contact areas of a glasslike carbon, soda-lime glass and a stainless steel at the indentation loads indicated in the respective figures

stabilized tetragonal zirconia polycrystals). Due to the inclined face angle $\beta = 22.0°$ of Vickers indentation, the technique of dark-field viewing was adopted. The numbers from 1 to 8 noted in the sequential images are corresponding to the numbered locations (the open circles) indicated in the associated $P - h$ loading path. Since the instrumented indentation microscope makes possible the synchronized observations between the $P - h$ hysteresis curve and the in situ contact images, it has very versatile potentials in the science and engineering of contact mechanics, not only in time-independent elastic/plastic deformations but also in time-dependent viscoelastic flows of homogeneous bulk materials as well as coating/substrate composites. The examples of its versatility are demonstrated in Figs. 1.24–1.26.

The P vs. A_c relationships of several engineering materials are shown in Fig. 1.24, directly providing the Meyer hardness as the slope of each linear plot through the relation of $H_M = P/A_c$. The dependence of the contact radius a_c on the penetration

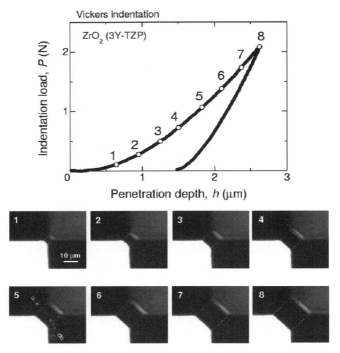

Fig. 1.23 The evolution of the contact areas (*dark-field images*) in a zirconia ceramic in situ observed along the loading path from 1 to 8

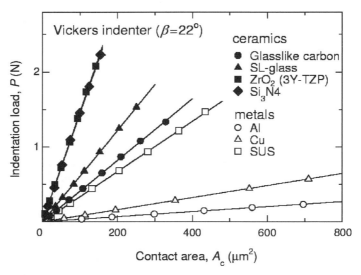

Fig. 1.24 Indentation load is plotted against the contact area that is in situ determined by the indentation microscope. The slope of each linear plot directly affords to the Meyer hardness without assuming/approximating the contact area

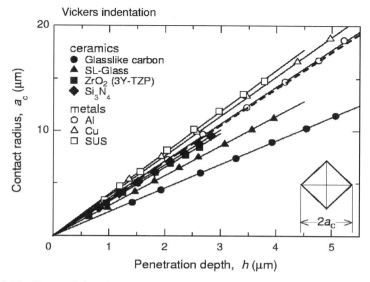

Fig. 1.25 The linear relations between the contact radius and the applied penetration depth for various engineering materials. The *dashed line* represents the relation given in (1.66). The contact profiles are piling-up for the materials (the *open symbols*) showing their linear plots above the *dashed line*, whereas the profiles are sinking-in for the materials (the *closed symbols*) having their linear plots are below the *dashed line*

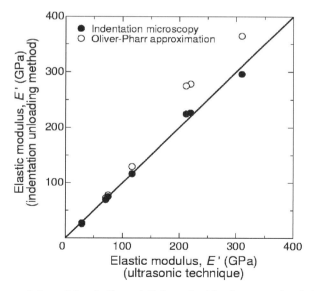

Fig. 1.26 The correlation of the elastic moduli determined by the conventional ultrasonic technique and the indentation unloading method. The *open circles* indicate the values estimated by the Oliver–Pharr approximation and the *closed circles* represent the values determined from the contact areas directly observed through the instrumented indentation microscope

depth h is plotted for Vickers indentation in Fig. 1.25, where the geometrical consideration on the indenter affords to the following a_c vs. h relation

$$a_c = \frac{\sqrt{2}}{\tan\beta}\left(\frac{h_c}{h}\right) \cdot h \qquad (1.66)$$

with $\beta = 22.0°$ for Vickers indentation. As intensively examined in the previous considerations, the normalized contact depth $h_c/h(\equiv 1/\gamma_1)$ dictates the details of the impression profile (sink-in or pileup) at the maximum penetration; the impression is sinking-in for $h_c/h < 1.0$, and piling-up for $h_c/h > 1.0$. The dashed line in Fig. 1.25 represents the relation of (1.66) with $h_c/h = 1.0$. Accordingly, the materials having their linear slopes larger than that of the dashed line imply the piling-up in the impression profiles, and the impression profile is sinking-in for the materials with their slopes smaller than that of the dashed line. Furthermore, it should be noted in Fig. 1.25 that we can easily determine in experiment the contact depth h_c or the γ_1-value from the slope of each linear line.

The indentation-determined elastic modulus values via (1.65) are correlated with those determined in the conventional ultrasonic method in Fig. 1.26. The closed circles represent the results obtained from the A_c-values directly determined in the indentation microscope, showing excellent coincidence with those in the ultrasonic test, while the open circles are the results calculated from the A_c-values estimated by the Oliver–Pharr approximation, always resulting in finite overestimates.

1.5 Concluding Remarks

The stress/strain field beneath the contact area of an indenter is extremely complicated even in an elastic contact. Such a complex mechanical environment inevitably results in undesirable complexities and ambiguities to measure contact parameters such as the contact hardness. To utilize the instrumented indentation test systems as efficient experimental tools for characterizing the material properties in micro/nanoregimes, thoroughly understanding the physical significances of the contact parameters is the most essential. In this chapter, therefore, the discussions have intensively focused on the indentation contact processes not only of homogeneous bulk materials, but also of coating/substrate composites in time-independent elastic and elastoplastic as well as in time-dependent viscoelastic regimes.

References

1. Gilman JJ (1973) In: Westbrook JH, Conrad H (eds) The Science of Hardness Testing and Its Research Applications. American Society for Metals, Metals Park, pp. 51–74
2. Hertz H (1881) J für die reine und angewandte Mathematik 92:156. (For English translation see: Jones DE, Schott GA (1896) Miscellaneous Papers by H. Hertz. Macmillan, New York, pp. 146–162)

3. Hertz H (1882) Verhandlungen des Vereins zur Beförderung des Gewerbefleisses, November, Leipzig. (For English translation see: Jones DE, Schott GA (1896) Miscellaneous Papers by H. Hertz. Macmillan, New York, pp. 163–183)
4. Boussinesq J (1885) Application des potentials à l'etude de l'équilibre et du mouvement des solides élastiques. Gauthier-Villars, Paris
5. Love AEH (1939) Q. J. Math. (Oxford series) 10:161
6. Sneddon IN (1965) Int. J. Eng. Sci. 3:47
7. Sneddon IN (1951) Fourier Transforms. McGraw-Hill, New York, chaps. 9–10
8. Meyer E (1908) Zeit. Ver. Deut. Ing. 52:645, 740
9. Tabor D (1951) Hardness of Metals. Clarendon, Oxford, chap. 3
10. Johnson KL (1985) Contact Mechanics. Cambridge University Press, Cambridge, chap. 6
11. Flory PJ (1953) Principles of Polymer Chemistry. Cornell University Press, New York
12. Ferry JD (1961) Viscoelastic Properties of Polymers. Wiley, New York
13. Radok JRM (1957) Q. Appl. Math. 15:198
14. Lee EH, Radok JRM (1960) J. Appl. Mech. 27:438
15. Hunter SC (1960) J. Mech. Phys. Solids 8:219
16. Yang WH (1966) J. Appl. Mech. 33:395
17. Ting TCT (1966) J. Appl. Mech. 33:845
18. Ting TCT (1968) J. Appl. Mech. 35:248
19. Shimizu S, Yanagimoto T, Sakai M (1999) J. Mater. Res. 14:4075
20. Cheng L, Xia X, Yu W, Scriven LE, Gerberich WW (2000) J. Polym. Sci. B Polym. Phys. 38:10
21. Sakai M, Shimizu S, Miyajima N, Tanabe Y, Yasuda E (2001) Carbon 39:605
22. Sakai M, Shimizu S (2001) J. Non-Cryst. Solids 282:236
23. Sakai M, Shimizu S (2002) J. Am. Ceram. Soc. 85:1210
24. Yang S, Zhang YW, Zeng K (2004) J. Appl. Phys. 95:3655
25. Stilwell NA, Tabor D (1961) Proc. Phys. Soc. London 78:169
26. Shorshorov MKh, Bulychev SI, Alekhim VA (1981) Sov. Phys. Doki. 26:769
27. Newey D, Wilkins MA, Pollock HM (1982) J. Phys. E 15:119
28. Frölich F, Grau P, Grellman W (1977) Phys. Stat. Sol. A 42:79
29. Lawn BR, Howes VR (1981) J. Mater. Sci. 16:2745
30. Pethica JB, Hutchings R, Oliver WC (1983) Philos. Mag. A 48:593
31. Loubet JL, Georges JM, Marchesini O, Meille G (1984) J. Tribol. 106:43
32. Loubet JL, Georges JM, Meille G (1986) In: Blau PJ, Lawn BR (eds) Microindentation Techniques in Materials Science and Engineering, ASTM STP 889. American Society for Testing and Materials, Philadelphia, pp. 72–89
33. Oliver WC, Hutchings R, Pethica JB (1986) In: Blau PJ, Lawn BR (eds) Microindentation Techniques in Materials Science and Engineering, ASTM STP 889. American Society for Testing and Materials, Philadelphia, pp. 90–108
34. Doerner MF, Nix WD (1986) J. Mater. Res. 1:601
35. Stone D, LaFontaine WR, Alexopoulos P, Wu TW, Li CY (1988) J. Mater. Res. 3:141
36. Joslin DL, Oliver WC (1990) J. Mater. Res. 5:123
37. Pharr GM, Cook RF (1990) J. Mater. Res. 5:847
38. Page TF, Oliver WC, McHargue CJ (1992) J. Mater. Res. 7:450
39. Pharr GM, Oliver WC, Brotzen FR (1992) J. Mater. Res. 7:613
40. Oliver WC, Pharr GM (1992) J. Mater. Res. 7:1564
41. Sakai M (1993) Acta Metall. Mater. 41:1751
42. Field JS, Swain MV (1993) J. Mater. Res. 8:297
43. Söderlund E, Rowcliffe DJ (1994) J. Hard Mater. 5:149
44. Cook RF, Pharr GM (1994) J. Hard Mater. 5:179
45. Zeng K, Söderlund E, Giannakopouloa AE, Rowcliffe DJ (1996) Acta Mater. 44:1127
46. Hainsworth SV, Chandler HW, Page TF (1996) J. Mater. Res. 11:1987
47. Gerberich WW, Yu W, Kramer D, Strojny A, Bahr D, Lilleodden E, Nelson J (1998) J. Mater. Res. 13:421

48. Bahr DF, Gerberich WW (1998) J. Mater. Res. 13:1065
49. Sakai M, Shimizu S, Ishikawa T (1999) J. Mater. Res. 14:1471
50. Bulychev SI (1999) Tech. Phys. 44:775
51. Hay JC, Bolshakov A, Pharr GM (1999) J. Mater. Res. 14:2296
52. Sakai M (1999) J. Mater. Res. 14:3630
53. Malzbender J, With G, Toonder J (2000) J. Mater. Res. 15:1209
54. Oliver WC (2001) J. Mater. Res. 16:3202
55. Sakai M, Nakano Y (2002) J. Mater. Res. 17:2161
56. Pharr GM, Bolshakov A (2002) J. Mater. Res. 17:2660
57. Sakai M (2003) J. Mater. Res. 18:1631
58. Oliver WC, Pharr GM (2004) J. Mater. Res. 19:3
59. Cheng YT, Cheng CM (2004) Mater. Sci. Eng. R Rep. 44:91
60. Zhang J, Sakai M (2004) Mater. Sci. Eng. A 381:62
61. Taljat B, Pharr GM (2004) Int. J. Solids Struct. 41:3891
62. Habbab H, Mellor BG, Syngellakis S (2006) Acta Mater. 54:1965
63. Dhaliwal RS (1970) Int. J. Eng. Sci. 8:273
64. Dhaliwal RS, Rau IS (1970) Int. J. Eng. Sci. 8:843
65. Chen WT, Engel PA (1972) Int. J. Solids Struct. 8:1257
66. King RB (1987) Int. J. Solids Struct. 23:1657
67. Yu HY, Sanday SC, Rath BB (1990) J. Mech. Phys. Solids 38:745
68. Wu TW (1991) J. Mater. Res. 6:407
69. Gao H, Chiu CH, Lee J (1992) Int. J. Solids Struct. 29:2471
70. Gerberich WW, Nelson JC, Lilleodden ET, Anderson P, Wyrobek JT (1996) Acta Mater. 44:3585
71. Moody NR, Hwang RQ, Venka-Traman S, Angelo JE, Norwood DP, Gerberich WW (1998) Acta Mater. 46:585
72. Menčík J, Munz D, Quandt E, Weppelmann ER, Swain MV (1997) J. Mater. Res. 12:2475
73. Gouldston A, Koh HJ, Zeng KY, Giannakopoulos AE, Suresh S (2000) Acta Mater. 48:2277
74. Malzbender J, With G, Toonder J (2000) Thin Solid Films 366:139
75. Chen X, Vlassak JJ (2001) J. Mater. Res. 16:2974
76. Toonder J, Balkenende R (2002) J. Mater. Res. 17:224
77. Saha R, Nix WD (2002) Acta Mater. 50:23
78. Tsui TY, Ross CA, Pharr GM (2003) J. Mater. Res. 18:1383
79. Jung YG, Lawn BR, Martyniuk M, Huang H, Hu KZ (2004) J. Mater. Res. 19:3076
80. Han SM, Saha R, Nix W (2006) Acta Mater. 54:1571
81. Johnson KL (1985) Contact Mechanics. Cambridge University Press, Cambridge, chaps. 2–5
82. Maugis D (2000) Contact, Adhesion and Rupture of Elastic Solids. Springer, Berlin, chap. 4
83. Shaw MC (1973) In: Westbrook JH, Conrad H (eds) The Science of Hardness Testing and Its Research Applications, American Society for Metals, Metals Park, pp. 1–74
84. Hill R (1950) The Mathematical Theory of Plasticity. Clarendon, Oxford, chaps. 5–6
85. Sakai M, Nowak R (1992) In: Bannister (ed) Ceramics, Adding the Value, vol. 2. The Australian Ceramic Society, Melbourne, pp. 922–931
86. Shames IH, Cozzarelli FA (1992) Elastic and Inelastic Stress Analysis. Prentice Hall, Englewood Cliffs, chap. 6
87. Sakai M (2002) Philos. Mag. A 82:1841
88. Sakai M, Zhang J, Matsuda A (2005) J. Mater. Res. 20:2173
89. Hsueh CH, Miranda P (2004) J. Mater. Res. 19:94
90. Timoshenko SP, Goodier JN (1951) Theory of Elasticity. McGraw-Hill, New York, chap. 12
91. Hsueh CH, Miranda P (2004) J. Mater. Res. 19:2774
92. Yang FQ (2003) Mater. Sci. Eng. A 358:226
93. Sakai M (2006) Philos. Mag. A 86:5607
94. Sakai M, Sasaki M, Matsuda A (2005) Acta Mater. 53:4455
95. Rekhson SM (1989) J. Am. Ceram. Soc. Bull. 68:1956
96. Briscoe BJ, Sebastian KS, Adams MN (1994) J. Phys. D Appl. Phys. 27:1156

97. Shimamoto A, Tanaka K, Akiyama Y, Yoshizaki H (1996) Philos. Mag. A 74:1097
98. Cheng YT, Cheng CM (1998) J. Mater. Res. 13:1059
99. Seitzman LE (1998) J. Mater. Res. 13:2936
100. Sawa T, Tanaka K (2002) Philos. Mag. A 82:1851
101. Miyajima T, Sakai M (2006) Philos. Mag. 86:5729
102. Sakai M, Hakiri N, Miyajima T (2006) J. Mater. Res. 21:2298

Chapter 2
Size Effects in Nanoindentation

Xue Feng, Yonggang Huang, and Keh-chih Hwang

2.1 Introduction

Microindentation hardness experiments have repeatedly shown that the indentation hardness increases with the decrease of indentation depth, i.e., the smaller the harder [3, 5, 9, 30, 31, 37, 43, 51]. On the basis of the Taylor dislocation model [48, 49] and a model of geometrically necessary dislocations (GND) underneath an indenter tip shown in the inset of Fig. 2.1, Nix and Gao established the relation between the microindentation hardness H and the indentation depth h

$$\left(\frac{H}{H_0}\right)^2 = 1 + \frac{h^*}{h}, \tag{2.1}$$

where h^* is a characteristic length on the order of micrometers that depends on the properties of indented material and the indenter angle, and H_0 is the indentation hardness for a large indentation depth (e.g., $h \gg h^*$) [33]. The above relation predicts a linear relation between H^2 and $1/h$, which agrees well with the microindentation hardness data for single crystal and polycrystalline copper [31] as shown in Fig. 2.1, as well as for single crystal silver [30].

The *nanoindentation hardness* data, however, do not follow (2.1) [10, 12, 13, 15, 23, 28, 29, 45, 46]. Here nanoindentation and microindentation typically refer to the indentation depth below and above 100 nm, respectively. As shown in Fig. 2.2, Lim and Chaudhri's nanoindentation hardness data for annealed copper start to deviate from (2.1) (the dotted straight line of the Nix-Gao model for annealed copper

X. Feng and K.-C. Hwang
FML, Department of Engineering Mechanics, Tsinghua University, Beijing 100084, China
fengxue@mail.tsinghua.edu.cn, huangkz@tsinghua.edu.cn

Y. Huang
Department of Civil and Environmental Engineering, Department of Mechanical Engineering, Northwestern University, Evanston, IL 60208
y-huang@northwestern.edu

F. Yang and J.C.M. Li (eds.), *Micro and Nano Mechanical Testing of Materials and Devices*,
doi: 10.1007/978-0-387-78701-5, © Springer Science+Business Media, LLC, 2008

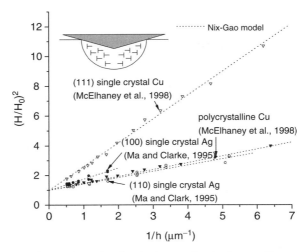

Fig. 2.1 Microindentation hardness data for single crystal and polycrystalline copper [31] as well as for single crystal silver [30]. Here h is the indentation depth, H is the microindentation hardness, and H_0 is the indentation hardness for large depths of indentation. The indentation hardness–depth relation of (2.1) is also shown for each set of experimental data, and it agrees well with the microindentation hardness data. The inset shows [33] model of geometrically necessary dislocations underneath a sharp indenter

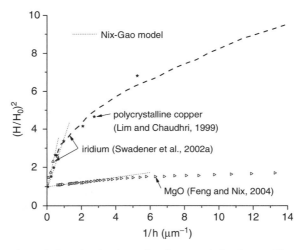

Fig. 2.2 Micro and nanoindentation hardness data for annealed polycrystalline copper [28], annealed iridium [45], and MgO [15]. The indentation hardness–depth relation of (2.1) is also shown for each set of experimental data, and it does *not* agree well with the nanoindentation hardness data

in Fig. 2.2) when the indentation depth h is on the order of 100 nm [28]. Even though the indentation hardness continues to increase as h decreases, the hardness data are significantly lower than the straight line predicted by [34]. Swadener et al. also showed that the nanoindentation hardness data for annealed iridium are smaller than that given by (2.1) when the indentation depth h becomes submicrometer (see the dotted straight line of the Nix-Gao model for iridium in Fig. 2.2) [45]. Recently, Feng and Nix and Elmustafa and Stone found that, once the indentation depth is less than 0.2 µm, (2.1) does not hold in MgO (see the dotted straight line of the Nix-Gao model for MgO in Fig. 2.2) and in annealed α-brass and aluminum, respectively [11, 15].

There are two main factors for the discrepancy between (2.1) and the nanoindentation hardness data.

(i) *Indenter tip radius.* Equation (2.1) holds only for "sharp," pyramid indenters since the effect of indenter tip radius (typically around 50 nm) has not been accounted for [23, 39]. Qu et al. studied pyramid indenters with spherical tips, and found that the finite tip radius indeed gives smaller indentation hardness than (2.1) [39]. Qu et al. also studied spherical indenters, and established an analytic relation between the indentation hardness and indentation depth that is very different from (2.1) [41]. On the basis of the maximum allowable GND density Huang et al. established a model for the effect of finite tip radius in nanoindentation [20, 21].

(ii) *Storage volume for GNDs.* Equation (2.1) assumes that all GNDs are stored in a hemisphere of radius a, where a is the contact radius of indentation. Such an assumption may not hold in nanoindentation. Swadener et al., Feng and Nix, and Durst et al. proposed to modify this storage volume for GNDs [10, 15, 45]. Huang et al. established a model of nanoindentation based on the maximum allowable GND density [20].

Other factors may also contribute to the discrepancy between (2.1) and nanoindentation hardness data, such as the intrinsic lattice resistance or friction stress [38], surface roughness [29, 52], and long-range stress associated with GNDs [13, 14].

The objective of this chapter is to review the analytic models based on the *maximum allowable GND density* to study the nanoindentation size effect, and validate them by comparing with the experiments. The analytical models give simple relations between the indentation hardness H and the indentation depth h or contact radius in nanoindentation.

This chapter is outlined as follows. The Taylor dislocation model [48, 49], which has been widely used to explain the indentation size effect, is summarized in Sect. 2.2. The conventional theory of mechanism-based strain gradient plasticity established from the Taylor dislocation model is reviewed in Sect. 2.3. Sections 2.4 and 2.5 review the nanoindentation models for the sharp, conical indenters and for spherical indenters, respectively.

2.2 Taylor Dislocation Model

The Taylor dislocation model [2, 48, 49] relates the shear flow stress τ to the dislocation density ρ by

$$\tau = \alpha\mu b\sqrt{\rho},\tag{2.2}$$

where μ, b, and α are the shear modulus, (magnitude of) Burgers vector and empirical coefficient around 0.3, respectively; the dislocation density ρ is composed of the densities ρ_S for statistically stored dislocations (SSD) [34] and ρ_G for geometrically necessary dislocations (GND) [1, 8, 34],

$$\rho = \rho_S + \rho_G.\tag{2.3}$$

The GND density ρ_G is related to the curvature of plastic deformation [33, 34], or effective plastic strain gradient η^p, by

$$\rho_G = \bar{r}\frac{\eta^p}{b},\tag{2.4}$$

where $\bar{r}$ is the Nye factor introduced by Arsenlis and Parks to reflect the effect of crystallography on the distribution of GNDs, and $\bar{r}$ is around 1.90 for fcc polycrystals [1, 25–27, 42].

The tensile flow stress σ_{flow} is related to the shear flow stress τ by

$$\sigma_{\text{flow}} = M\tau,\tag{2.5}$$

where M is the Taylor factor, and $M = 3.06$ for fcc metals [4, 6, 7, 24] as well as for bcc metals that slip on $\{110\}$ planes [4]. The substitution of (2.2)–(2.4) into (2.5) yields

$$\sigma_{\text{flow}} = M\alpha\mu b\sqrt{\rho_S + \bar{r}\frac{\eta^p}{b}}.\tag{2.6}$$

For uniaxial tension, the flow stress is related to the plastic strain ε^p by $\sigma_{\text{flow}} = \sigma_{\text{ref}}\, f(\varepsilon^p)$, where σ_{ref} is a reference stress and f is a nondimensional function determined from the uniaxial stress-strain curve. As the plastic strain gradient η^p vanishes in uniaxial tension, the SSD density ρ_S [33] is determined from (2.6) as $\rho_S = [\sigma_{\text{ref}}\, f(\varepsilon^p)/(M\alpha\mu b)]^2$. The flow stress in (2.6) then becomes

$$\sigma_{\text{flow}} = \sqrt{[\sigma_{\text{ref}}\, f(\varepsilon^p)]^2 + M^2\bar{r}\alpha^2\mu^2 b\eta^p} = \sigma_{\text{ref}}\sqrt{f^2(\varepsilon^p) + l\eta^p},\tag{2.7}$$

where

$$l = M^2\bar{r}\alpha^2 \left(\frac{\mu}{\sigma_{\text{ref}}}\right)^2 b = 18\alpha^2 \left(\frac{\mu}{\sigma_{\text{ref}}}\right)^2 b\tag{2.8}$$

is the intrinsic material length in strain gradient plasticity, $M = 3.06$ and $\bar{r} = 1.90$. This intrinsic material length represents a natural combination of elasticity (via the shear modulus, μ), plasticity (via the reference stress σ_{ref}) and atomic spacing (via the Burgers vector b). Even though this intrinsic material length l depends on the choice of the reference stress σ_{ref} in uniaxial tension, the flow stress σ_{flow} in (2.7) *does not* because the strain gradient term inside the square root in (2.7) becomes $\sigma_{\text{ref}}^2 l \eta^p = 18\alpha^2 \mu^2 b \eta^p$, and is independent of σ_{ref}.

2.3 The Conventional Theory of Mechanism-Based Strain Gradient Plasticity

The conventional theory of mechanism-based strain gradient plasticity (CMSG) has been established from the Taylor dislocation model to study the size-dependent material behavior at the micro and submicrometer scale. Following [25–27], Huang et al. rewrote the uniaxial stress-plastic strain relation $\sigma = \sigma_{\text{ref}} f(\varepsilon^p)$ to the following expression [19]

$$\dot{\varepsilon}^p = \dot{\varepsilon} \left[\frac{\sigma}{\sigma_{\text{ref}} f(\varepsilon^p)} \right]^m , \tag{2.9}$$

where $\dot{\varepsilon}^p$ is the rate of plastic strain, $\dot{\varepsilon}$ is the effective strain rate, and m is the rate-sensitivity exponent, which usually takes a large value ($m \geqslant 20$). For $m \to \infty$, (2.9) degenerates to $\sigma = \sigma_{\text{ref}} f(\varepsilon^p)$. For any finite m, the constitutive relation (2.9) is still rate independent as it has strain rates on both sides. In fact, $m \geqslant 20$ gives an excellent representation of the uniaxial stress-strain relation $\sigma = \sigma_{\text{ref}} f(\varepsilon^p)$ [19].

The volumetric strain rate $\dot{\varepsilon}_{kk}$ and deviatoric strain rate $\dot{\varepsilon}'_{ij}$ in CMSG are related to the stress rate in the same way as in classical plasticity, i.e.,

$$\dot{\varepsilon}_{kk} = \frac{\dot{\sigma}_{kk}}{3K} , \tag{2.10}$$

$$\dot{\varepsilon}'_{ij} = \dot{\varepsilon}'^e_{ij} + \dot{\varepsilon}^p_{ij} = \frac{\dot{\sigma}'_{ij}}{2\mu} + \frac{3\dot{\varepsilon}^p}{2\sigma_e} \sigma'_{ij} , \tag{2.11}$$

where K is the elastic bulk modulus, $\dot{\varepsilon}'^e_{ij}$ and $\dot{\varepsilon}^p_{ij}$ are the elastic deviatoric and plastic strain rates, respectively, σ'_{ij} is the deviatoric stress, and $\sigma_e = \left(3\sigma'_{ij}\sigma'_{ij}/2 \right)^{1/2}$ is the effective stress. The effective plastic strain rate $\dot{\varepsilon}^p = \left(2\dot{\varepsilon}^p_{ij}\dot{\varepsilon}^p_{ij}/3 \right)^{1/2}$ is obtained from (2.9) except that the tensile flow stress $\sigma_{\text{ref}} f(\varepsilon^p)$ in the denominator is replaced by the flow stress in (2.7) as established from the Taylor dislocation model accounting for the strain gradient effect, i.e.,

$$\dot{\varepsilon}^p = \dot{\varepsilon} \left(\frac{\sigma_e}{\sigma_{\text{flow}}} \right)^m = \dot{\varepsilon} \left[\frac{\sigma_e}{\sigma_{\text{ref}} \sqrt{f^2(\varepsilon^p) + l\eta^p}} \right]^m , \tag{2.12}$$

where $\dot{\varepsilon} = \left(2\dot{\varepsilon}'_{ij}\dot{\varepsilon}'_{ij}/3 \right)^{1/2}$.

It is noted that $\dot{\varepsilon}$ in (2.12) is equivalent to $\sigma'_{kl}\dot{\varepsilon}_{kl}/\sigma_e$ in uniaxial tension. Equation (2.12) can also be written as

$$\dot{\varepsilon}^p = \frac{\sigma'_{kl}}{\sigma_e}\dot{\varepsilon}_{kl}\left[\frac{\sigma_e}{\sigma_{ref}\sqrt{f^2(\varepsilon^p)+l\eta^p}}\right]^m. \tag{2.13}$$

Equations (2.10)–(2.12) give the stress rate in terms of the strain rate as

$$\dot{\sigma}_{ij} = K\dot{\varepsilon}_{mm}\delta_{ij} + 2\mu\left\{\dot{\varepsilon}'_{ij} - \frac{3\sigma'_{ij}\sigma'_{kl}}{2\sigma_e^2}\left[\frac{\sigma_e}{\sigma_{ref}\sqrt{f^2(\varepsilon^p)+l\eta^p}}\right]^m\dot{\varepsilon}'_{kl}\right\}, \tag{2.14}$$

which can be inverted to write the strain rate in terms of the stress rate as

$$\dot{\varepsilon}_{ij} = \frac{1}{9K}\dot{\sigma}_{mm}\delta_{ij} + \frac{1}{2\mu}\dot{\sigma}'_{ij} + \frac{\sigma'_{ij}}{\sigma_e}\frac{\left[\frac{\sigma_e}{\sigma_{ref}\sqrt{f^2(\varepsilon^p)+l\eta^p}}\right]^m}{2\mu\left\{1-\left[\frac{\sigma_e}{\sigma_{ref}\sqrt{f^2(\varepsilon^p)+l\eta^p}}\right]^m\right\}}\dot{\sigma}_e. \tag{2.15}$$

The constitutive relation (2.14) [or (2.15)] for CMSG is identical to that for classical J_2-flow theory, $\dot{\varepsilon}_{ij} = \frac{1}{9K}\dot{\sigma}_{mm}\delta_{ij} + \frac{1}{2\mu}\dot{\sigma}'_{ij} + \frac{\sigma'_{ij}}{\sigma_e}\frac{3}{2h_p}\dot{\sigma}_e$, except that the incremental plastic modulus $h_p = \sigma_{ref}\,f'(\varepsilon^p)$ in uniaxial tension is replaced by $3\mu\left\{\left[\sigma_{ref}\sqrt{f^2(\varepsilon^p)+l\eta^p}/\sigma_e\right]^m - 1\right\}$. The plastic strain gradient increases the incremental plastic modulus, similar to some of early strain gradient plasticity theories [?, 35].

The effective plastic strain gradient η^p is given by [16–18]

$$\eta^p = \int\dot{\eta}^p dt, \ \dot{\eta}^p = \sqrt{\frac{1}{4}\dot{\eta}^p_{ijk}\dot{\eta}^p_{ijk}}, \ \dot{\eta}^p_{ijk} = \dot{\varepsilon}^p_{ik,j} + \dot{\varepsilon}^p_{jk,i} - \dot{\varepsilon}^p_{ij,k}, \tag{2.16}$$

where $\dot{\varepsilon}^p_{ij}$ is the tensor of the plastic strain rate. The equilibrium equations in CMSG are identical to those in conventional continuum theories. There are no extra boundary conditions beyond those in conventional continuum theories.

2.4 Nanoindentation Model for Sharp, Conical Indenters

2.4.1 Indentation Model

Figure 2.3 shows a schematic diagram of GNDs underneath the indenter [33]. For simplicity, the indenter is assumed to be a cone, and the indentation is accommodated by circular loops of GNDs with Burgers vector normal to the plane of the surface. Such an axisymmetric model neglects crystalline symmetries, and does not account for the indenter pile-up or sink-in either, though such effect is considered in

Fig. 2.3 A schematic diagram
of geometrically necessary
dislocations underneath the
indenter [33]

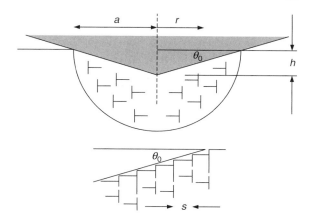

the finite element analysis in Sect. 2.4.2. As the indenter is forced into the surface
of a single crystal, GNDs are required to account for the permanent shape change at
the surface.

Let θ_0 denote the angle between the surfaces of the plane and conical indenter,
a the contact radius and h the indentation depth. It is easy to show that

$$\tan \theta_0 = \frac{h}{a} = \frac{b}{s}, \tag{2.17}$$

where b is the Burgers vector and s is the spacing between individual slip steps
on the indentation surface. The length $d\lambda$ of injected dislocation loops with radii
between r and $r + dr$ is given by [33]

$$d\lambda = 2\pi r \frac{dr}{s} = 2\pi r \frac{h}{ba} dr. \tag{2.18}$$

The total length of injected dislocation loops is obtained by integrating (2.18) as
$\lambda = \pi h a / b$. The average GND density can be obtained from $\rho_G = \lambda / V$, where
$V = 2\pi a^3 / 3$ is the volume of the hemisphere [33]. This gives $\rho_G = 3\tan^2 \theta_0 / 2bh$,
which becomes very large at small indentation depth h (as in nanoindentation).

In reality, the GND density ρ_G cannot be very large because of the strong
repulsive forces between GNDs, which push dislocations to spread beyond the
hemisphere at small indentation depth [15, 45]. Feng and Nix developed a phenomenological model to calculate the enlarged storage volume for GNDs [15]. Huang
et al. introduced a maximum allowable GND density ρ_G^{max} such that ρ_G can never
exceed ρ_G^{max}, i.e.,

$$\rho_G \leqslant \rho_G^{max}, \tag{2.19}$$

where ρ_G^{max} is to be determined from experiments, and as to be shown later it is
on the order of $10^{16} \, \text{m}^{-2}$ [20]. For indentation depth $h \geqslant 3\tan^2 \theta_0 / 2b\rho_G^{max}$, the
average GND density $\rho_G = 3\tan^2 \theta_0 / 2bh$ satisfies (2.19) such that the relation (2.1)
still holds. For small indentation depth $h < 3\tan^2 \theta_0 / 2b\rho_G^{max}$, ρ_G exceeds ρ_G^{max}.

However, if ρ_G were simply replaced by ρ_G^{max}, the resulting indentation hardness in (2.1) would have become a constant $H = H_0 \sqrt{1 + \frac{2bh^* \rho_G^{max}}{3\tan^2 \theta_0}}$ (independent of the indentation depth!), which would not agree with the experimentally observed depth-dependent indentation hardness in Fig. 2.2.

Huang et al. defined the local GND density at radius r as [20]

$$(\rho_G)_{local} = \frac{d\lambda}{dV} = \frac{\tan \theta_0}{br}, \tag{2.20}$$

where $d\lambda$ is the length of injected dislocation loops with radii between r and $r + dr$ in (2.18), and $dV = 2\pi r^2 dr$ is the volume of material between r and $r + dr$. For the maximum allowable GND density criterion (2.19), $(\rho_G)_{local}$ is replaced by ρ_G^{max} for $r < \tan \theta_0 / b\rho_G^{max}$, and (2.20) then becomes

$$(\rho_G)_{local} = \rho_G^{max} \quad \text{if } h < h_{nano}, \tag{2.21a}$$

$$(\rho_G)_{local} = \begin{cases} \rho_G^{max} & \text{for } r < \dfrac{h_{nano}}{\tan \theta_0} \\[2mm] \dfrac{\tan \theta_0}{br} & \text{for } \dfrac{h_{nano}}{\tan \theta_0} \leqslant r \leqslant a = \dfrac{h}{\tan \theta_0} \end{cases} \quad \text{if } h \geqslant h_{nano}, \tag{2.21b}$$

where

$$h_{nano} = \frac{\tan^2 \theta_0}{b\rho_G^{max}} \tag{2.22}$$

is a new characteristic length for nanoindentation. It is 51 nm for $\rho_G^{max} = 10^{16} \, \text{m}^{-2}$ (and $b = 0.25 \, \text{nm}$ and $\theta_0 = 19.7°$ [33]), and is much smaller than the characteristic length h^* in (2.1) for microindentation, which is typically on the order of micrometers [33].

Strictly speaking, a constant local GND density (in (2.21a) and the first line of (2.21b)) is not consistent with [33] GND distribution underneath a "sharp" indenter shown in Fig. 2.3. However, the constant local GND density in (2.21) only holds for small radius $r < h_{nano} / \tan \theta_0$, which is on the order of 100 nm. For such small r, the indenter tip is never sharp because the tip radius is also around 100 nm (or less).

The average GND density ρ_G is obtained by averaging (2.21) over the hemisphere, i.e., $\rho_G = \frac{1}{V} \int_V (\rho_G)_{local} dv$, where $V = \frac{2}{3}\pi a^3$ is the volume of the hemisphere. This gives

$$\rho_G = \rho_G^{max} \begin{cases} 1 & \text{if } h < h_{nano} \\[2mm] \dfrac{3h_{nano}}{2h} - \dfrac{h_{nano}^3}{2h^3} & \text{if } h \geqslant h_{nano} \end{cases}. \tag{2.23}$$

Figure 2.4 shows the normalized GND density, ρ_G / ρ_G^{max}, vs. the normalized reciprocal of indentation depth, h_{nano} / h. The relation (2.1) is also shown by the straight line with the slope 3/2. For large indentation depth (i.e., small h_{nano}/h), the present result indeed approaches (2.1). The GND density deviates from (2.1) as the indentation depth decreases, and eventually reaches a constant of ρ_G^{max} when the indentation depth is less than h_{nano}.

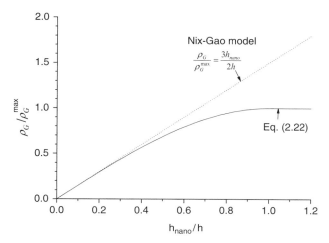

Fig. 2.4 The normalized density of geometrically necessary dislocations (GND), $\rho_G/\rho_G^{\max}$, vs. the normalized reciprocal of indentation depth, h_{nano}/h, where $\rho_G^{\max}$ is the maximum allowable GND density, and h_{nano} is a characteristic length in nanoindentation given in (2.22)

The volume dV in (2.20) is for the *spherical shell* of radius r and thickness dr. This is not completely consistent with the dislocation loops of radius r which are all within the *cylindrical shell* (of radius r and thickness dr). Huang et al. calculated the dislocation density on the basis of the cylindrical shell model and found that, even though the local GND densities in (2.20) and (2.21) change, the average GND density in (2.23) remains the same [20]. In other words, the choice of spherical or cylindrical shells does not affect the average GND density based on which the indentation hardness is calculated. In fact, the finite element analysis in Sect. 2.4.2 involves neither spherical nor cylindrical shells.

The indentation hardness H is related to the tensile flow stress σ_{flow} by the Tabor factor of between 2.8 and 3 [47]; also ([33, 40]),

$$H = 3\sigma_{\text{flow}}, \qquad (2.24)$$

where σ_{flow} is given in (2.5). The average plastic strain in indentation is related to the indenter shape ($\tan\theta_0$) instead of indentation depth h [33] such that the term $\sigma_{\text{ref}}\, f(\varepsilon^{\text{p}})$ in (2.5) gives the classical indentation hardness H_0 (for very large indentation depth) as in (2.1). The indentation hardness is then found from (2.24), (2.5), and (2.23) as

$$\left(\frac{H}{H_0}\right)^2 = 1 + \begin{cases} \dfrac{2h_{\text{micro}}}{3h_{\text{nano}}} & \text{if} \quad h < h_{\text{nano}} \\[2ex] \dfrac{h_{\text{micro}}}{h} - \dfrac{h_{\text{micro}}h_{\text{nano}}^2}{3h^3} & \text{if} \quad h \geqslant h_{\text{nano}} \end{cases}, \qquad (2.25)$$

where

$$h_{\text{micro}} = \frac{27M^2}{2} b\alpha^2 \tan^2 \theta_0 \left(\frac{\mu}{H_0}\right)^2 \tag{2.26}$$

is the characteristic length for microindentation, and is in general on the order of
micrometers. It degenerates to h^* in (2.1) for isotropic materials with von Mises
yield criterion ($M = \sqrt{3}$). The indentation hardness in (2.25) involves two lengths,
namely h_{micro} on the order of micrometers and h_{nano} on the order of 50 nm. For
indentation depth around h_{micro} (and therefore much larger than h_{nano}), (2.25)
degenerates to (2.1), $\left(\frac{H}{H_0}\right)^2 = 1 + \frac{h_{\text{micro}}}{h}$, which depends linearly on $1/h$. For indentation depth between h_{nano} and h_{micro}, (2.25) becomes a nonlinear function
$\left(\frac{H}{H_0}\right)^2 = 1 + \frac{h_{\text{micro}}}{h} - \frac{h_{\text{micro}}h_{\text{nano}}^2}{3h^3}$. For small indentation depth $h < h_{\text{nano}}$, the indentation hardness becomes a constant $\left(\frac{H}{H_0}\right)^2 = 1 + \frac{2h_{\text{micro}}}{3h_{\text{nano}}}$.

Figure 2.5 shows the square of indentation hardness, H^2, vs. the reciprocal of
indentation depth, $1/h$, for iridium. The experimental data [45] are also shown. The
classical indentation hardness (for very large indentation depth) is $H_0 = 2.52\,\text{GPa}$
determined from the intercept with the vertical axis. The characteristic length for microindentation is $h_{\text{micro}} = 2.42\,\mu\text{m}$ determined from the slope of the straight segment
at small $1/h$, which corresponds to the empirical coefficient in the Taylor disloca-
tion $\alpha = 0.273$ for iridium ($M = 3.06$, $b = 0.271\,\text{nm}$ and $\mu = 217\,\text{GPa}$, see [40,45])
and $\theta_0 = 19.7°$. The coefficient α has the correct order of magnitude (around 0.3).
The characteristic length for nanoindentation is $h_{\text{nano}} = 286\,\text{nm}$, which is indeed
much smaller than h_{micro}, and corresponds to the maximum allowable GND density
$\rho_G^{\text{max}} = 1.66 \times 10^{15}\,\text{m}^{-2}$ for iridium ($b = 0.271\,\text{nm}$) and $\theta_0 = 19.7°$. Equation (2.25)

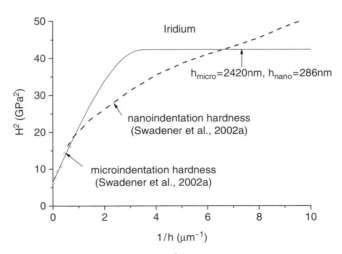

Fig. 2.5 The square of indentation hardness, H^2, vs. the reciprocal of indentation depth, $1/h$,
given by the analytic relation of (2.25) with the maximum allowable GND density $\rho_G^{\text{max}} = 1.66 \times 10^{15}\,\text{m}^{-2}$ for iridium. The experimental data [45] are also shown

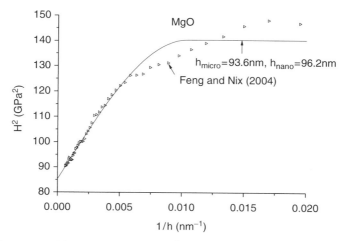

Fig. 2.6 The square of indentation hardness, H^2, vs. the reciprocal of indentation depth, $1/h$, given by the analytic relation of (2.25) with the maximum allowable GND density $\rho_G^{max} = 4.48 \times 10^{15}\,\text{m}^{-2}$ for MgO. The experimental data [15] are also shown

agrees very well with microindentation hardness data, and captures the trend in nanoindentation but has some discrepancies.

It is important to note that the characteristic length h_{micro} for microindentation is inversely proportional to the square of classical indentation hardness H_0^2, and therefore becomes very small for ultra hard materials such as MgO. We show in Fig. 2.6 the square of indentation hardness, H^2, vs. the reciprocal of indentation depth, $1/h$, for MgO that has the classical indentation hardness $H_0 = 9.23\,\text{GPa}$. The experimental data [15] are also shown. The characteristic length for microindentation is indeed very small, $h_{micro} = 93.6\,\text{nm}$, determined from the slope of straight segment at small $1/h$, which corresponds to the empirical coefficient in the Taylor dislocation $\alpha = 0.322$ for MgO ($M = 3.06$, $b = 0.298\,\text{nm}$ and $\mu = 126\,\text{GPa}$, see [15]) and $\theta_0 = 19.7°$, and once again has the correct order of magnitude (around 0.3). The characteristic length for nanoindentation is $h_{nano} = 96.2\,\text{nm}$, which is about the same as h_{micro}. This corresponds to the maximum allowable GND density $\rho_G^{max} = 4.48 \times 10^{15}\,\text{m}^{-2}$ for MgO ($b = 0.298\,\text{nm}$) and $\theta_0 = 19.7°$. Equation (2.25) agrees very well with microindentation hardness data, and captures the trend in nanoindentation but has little discrepancies. This explains why the indentation hardness does not continue to rise at small depths of indentation.

2.4.2 Finite Element Analysis for Nanoindentation

In this section, we use the finite element method for CMSG to examine the effect of maximum allowable GND density in micro and nanoindentation. Unlike [33] and Sect. 2.4.1, it is not necessary to assume GND distributions underneath the indenter in the finite element analysis, or the storage volume for GNDs.

On the basis of the maximum allowable GND density ρ_G^{max}, the flow stress in (2.7) becomes

$$
\sigma_{flow} = \begin{cases} \sigma_{ref}\sqrt{f^2(\varepsilon^p) + l\eta^p} & \text{if } \eta^p < \dfrac{b\rho_G^{max}}{\bar{r}} \\[2ex] \sigma_{ref}\sqrt{f^2(\varepsilon^p) + \dfrac{l}{\bar{r}}b\rho_G^{max}} & \text{if } \eta^p \geq \dfrac{b\rho_G^{max}}{\bar{r}} \end{cases}, \tag{2.27}
$$

which imposes a limit on the flow stress increase due to GNDs. The maximum allowable GND density ρ_G^{max} is to be determined from nanoindentation experiments.

A sharp, conical indenter with the included angle of 140.6° is studied. Such an included angle gives the same contact area $A = 24.5\,h^2$ as the Berkovich indenter at the same indentation depth h [31]. The indentation hardness is defined as the mean pressure exerted by the indenter at the maximum load. Indentation can be represented by the contact model between the rigid indenter and the indented material in the finite element analysis. The finite sliding, hard contact model in the finite element program of ABAQUS is used, which allows the sliding between two contact surfaces but no interpenetration. The normal and shear tractions are continuous within the contact zone, and they vanish outside the contact zone. The effect of indenter pile-up or sink-in is automatically accounted for in the contact model. For the friction coefficient less than or equal to 0.2, there are essentially no differences between the indentation hardness predicted by the frictional and frictionless contact models. Only the results for frictionless contact are presented.

The relation between flow stress σ_{flow} and plastic strain ε^p in uniaxial tension can be generally expressed via the power law as

$$
\sigma_{flow} = \sigma_{ref}f(\varepsilon^p) = \sigma_Y\left(1 + \frac{E\varepsilon^p}{\sigma_Y}\right)^N, \tag{2.28}
$$

where E is the Young's modulus, σ_Y is the (initial) yield stress, and N (<1) is the plastic work hardening exponent. For vanishing plastic strain $\varepsilon^p \to 0$, the flow stress approaches the yield stress σ_Y. For large plastic strain, (2.28) gives a power law relation between the flow stress and plastic strain. The reference stress in (2.28) can be taken as $\sigma_{ref} = \sigma_Y(E/\sigma_Y)^N$, and the function f then becomes $f(\varepsilon^p) = (\sigma_Y/E + \varepsilon^p)^N$.

2.4.2.1 Nanoindentation Hardness for MgO

For MgO, Young's modulus $E = 297\,$GPa, Poisson's ratio $\nu = 0.177$, shear modulus $\mu = 126\,$GPa [15], Burgers vector $b = 0.298\,$nm, and the plastic work hardening exponent $N = 0.1$. The (initial) yield stress is $\sigma_Y = 2.82\,$GPa such that the classical indentation hardness (for very large indentation depth) agrees with the experimental value $H_0 = 9.24\,$GPa [15]. This (initial) yield stress is slightly less than the flow stress obtained from the Tabor relation $\sigma_{flow} = H_0/3 = 3.08\,$GPa.

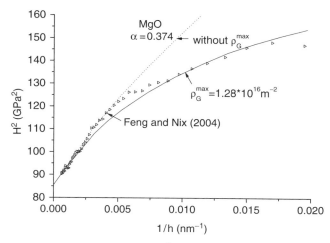

Fig. 2.7 The square of indentation hardness, H^2, vs. the reciprocal of indentation depth, $1/h$, given by the finite element analysis for MgO. The maximum allowable GND density is $\rho_G^{max} = 1.28 \times 10^{16}\,\mathrm{m}^{-2}$. The experimental data [15] and the finite element results without accounting for ρ_G^{max} are also shown

The empirical coefficient in the Taylor dislocation model is determined from the slope of the $H^2 \sim 1/h$ curve for large indentation depth $h > 300\,\mathrm{nm}$ (i.e., microindentation) as $\alpha = 0.374$, which is slightly larger than 0.322 reported in Sect. 2.4.1 but still has the correct order of magnitude (around 0.3).

Figure 2.7 shows the square of indentation hardness H^2 vs. the reciprocal of indentation depth $1/h$ given by the finite element analysis for MgO. The experimental data of [15] are also shown. The maximum allowable GND density is $\rho_G^{max} = 1.28 \times 10^{16}\,\mathrm{m}^{-2}$, which is a little less than three times of that in the dislocation model in Sect. 2.4.1. The results based on ρ_G^{max} agree well with the experimental data. For comparison Fig. 2.7 also shows the results without accounting for ρ_G^{max}, which agree well with the microindentation hardness data (at small $1/h$), but not with the nanoindentation hardness data (at large $1/h$). Therefore, the strain gradient plasticity theory based on Taylor dislocation model accounting for the maximum allowable GND density can capture the indentation size effect in micro and nanoindentation.

2.4.2.2 Nanoindentation Hardness for Iridium

For iridium, the Young's modulus $E = 540\,\mathrm{GPa}$, Poisson's ratio $\nu = 0.246$, shear modulus $\mu = 217\,\mathrm{GPa}$ [45], and Burgers vector $b = 0.271\,\mathrm{nm}$. Swadener et al. [45] reported the stress-strain curve of iridium, which can be approximated by the power law hardening at small strain and linear hardening at relatively large strain

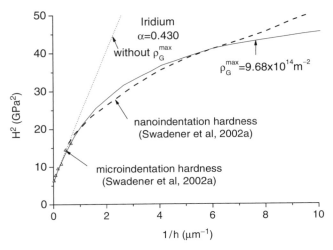

Fig. 2.8 The square of indentation hardness, H^2, vs. the reciprocal of indentation depth, $1/h$, given by the finite element analysis for iridium. The maximum allowable GND density is $\rho_G^{max} = 0.968 \times 10^{15}\,\mathrm{m}^{-2}$. The experimental data [45] and the finite element results without accounting for ρ_G^{max} are also shown

$$\sigma = \begin{cases} \left[121 + 2575\,(\varepsilon^p)^{0.638}\right]\ \mathrm{MPa} & \text{for}\quad \varepsilon^p < 0.02 \\[2mm] (200 + 6671\varepsilon^p)\qquad \mathrm{MPa} & \text{for}\quad \varepsilon^p \geqslant 0.02 \end{cases} \tag{2.29}$$

On the basis of the above relation, the classical indentation hardness (for very large indentation depth) is found as $H_0 = 2.38\,\mathrm{GPa}$, which agrees reasonably well with experiments [45]. The empirical coefficient in the Taylor dislocation model is determined from the slope of the $H^2 \sim 1/h$ curve for large indentation depth $h > 2\,\mu\mathrm{m}$ (i.e., microindentation) as $\alpha = 0.430$, which is larger than 0.273 reported in Sect. 2.4.1 but still has the correct order of magnitude (around 0.3).

Figure 2.8 shows the square of indentation hardness H^2 vs. the reciprocal of indentation depth $1/h$ given by the finite element analysis for iridium. The experimental data of [45] are also shown. The maximum allowable GND density is $\rho_G^{max} = 0.968 \times 10^{15}\,\mathrm{m}^{-2}$, which is more than one half of that in the dislocation model in Sect. 2.4.1. Once again the results based on ρ_G^{max} agree well with the experimental data, while those without accounting for ρ_G^{max} agree only with the microindentation and not the nanoindentation hardness data.

2.4.2.3 Distribution of GND Density

The finite element analysis in this section has better agreement with the experimental data than the simple analytic model in Sect. 2.4.1. This is partially because the analytic model assumes a hemisphere for the storage volume of GNDs, while

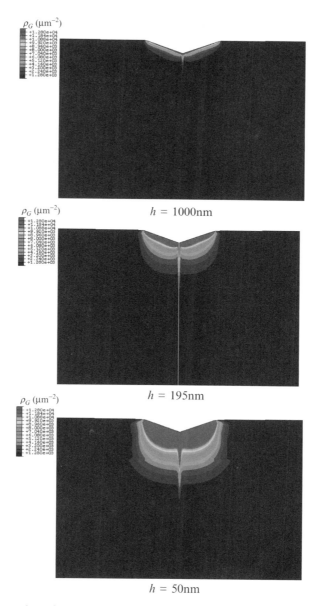

Fig. 2.9 Contour plots of geometrically necessary dislocation density in a region of $18h \times 12h$ underneath the indenter in MgO, where the indentation depth $h = 1,000$, 195, and 50 nm. The maximum allowable density of geometrically necessary dislocations is $\rho_G^{max} = 1.28 \times 10^{16}\,\text{m}^{-2}$. The contour values range from ρ_G^{max} to its one tenth, $\rho_G^{max}/10$

the finite element analysis avoids such an assumption. Figure 2.9 shows the contour plots of GND density ρ_G in MgO, where ρ_G is obtained by $\bar{r}\eta^p/b$ (2.27) from the finite element analysis or from the maximum allowable GND density

$\rho_G^{max} = 1.28 \times 10^{16}\,\text{m}^{-2}$, whichever is smaller. The contour values range from ρ_G^{max} to its one tenth, $\rho_G^{max}/10$. The indentation depths are $h = 1,000\,\text{nm}$ (microindentation), 195 nm, and 50 nm (nanoindentation). Each contour is shown for a region of $18h \times 12h$ underneath the indenter, which scales with the indentation depth h. For $h = 1,000\,\text{nm}$, the GND density is less than ρ_G^{max} except in a very narrow layer immediately underneath the indenter. The GND density contours are different from hemispheres. For $h = 195\,\text{nm}$ and $h = 50\,\text{nm}$, the GND density reaches ρ_G^{max} in relatively large regions. The GND density contours are somewhat like hemispheres. This implies that the assumption of a hemisphere storage volume for GNDs in the analytic model cannot be accurate for both micro and nanoindentation.

2.5 Nanoindentation Model for Spherical Indenter

The inset of Fig. 2.10 shows a schematic diagram of a spherical indenter with the indenter radius R. The contact radius of indentation is a. As the indenter is forced into the surface of the material, geometrically necessary dislocations (GNDs) are required to account for the permanent shape change at the surface. Figure 2.10 shows the indentation hardness vs. the ratio a/R for iridium [45, 46], which clearly displays the size effect in spherical indentation, i.e., the smaller the indenter radius, the larger the indentation hardness.

In a manner similar to Nix and Gao [33], Swadener et al. calculated the total length of GND loops in spherical indentation as $2\pi a^3/(3bR)$, where b is the Burgers vector [45, 46]. If all GNDs are confined in a hemisphere with the radius equal to the contact radius a [33, 45, 46], the average density of GND ρ_G underneath the indenter is

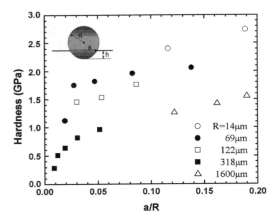

Fig. 2.10 The indentation hardness data of iridium vs. the ratio of contact radius a to the indenter radius R. The inset shows a *schematic diagram* of a spherical indenter with the indenter radius R

$$\rho_G = \bar{r} \frac{2\pi a^3/(3bR)}{2\pi a^3/3} = \frac{\bar{r}}{bR}, \tag{2.30}$$

where $\bar{r}$ is the Nye factor introduced by Arsenlis and Parks to reflect the effect of crystallography on the distribution of GNDs, and $\bar{r}$ is around 1.90 for fcc polycrystals [1, 42]. Similar expressions of ρ_G have also been obtained by [35] and [50], which all show that the average density of GNDs underneath a spherical indenter is independent of the contact radius a, and is inversely proportional to the indenter radius R.

For small indenter radius R, the GND density in (2.30) becomes large. In reality, the GND density ρ_G cannot be very large because of the strong repulsive forces between GNDs, which push dislocations to spread beyond the hemisphere at small indenter tip radius [15, 20, 45, 46]. Huang et al. introduced a maximum allowable GND density ρ_G^{max} such that ρ_G can never exceed ρ_G^{max}, i.e., (2.19) [20]. For indenter radius $R \geqslant \bar{r}/b\rho_G^{max}$, the average GND density $\rho_G = \bar{r}/bR$ satisfies (2.19) such that the relation (2.30) still holds. For small indenter radius $R < \bar{r}/b\rho_G^{max}$, ρ_G is replaced by ρ_G^{max}, and (2.30) then becomes [21]

$$\rho_G = \begin{cases} \rho_G^{max} & \text{for} \quad R < \dfrac{\bar{r}}{b\rho_G^{max}} \\[2ex] \dfrac{\bar{r}}{bR} & \text{for} \quad R \geqslant \dfrac{\bar{r}}{b\rho_G^{max}} \end{cases}. \tag{2.31}$$

These two cases are called the nano-indentation $(R < \bar{r}/b\rho_G^{max})$ and micro-indentation $(R \geqslant \bar{r}/b\rho_G^{max})$, respectively, for spherical indenters.

The average effective plastic strain underneath a spherical indenter of radius R can be estimated as [22, 45, 46, 50]

$$\varepsilon^p \approx 0.2 \frac{a}{R} = \frac{a}{5R}, \tag{2.32}$$

where a is the contact radius. The average flow stress underneath a spherical indenter can then be obtained as

$$\sigma_{flow} = \begin{cases} \sigma_{ref}\sqrt{f^2\left(\dfrac{a}{5R}\right) + \dfrac{lb\rho_G^{max}}{\bar{r}}} & \text{for} \quad R < \dfrac{\bar{r}}{b\rho_G^{max}} \\[3ex] \sigma_{ref}\sqrt{f^2\left(\dfrac{a}{5R}\right) + \dfrac{l}{R}} & \text{for} \quad R \geqslant \dfrac{\bar{r}}{b\rho_G^{max}} \end{cases}, \tag{2.33}$$

where l is the intrinsic material length in (2.8).

The indentation hardness is then related to the flow stress via the Tabor relation [47],

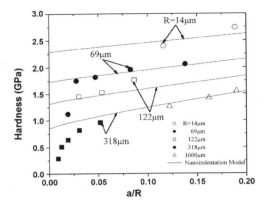

Fig. 2.11 The spherical indentation hardness H vs. the radius ratio a/R given by the analytic model in (2.34) (*solid line*). The experimental results are also shown for comparison ($R = 14\,\mu$m, *open circle*; $R = 69\,\mu$m, *filled circle*; $R = 122\,\mu$m, *open square*; $R = 318\,\mu$m, *filled square*; $R = 1{,}600\,\mu$m, *open triangle*) [45, 46]. The material properties of iridium are Young's modulus $E = 540\,$GPa, Poisson's ratio $\nu = 0.246$, Burgers vector $b = 0.271\,$nm, uniaxial stress-plastic strain relation $\sigma = 108 + 2850\,(\varepsilon^{\mathrm{p}})^{0.638}$ MPa, coefficient in the Taylor dislocation model of (2.2) $\alpha = 1/3$, and maximum allowable GND density $\rho_{\mathrm{G}}^{\max} = 1.8 \times 10^{14}\,\mathrm{m}^{-2}$

$$
H = 2.8\sigma_{\mathrm{flow}} =
\begin{cases}
2.8\sigma_{\mathrm{ref}}\sqrt{f^2\left(\dfrac{a}{5R}\right) + \dfrac{lb\rho_{\mathrm{G}}^{\max}}{\bar{r}}} & \text{for} \quad R < \dfrac{\bar{r}}{b\rho_{\mathrm{G}}^{\max}} \\[3mm]
2.8\sigma_{\mathrm{ref}}\sqrt{f^2\left(\dfrac{a}{5R}\right) + \dfrac{l}{R}} & \text{for} \quad R \geqslant \dfrac{\bar{r}}{b\rho_{\mathrm{G}}^{\max}}
\end{cases}
. \qquad (2.34)
$$

This provides a very simple way to estimate the effect of indenter radius R in spherical indentation.

Figure 2.11 compares the spherical indentation hardness in (2.34) with [45, 46] experimental data for iridium. The coefficient α in the Taylor dislocation model of (2.2) is fixed at $\alpha = 1/3$, which has the correct order of magnitude (around 0.3). The radii of spherical indenters are $R = 14, 69, 122, 318$ and $1{,}600\,\mu$m. (The curve for $R = 1{,}600\,\mu$m is very close to that for $318\,\mu$m and is therefore not presented here.) For all indenter radii except the smallest one $R = 14\,\mu$m, the microindentation hardness $H = 2.8\sigma_{\mathrm{ref}}\sqrt{f^2\,(a/5R) + l/R}$ for spherical indenters agrees well with the experimental data. The discrepancy at small a/R is due to the neglect of elasticity effect in (2.34). For very small a/R, the deformation underneath the indenter is elastic, i.e., the elastic Hertzian contact between the indenter and iridium. The average contact pressure in elastic contact is proportional to the contact radius a, and therefore approaches zero as $a \to 0$. As the hardness H is defined as the average contact pressure in the present study, H also approaches zero as $a \to 0$. As the indentation depth increases, the deformation underneath the indenter is elastic-plastic, but the effect of elasticity is still significant, which gives experimentally measured hardness smaller than that given by (2.34) at small a/R.

For the smallest indenter $R = 14\,\mu\mathrm{m}$, the microindentation hardness in (2.34) would significantly overestimate the indentation hardness. Instead we use the nanoindentation hardness $H = 2.8\sigma_{\mathrm{ref}}\sqrt{f^2\left(a/5R\right) + lb\rho_G^{\mathrm{max}}/\bar{r}}$ for spherical indenters, which agrees reasonably well with the experimental hardness data for the smallest indenter $R = 14\,\mu\mathrm{m}$ with the maximum allowable GND density $\rho_G^{\mathrm{max}} = 1.8 \times 10^{14}\,\mathrm{m}^{-2}$. This ρ_G^{max} is smaller than $\rho_G^{\mathrm{max}} = 0.968 \times 10^{15}\,\mathrm{m}^{-2}$ in Sect. 2.4 for iridium with sharp, conical indenters, which suggests that the maximum allowable GND density depends not only on materials but also on the indenter tip geometry (conical vs. spherical indenters). The corresponding critical radius separating the nano and microindentation is $39\,\mu\mathrm{m}$ for iridium. Therefore, (2.34) can be used to estimate the indentation hardness with spherical indenters for both nano and microindentation.

2.6 Concluding Remarks

(1) Analytic models are established for nanoindentation hardness with sharp, conical indenters, and spherical indenters on the basis of the maximum allowable GND density. The models give simple relations between indentation hardness and depth/contact radius. For the sharp, conical indenters, the model degenerates to [33] at relatively large indentation depth as in microindentation. For spherical indenters, the model degenerates to [41] for relatively large indenters as in microindentation. The models agree reasonably well with both micro and nanoindentation hardness data of MgO and iridium.

(2) The finite element analysis for the strain gradient plasticity theory based on the Taylor dislocation model and the maximum allowable GND density agrees very well with the experimental data.

(3) Without accounting for the maximum allowable GND density, the indenter tip radius effect alone cannot explain the nanoindentation size effect.

It is important to note that the present nanoindentation model has a lower limit because, once the indentation depth reaches the order of nanometers, discrete dislocation events dominate and the indentation hardness results are scattered. The present model should not be used in this regime.

Acknowledgments Y.H. acknowledges the support from NSF (grant CMS-0084980) and ONR (grant N00014-01-1-0205, program officer Dr. Y.D.S. Rajapakse). The support from NSFC is also acknowledged. Y.H and WDN acknowledge the support from an NSF-NIRT project "Mechanism Based Modeling and Simulation in Nanomechanics," through grant No. NSF CMS-0103257, under the direction of Dr. Ken Chong. Research at the ORNL SHaRE User Facility (GMP) was sponsored by the Division of Materials Sciences and Engineering, U.S. Department of Energy, under contract DE-AC05-00OR22725 with UT-Battelle, LLC.

References

1. Arsenlis A, Parks DM (1999) Acta Mater. 47:1597
2. Ashby MF (1970) Philos. Mag. 21:399
3. Atkinson M (1995) J. Mater. Res. 10:2908
4. Bailey JE, Hirsch PB (1960) Philos. Mag. 5:485
5. Begley MR, Hutchinson JW (1998) J. Mech. Phys. Solids 46:2049
6. Bishop JFW, Hill R (1951) Philos. Mag. 42:414
7. Bishop JFW, Hill R (1951) Philos. Mag. 42:1298
8. Cottrell AH (1964) The Mechanical Properties of Materials. Wiley, New York
9. De Guzman MS, Neubauer G, Flinn P, Nix WD (1993) Mater. Res. Symp. Proc. 308:613
10. Durst K, Backes B, Goken M (2005) Scr. Mater. 52:1093
11. Elmustafa AA, Stone DS (2002) Acta Mater. 50:3641
12. Elmustafa AA, Stone DS (2003) J. Mech. Phys. Solids 51:357
13. Elmustafa AA, Ananda AA, Elmahboub WM (2004) J. Mater. Res. 19:768
14. Elmustafa AA, Ananda AA, Elmahboub WM (2004) J. Eng. Mater. Technol. 126:353
15. Feng G, Nix WD (2004) Scr. Mater. 51:599
16. Gao H, Huang Y, Nix WD, Hutchinson JW (1999) J. Mech. Phys. Solids 47:1239
17. Huang Y, Gao H, Nix WD, Hutchinson JW (2000) J. Mech. Phys. Solids 48:99
18. Huang Y, Xue Z, Gao H, Nix WD, Xia ZC (2000) J. Mater. Res. 15:1786
19. Huang Y, Qu S, Hwang KC, Li M, Gao H (2004) Int. J. Plast. 20:753
20. Huang Y, Zhang F, Hwang KC, Nix WD, Pharr GM, Feng G (2006) J. Mech. Phys. Solids 54:1668
21. Huang Y, Feng X, Pharr GM, Hwang KC (2007) Model. Simul. Mater. Sci. Eng. 15:255
22. Johnson KL, (1970) J. Mech. Phys. Solids 18:115
23. Kim JY, Lee BW, Read DT, Kwon D (2005) Scr. Mater. 52:353
24. Kocks UF (1970) Metall. Mater. Trans. 1:1121
25. Kok S, Beaudoin AJ, Tortorelli DA (2002) Int. J. Plast. 18:715
26. Kok S, Beaudoin AJ, Tortorelli DA (2002) Acta Mater. 50:1653
27. Kok S, Beaudoin AJ, Tortorelli DA, Lebyodkin M (2002) Model. Simul. Mater. Sci. Eng. 10:745
28. Lim YY, Chaudhri MM (1999) Philos. Mag. A 79:2979
29. Liu Y, Ngan AHW (2001) Scr. Mater. 44:237
30. Ma Q, Clarke DR (1995) J. Mater. Res. 10:853
31. McElhaney KW, Vlasssak JJ, Nix WD (1998) J. Mater. Res. 13:1300
32. Nix WD (1989) Metall. Trans. A 20A:2217
33. Nix WD, Gao H (1998) J. Mech. Phys. Solids 46:411
34. Nye JF (1953) Acta Metall. Mater. 1:153
35. Tymiak NI, Kramer DE, Bahr DF, Wyrobek TJ, Gerberich WW (2001) Acta Mater. 49:1021
36. Oliver WC, Pharr GM (1992) J. Mater. Res. 7:1564
37. Poole WJ, Ashby MF, Fleck NA (1996) Scr. Mater. 34:559
38. Qiu X, Huang Y, Nix WD, Hwang KC, Gao H (2001) Acta Mater. 49:3949
39. Qu S, Huang Y, Nix WD, Jiang H, Zhang F, Hwang KC (2004) J. Mater. Res. 19:3423
40. Qu S, Siegmund T, Huang Y, Wu PD, Zhang F, Hwang KC, (2005) Comp. Sci. Tech. 65:1244
41. Qu S, Huang Y, Pharr GM, Hwang KC (2006) Int. J. Plast. 22:1265
42. Shi MX, Huang Y, Gao H (2004) Int. J. Plast. 20:1739
43. Stelmashenko NA, Walls AG, Brown LM, Milman YV (1993) Acta Metall. Mater. 41:2855
44. Suresh S, Nieh TG, Choi BW (1999) Scr. Mater. 41:951
45. Swadener JG, George EP, Pharr GM (2002) J. Mech. Phys. Solids 50:681
46. Swadener JG, Misra A, Hoagland RG, Nastasi A (2002) Scr. Mater. 47:343
47. Tabor D (1951) The Hardness of Metal. Clarendon, Oxford, p. 44

48. Taylor GI (1934) Proc. R. Soc. Lond. A 145:362
49. Taylor GI (1938) J. Inst. Met. 62:307
50. Xue Z, Huang Y, Hwang KC, Li M (2002) J. Eng. Mater. Technol. 124:371
51. Zagrebelny AV, Lilleodden ET, Gerberich WW, Carter BC (1999) J. Am. Ceram. Soc. 82:1803
52. Zhang TY, Xu WH, Zhao MH (2004) Acta Mater. 52:57

Chapter 3
Indentation in Shape Memory Alloys

Yang-Tse Cheng and David S. Grummon

3.1 Introduction

Research on shape memory alloys (SMAs) has been broadly active since the discovery of shape memory in the compound NiTi in 1963, a decade after first reports of the effect in Au–Cd. For general reviews, see [1–4]. Early work on NiTi-based SMAs (primarily NiTi, and NiTiX, where X = Pt, Pd, Au, Cu, Hf, Zr, or Nb, and others) led to applications such as the NiTi hydraulic tube couplings developed by the Raychem Corporation. Today, a wide variety of new ideas have emerged [1–5] for applications such as sensors, actuators, damping materials, MEMS, biomedical devices, and hydro/aerodynamic control at surfaces. A noticeable resurgence of interest in SMAs has occurred, largely in response to recent advances in alloy preparation techniques (including physical vapor deposition routes), machining and joining technologies, and modeling capabilities.

It is well known that NiTi alloys can exhibit either the shape memory effect (SME) or the superelastic effect (SE, often called pseudoelasticity). These are, respectively, isobaric and isothermal forms of shape memory, and a single alloy may exhibit either, depending on stress and temperature. Real behavior often combines aspects of each effect in a complex hysteretic fashion.

Both shape memory and superelasticity depend on martensitic transformations, which, for NiTi, involve athermal first-order displacive transformations from a CsCl parent to a monoclinic, orthorhombic, or rhombohedral low-temperature allotrope, sometimes in succession. The heats of transformation may exceed $20\,\mathrm{J\ g^{-1}}$ and the

Y.-T. Cheng
Materials and Processes Laboratory, General Motors R&D Center, Warren, Michigan 48090, USA
Yangtcheng@aol.com

D.S. Grummon
Department of Chemical Engineering and Materials Science, Michigan State University, East Lansing, Michigan 48823, USA
grummon@egr.msu.edu

F. Yang and J.C.M. Li (eds.), *Micro and Nano Mechanical Testing of Materials and Devices*,   71
doi: 10.1007/978-0-387-78701-5, © Springer Science+Business Media, LLC, 2008

reactions may have large temperature and stress hysteresis. The maximum practical strain energy storage density in NiTi ($>10^6$ MJ m^{-3}) [6] is extremely high.

Displacive transformations are found in many metals and ceramics, but robust shape memory can usually only be obtained when:

(a) The parent phase is a high-symmetry ordered compound.
(b) The transformational volume change is negligible.
(c) Twin boundaries in the martensite are glissile (i.e., mobile under the influence of shear stresses).

Practical SMAs must usually also possess classical metallic toughness, and good strength against conventional slip plasticity. Furthermore, the martensite phase must be thermally stable below M_f. It is not surprising then that only a select few alloy systems have found wide application. With regard to the above criteria, the NiTi system is exceptionally robust.

Figure 3.1a schematically illustrates the nature of the martensite microstructure that forms on cooling the austenite, "A," below the M_s temperature. The austenite-

Fig. 3.1 (**a**) Self-accommodated formation of monoclinic martensite from CsCl austenite in NiTi. *Top*: cooperative formation of lattice correspondence variants L_1 and L_2 results in the appearance of the habit-plane variant H. This internal twinning gives coherency to the habit plane and results in a highly mobile $M \parallel A$ boundary. At a slightly larger scale (*middle*), multiple habit-plane variants assemble cooperatively via "self-accommodation." In the idealized case, the net shear strain is zero. (**b**) *Bottom*: external stress can drive shear deformation of the martensite phase by intervariant boundary motion ("detwinning"). It is this shear that can recover during the M-to-A transformation (the martensite may have the approximate appearance of the 5% sheared triangle if it forms in the presence of external stress, as is the case in transformational superelasticity)

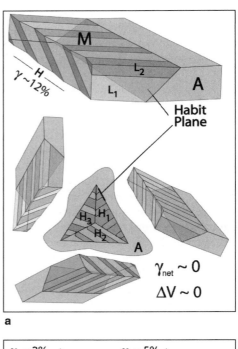

a

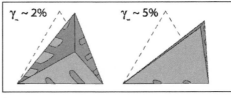

b

to-martensite transformation – in the absence of stress – produces no macroscopic shape change, because the large transformational shear strain is locally accommodated by formation of multiple crystal orientation variants of the martensite. This is possible because of the symmetry change: any one of 12 differently oriented monoclinic martensite domains ("lattice correspondence variants," L) can displacively arise from a given orientation of the cubic austenite grain. It is observed that, at the nano- and microscales, transformation proceeds by the formation of twin-related pairs of these orientation variants (labeled L_1 and L_2 in the top sketch Fig. 3.1a), which nucleate cooperatively to produce a single, coherent *habit-plane variant* (labeled H). It is further observed that multiples of distinct habit-plane variants (labeled H_1, H_2, and H_3) – of which there are 24 possibilities – evolve cooperatively, such that the *net* shear strain is near zero. The resulting microstructure is generally complex and heavily twinned at a very fine scale, and can frequently contain martensite–austenite phase mixtures.

Shape memory is possible because of the mobility of twin boundaries in the martensite. When shear stresses favor one martensite variant orientation over another, twinning reactions (sometimes called "detwinning") can produce a large (but not unlimited) macroscopic plastic strain (see Fig. 3.1b). Under stress, some variants shrink as others grow; but since each must later revert to the same original austenite orientation (if heated), the apparent plastic deformation of the martensite can be reversed as the austenite is "recovered." The "remembered shape" is that which is "set" in the austenite. The cubic phase defines the reference condition and it is only deformation of the martensite that can be thermally reversed, at least against nontrivial reaction stresses.

A second form of shape memory occurs under isothermal conditions at temperatures above the austenite finish (A_f). When the austenite is stressed at temperatures just above A_f, martensite can form spontaneously because shear stress effectively increases the martensite start (M_s) temperature. When martensite variants form under stress, macroscopic shear occurs because the (shear) stress favors certain variants over others, and the resulting microstructure can be envisioned as being similar to that of mechanically deformed martensite suggested in Fig. 3.1b (right). But when the stress is removed, the phase equilibrium temperatures drop back down and the martensite *isothermally* reverts to austenite. Shape recovery occurs much as in the case of thermal shape memory, except that the process occurs on falling stress at constant temperature. Very large amounts of strain energy can be stored and released via this mechanism and superelastic objects are highly robust against damage and distortion.

It should be noted that, if the martensite is strained further than can be accommodated by the detwinning process illustrated in Fig. 3.1b (or if stress on an austenitic superelastic alloy is high enough), conventional slip plasticity – and consequent dislocation production – may occur that will inhibit complete return to an original shape. The maximum recoverable strain for practical NiTi is 4–8% in tension.

Returning to the first form of shape memory, thermal SME, our description so far describes a "one-way" shape recovery on heating. If the return to set shape is mechanically restrained, very large forces can be developed, as is done in the case

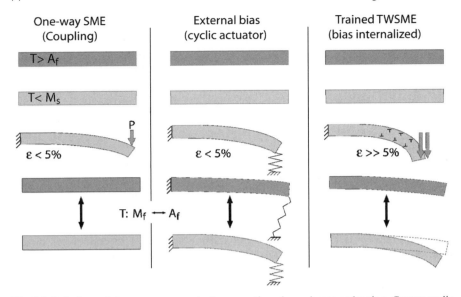

Fig. 3.2 *Left*: thermal shape memory producing a one-time shape change on heating. *Center*: cyclic actuators require some external agency (force) to cause the martensite to deform on cooling. *Right*: the "biasing force" for actuator resetting can be developed by internal defects created by loading the martensite into the dislocation slip regime

of the Raychem tube connector. As shown schematically in Fig. 3.2, when SMAs are to be used as *cyclic* actuators, however, some kind of force-producing element, such as a spring or pendant mass, must be mechanically linked so as to deform the martensite phase during the cooling part of the actuation cycle. When the actuator is warm, the austenite appears "hard" to the biasing force and responds as a stiff elastic element that remains contracted along the bias axis. When cooling occurs in the presence of the bias force (think of the pendant mass), the resultant martensite looks plastically *soft* to the bias stress, and plasticity occurs up to the strain limit for transformational shear (i.e., 4–8%). In the pendant mass example, cyclic heating and cooling alternately raises and lowers the hanging weight, whereas the spring shown in the sketch stretches and contracts. To reiterate, in the absence of the bias forces, no shape change is expected on cooling – there is no strong memory of previous martensite shapes.

SMAs may, however, be specially processed to possess some degree of memory of *both* the shapes of the parent and martensitic phases, which gives a two-way SME or TWSME (see Fig. 3.2, right-hand sketch). TWSME is usually obtained via thermomechanical "training" that involves heating and cooling under controlled load, displacement, and temperature conditions, often using stresses that will cause generation of dislocations. Two-way SME is believed to depend on nucleation of preferred martensite variants under the influence of stress fields around dislocations introduced during the training process. Though two-way alloys provide an internalized "reset" mechanism for actuators, significant work can only be extracted during the heating part of the cycle. Also, tighter limits on strain recovery and thermal stability apply when TWSME is required.

Although there have been extensive studies of bulk SME and SE behavior using macroscopic tensile, shear, and compression loading, until recently, few studies have been devoted to the microscopic behavior of these alloys under contact loading conditions. To help better understand shape memory and superelastic effects at small scales and under complex loading conditions, we and several other groups have studied SMAs using micro- and nanoindentation techniques [7–32]. Unlike in simple tension, compression, or shear experiments, the stress distribution under an indenter is highly nonuniform and strains can easily exceed the maximum that can be recovered by SMEs.

The shape of the indenter has a strong effect on both the magnitude and distribution of these stresses [33–35]. If one ignores natural tip imperfection, sharp pyramids (i.e., Vickers and Berkovich indenters) do not possess a characteristic length scale, so the indentation response in SMAs should be independent of indentation depth. Spherical indenters do have a natural length scale, the ball radius R, and indentation response is thus expected to have some form of depth dependence. Pyramids do too if the displacements are on the order of the (imperfect) tip radius.

Below, we provide a brief overview of recent work on indentation in SMAs. Together, they demonstrate the robust operation of shape memory and superelastic mechanisms at both small length scales and under complex (highly anisotropic and nonuniform) loading conditions. Some of these results are suggestive of several new kinds of applications for SMAs.

3.2 Indentation-Induced Shape Memory Effect

Figure 3.3a shows a 3D profile of a spherical indent made in martensitic NiTi. The initial indent profile is shown together with the altered profile after heating past A_f [10]. The data show that nearly perfect recovery of a spherical indenter impression is

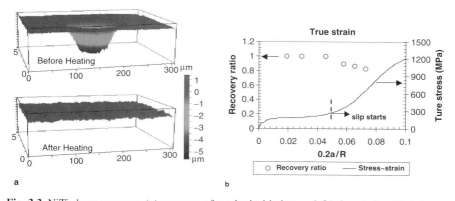

Fig. 3.3 NiTi shape memory: (**a**) recovery of a spherical indent and (**b**) the relationship between thermally activated recovery ratio δ and representative strain $\varepsilon_r = 0.2a/R$ and the relationship between true stress and true tensile strain measured for the same alloy

possible by the SME. Further experiments showed, however, that if deeper spherical indents were made, recovery would not be complete, as discussed further below.

The magnitude of "indent recovery" (a normalized change in the depth of an indent corresponding to a temperature change) may be quantified by defining a thermally induced recovery ratio:

$$\delta = (h_f - h'_h)/h_f. \tag{3.1}$$

Here, h_f is the residual indentation depth recorded immediately after unloading and h'_h is the final indentation depth after the completion of thermal recovery. Both h_f and h'_h for the present experiment were measured at room temperature, which was below M_f, under the assumption that no significant displacement occurs when cooling from above A_f.

Experiments using increasing indentation load showed that, for spherical indentation, δ depends on both the indenter radius and the indentation depth in a way that can be rationalized with the help of the concept of representative strain ε_r [33]. This is defined as

$$\varepsilon_r = 0.2a/R, \tag{3.2}$$

where a and R are the in-plane contact radius and the radius of the indenter, respectively. Figure 3.3b shows the relationship between the thermally induced recovery ratio and ε_r (open circles), with the tensile stress–strain relationship obtained for the same NiTi martensite [16] superimposed as a solid line. It shows that δ is about unity (perfect recovery) until a critical strain is reached. This strain coincides with the end of the stress plateau in the stress–strain curve, which is further associated with the maximum recoverable strain for the martensitic NiTi alloy used in the experiment.

Recovery has also been observed for Berkovich and Vickers indents made in SMAs. Figure 3.4 shows the cross-section profiles of two Vickers indent on a martensitic NiTi [10]. The thermally activated recovery ratio for the Berkovich and Vickers indents is plotted as a function of maximum indenter displacement in

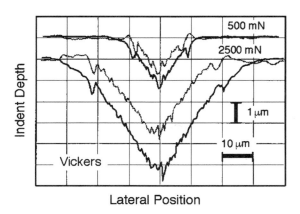

Fig. 3.4 Cross sections of surface profiles of Vickers indents on a NiTi shape memory alloy before and after heating for two different indent loads

Fig. 3.5 Relationship between recovery ratio and initial indentation depth for sharp indents in a NiTi shape memory alloy

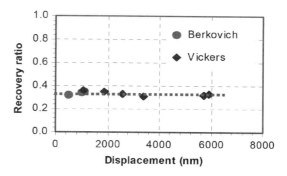

Fig. 3.5 [16]. The recovery ratio of both Berkovich and Vickers indents in martensitic NiTi is about 0.34 and it is nearly depth independent from about 100 nm up to 6 μm, setting an upper limit on length scale effects in NiTi shape memory of ~100 nm.

Other groups have also observed indentation-induced SMEs. For example, Shaw et al. [12] observed the recovery of Berkovich indents in NiTi thin films on heating, showing that shape memory persists down to the 10-nm scale. They proposed a modified spherical cavity model to rationalize the magnitude of recovery of these indents. Indent recovery has also been observed in shape memory CuAlNi single crystal samples by Liu et al. [17, 18], using a nanoindenter and an atomic force microscope with in situ heating capabilities. The authors observed a much higher recovery ratio (0.7–0.8) when the indentation depth was shallow. In addition to observing partial depth recovery, Huang et al. [22] observed a change in the surface profile from sinking-in (for the initial indent) to piling-up for the recovered profile.

It has consistently been reported by various groups [12, 22, 26] that Berkovich indents made to depths less than 10 nm recover completely on heating and that the normalized degree of recovery begins to increase for indents less than about 100-nm deep. This observation has generally been attributed to tip imperfections, which some investigators have measured. Few Berkovich indenters have tip radii sharper than 50 nm. Thus, Berkovich indents at the nanoscale are expected to behave like spherical indents.

Indentation can also induce a two-way SME. This was first observed in a NiTi SMA sample after it had been subjected to a thermomechanical training using fixed displacement [24]. A 3.175-mm diameter tungsten carbide ball was forced into the sample and held in place for subsequent thermal cycling. The fixtured specimen was then heated and cooled for 30 cycles from above A_f to below M_f. The 3D image of the indent (Fig. 3.6a) was then captured at high and low temperature. Figure 3.6b shows the depth of the indent as it experienced several heating–cooling cycles.

In these plots, D_A is the depth at high temperature when the sample was in the austenitic phase. D_M is the depth at low temperature when the sample was in the martensitic phase. After the first heating step, the indent depth decreased from the original indent depth of about 170 to 80 μm – an approximately 47% decrease indent depth recovery. After cooling the sample to obtain the martensite, the indent

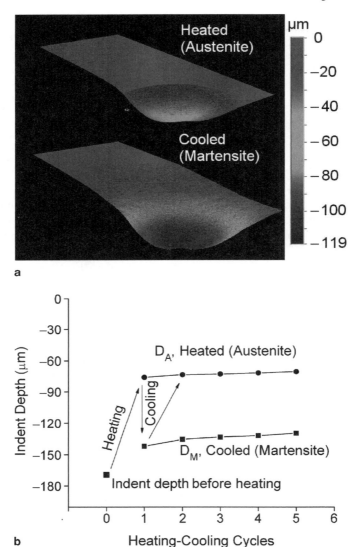

a

b

Fig. 3.6 3D profile of a heated and cooled indent made by thermal cycling of a fixed displacement indenter: (**a**) first-cycle change in profile and (**b**) two-way indent depth changes over the first few thermal cycles

depth increased from about 80 to 140 μm – an approximately 75% increase in the depth of indent. Subsequent heating–cooling cycles produced nearly constant indent depth ratio, $(D_M - D_A)/D_M$, of about 45%. In summary, as shown in Fig. 3.6, we have observed two-way depth recovery for spherical indentations in a NiTi alloy, using a simple constrained-indenter thermomechanical training method.

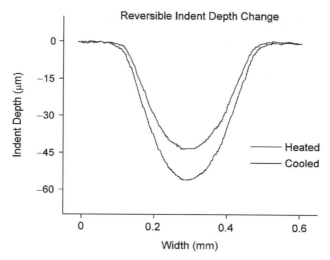

Fig. 3.7 Cross-sectional profiles of reversible indent depth change for a single-indent training regime (no thermal cycling during indentation). The indent in the austenitic phase (*heated*) is shallower than in the martensitic phase (*cooled*)

While this training method was effective in inducing a two-way SME, we have also found that significant two-way SME can be obtained in the martensitic NiTi via a single isothermal indentation without additional thermal training cycles [25]. A reversible depth change of about 24% of the total indent depth was observed (see Fig. 3.7). Though this is smaller than the recovery produced by thermally cycling a fixed indenter, it is significant that robust two-way shape memory can be induced by single-indent events. Recently, Su et al. [29] reported that Berkovich indentations could also impart two-way SME.

An interesting variation on the indentation-induced two-way shape memory can be observed if postindentation planarization is used. We have recently found that reversible surface *protrusions* can be made by mechanically polishing down to the base of deep spherical indents ("planarizing") while in the as-indented martensitic condition. Figure 3.8 shows a 3×3 matrix of circular protrusions after this indented-and-planarized martensitic NiTi sample was heated to the austenitic phase [25]. Remarkably, these protrusions disappeared when the sample was cooled back to the martensite, and the cycle could be repeated many times. Indents and protrusions can also be made at the nanoscale using the same method on sputtered thin films [25].

Figure 3.9 shows one proposed mechanism for the observed two-way indent recovery and reversible surface protrusions [25]. The creation of dislocations in a zone immediately under an indent is believed to be necessary for the observation of two-way reversibility. Indeed, the two-way effect is found to disappear if the plastically deformed zone is removed by polishing deeper below the base of the indent [32]. At present, however, the role of dense dislocation structures produced by deep spherical indentation is yet to be verified. It is also possible that elastic strain

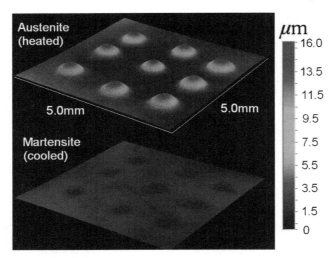

Fig. 3.8 3D profiles of a 3 × 3 matrix of circular protrusions on the surface of NiTi that can be made to cyclically appear and disappear when warmed and cooled, respectively

energy storage in a much larger volume of material is the driver of the effect. In this view, the role of dislocation activity near the surface is to create local strain gradients that give rise to extensive residual stress fields following the withdrawal of the indenter load.

These experiments show that indentation is not only a useful tool for studying microscopic shape memory behavior, but also a means to create reversible surface features with many potential applications.

3.3 Indentation in Superelastic Austenite

Recently, several groups have demonstrated microscopic superelastic effects using instrumented micro- and nanoindentation techniques [7, 8, 11, 13, 14]. For example, using Berkovich and spherical diamond indenters, we obtained load–displacement curves for both an austenitic NiTi alloy at $T > A_f$ and pure copper (Figs. 3.10 and 3.11) [11]. The significant differences between the load–displacement curves in NiTi and Cu lie in the magnitude of final indentation depth h_f relative to the maximum indentation depth h_m, and the recoverable work W_e relative to the total work W_t. The total work W_t required to move an indenter into a solid is the area under the loading curve. The reversible work W_e is the area under the unloading curve. The area between the loading and unloading curves is the dissipated energy. The energy recovery ratio η_w may be defined as the ratio of reversible work to total work W_e/W_t. The depth recovery ratio η_h may be defined as $\eta_h = (h_m - h_f)/h_m$.

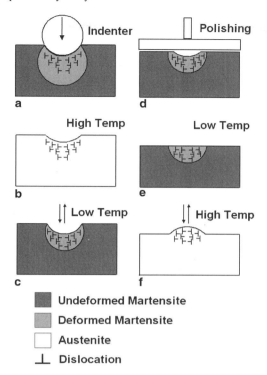

Fig. 3.9 Experimental procedures and mechanisms for achieving two-way reversible indentations and protrusions. (**a**) Indentation is accommodated both by detwinning deformation of the martensite and by slip plasticity with defect production. (**b**) When heated to recover the austenite, the indent displacement is partially recovered. (**c**) When cooled below M_f, the residual indent becomes deeper and thermal indent depth changes are reversible on heating and cooling. The dislocation structure and its associated stress field may facilitate the growth of oriented martensite variants during austenite-to-martensite transformation, leading to two-way indent depth change. Alternatively, longer-range residual stress fields arising from indent plasticity gradients may drive shape changes. (**d**) The residual indents are removed by planar grinding while in the martensitic condition. (**e**) The oriented martensite region, near-surface dislocations, and long-range residual stress fields are still present after the planarization step. (**f**) A surface protrusion is formed upon heating which is found to be perfectly reversible

For Berkovich indentation in the austenitic NiTi, η_w and η_h are about 45%, which is significantly larger than about 8% for copper for indentation depths between 250 and 2,000 nm. For elastic–plastic solids such as copper, the reversible and irreversible work result from elastic and plastic deformation, respectively. It has been shown that η_w is proportional to the ratio of the hardness to the reduced Young's modulus H/E^*. Since H/E^* is less than 2% for most metals, η_w is less than 10% [35]. In contrast, a much greater fraction of the total work of indentation is seen to be reversible for austenitic NiTi. We conclude that a very high degree of superelastic recovery is possible under Berkovich nanoindenter loading of austenitic NiTi SMAs.

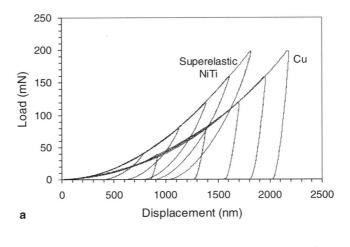

a

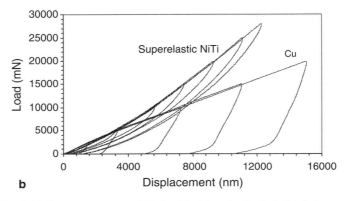

b

Fig. 3.10 Load–displacement curves of Berkovich (**a**) and spherical (**b**) indents made in an austenitic NiTi alloy and Cu

The indentation-induced superelastic effect is significantly more pronounced for spherical indentation conditions than that for pyramidal indentation. For spherical indentation, η_h has a weak depth dependence but declines as a function of the representative strain for strains above 0.05. This can be rationalized by comparing the dependence of recovery ratio on representative strain with the tensile true stress–true strain curve of the same austenitic alloy, as is shown in Fig. 3.11 [16]. Depth recovery is nearly complete when the representative strain is less than 0.05, corresponding closely with the end of the stress–strain plateau. This is in turn linked to "exhaustion" of the capability for stress-induced martensite formation. Incomplete indent reversibility increases with straining past 5% as slip dislocation activity ramps up.

The large difference in superelastic recovery between Berkovich and spherical indenters is inevitable given the differences in expected indentation stress

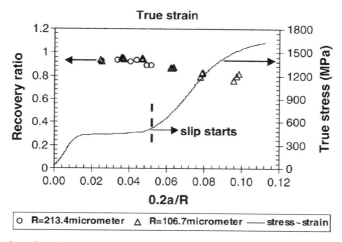

Fig. 3.11 Indentation depth recovery ratio as a function of the representative strain for spherical indenters (*open symbols*), superimposed on the true stress–true strain curve for the same austenitic NiTi material

distribution between the two. The stress at the apex of a perfectly sharp pyramidal or conical indenter is singular, so plastic deformation begins immediately with contact. A large volume of material directly below pyramidal indenters therefore becomes highly strained, and significant deformation occurs by dislocation motion since indentation-induced transformational strain mechanisms are quickly exhausted. This volume does not subsequently contribute to superelastic transformational recovery and in fact probably inhibits it by forming elastic back stresses.

In contrast, the maximum stress under a spherical indenter remains finite and is moderate for low a/R ratios. In austenitic NiTi, the smaller maximum strain caused by the stress distribution under the sphere can evidently be accommodated by stress-induced martensitic formation, allowing for more complete superelastic recovery on unloading.

Indentation-induced superelastic effects have also been demonstrated using temperature-controlled micro- and nanoindentation techniques [20, 28]. Here, the indentation response of a SMA above A_f, where SE response is expected, was compared with that below M_f. In [20], we used a temperature-controlled Berkovich indentation instrument to show that indentation-induced superelastic effects exist at temperatures up to $A_f + 100\,\mathrm{K}$, higher than is usually observed in uniaxial tension or compression tests in bulk and thin film NiTi. A model that considers the high hydrostatic pressure under the Berkovich indenter has been proposed to explain the observed elevated temperature limit.

Recently, Wood and Clyne [28] used a variable temperature indenter to examine superelastic behavior in NiTi. Their spherical indentation results clearly demonstrated superelastic behavior when the sample is austenitic, but not in the martensitic state, as shown in Fig. 3.12 [28].

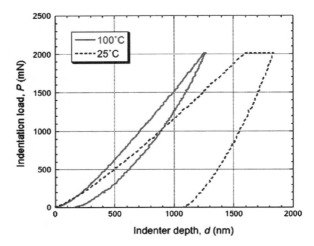

Fig. 3.12 Load–displacement curves obtained during indentation carried out at temperatures, such that the starting material would primarily be either in the parent phase (100 °C) or in the martensitic state (25 °C), using a large spherical ($R = 650\,\mu$m) indenter tip. The dwell period at maximum load was 120 s

3.4 Summary and Outlook

We have reviewed some recent work on indentation in SMAs that establishes the microscopic shape memory and superelastic effects, both under complex loading conditions and over a wide range of depths ranging from 10 nm to hundreds of micrometers. Specifically:

1. Indentation can induce both one- and two-way SMEs when an indentation is made in the martensitic phase of NiTi. The magnitude of this indentation-induced SME is depth independent for sharp Berkovich and Vickers indenters, unless tip blunting is present. With spherical indenters, indentation-induced one-way shape memory is almost perfect when the initial indent produces a representative strain less than 0.05. This is close to the maximum recoverable strain measured for the same SMA under uniaxial tension. One-way shape memory recovery decreases with increasing representative strain beyond 0.05.

2. When loading spherical indenters to reach representative strains beyond 0.05, one-way shape memory declines, but a potentially useful two-way indentation-induced SME begins to appear. This "trained-indent" can drive thermally reversible displacements that are still an appreciable fraction of the initial indent depth. Furthermore, by subsequently applying a precision surface-grinding step to remove visible trace of the indent, it is possible to form reversible *protrusions* on the surface. The method is easily extended to arrays, or alternate indenter forms such as punches and coining dies, or to impact methods.

3. Indentation can induce a superelastic effect when indents are made in the austenitic phase of a SMA. The magnitude of this indentation-induced

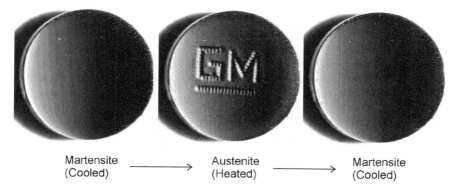

Martensite (Cooled) ——————> Austenite (Heated) ——————> Martensite (Cooled)

Fig. 3.13 This temperature-controlled reversible sign shows how images can appear on a surface simply by heating

superelastic effect appears to be depth independent for sharp indenters if tip rounding is negligible. Under spherical indenters, the magnitude of the indentation-induced superelastic effect is almost complete when the representative strain is less than the maximum recoverable strain of the SMA under tension ($\sim$5%), and decreases with increasing representative strain above this limit.

These effects may be exploited for a number of applications. For example, the one-way SME has been studied for tribological applications, where surface damage in the form of indents and scratches can be healed upon heating [36]. The one-way SME has also been studied for high-density data storage, where bits of information can be stored and erased as recoverable indents [19]. The two-way SME can also be used for information storage, as demonstrated in Fig. 3.13.

The microscopic superelastic effect can also be expressed by sputtered thin films of NiTi. It has, for example, recently been suggested that interlayers of vapor-deposited NiTi can function as a "metallic adhesive" for improving the bonding between hard coatings and ductile substrates [37]. The superelastic effect can also be used for reducing friction [38] and for improving resistance to both wear [37–39] and cavitation erosion [40].

We hope this chapter will stimulate future work on indentation in SMAs, especially in areas such as fundamental understanding of the mechanisms, numerical modeling [28,31,32,41], and novel applications of indentation-induced shape memory and superelastic effects.

Acknowledgments We would like to thank former Ph.D. students, Drs. Wangyang Ni and Yijun Zhang, for their contributions to some of the work reviewed in this chapter. We would also like to thank the U.S. National Science Foundation for partial support of this work under SGER Contract No. CMS0336810 and GOALI Contract No. CMS0510294.

References

1. Otsuka K, Wayman CM (1998) Shape Memory Alloys. Cambridge University Press, Cambridge
2. Duerig T, Melton KN, Stockel D, Wayman CM (1990) Engineering Aspect of Shape Memory Alloys. Butterworth-heinemann, Boston
3. Otsuka K, Kakeshita T (2002) MRS Bull. 27:91
4. Wayman CM (1993) MRS Bull. 18:49
5. Rice C (2002) In: Schwartz M (ed) Encyclopedia of Smart Materials. Wiley, New York, pp. 921–936
6. Wolf RH, Heuer AH (1995) J. Microelectromech. Syst. 4:206
7. Liu R, Li DY, Xie YS, Llewellyn R, Hawthorne HM (1999) Scr. Mater. 41:691
8. Cheng FT, Shi P, Man HC (2001) Scr. Mater. 45:1083
9. Gall K, Juntunen K, Maier HJ, Sehitoglu H, Chumlyakov YI (2001) Acta Mater. 49:3205
10. Ni WY, Cheng YT, Grummon DS (2002) Appl. Phys. Lett. 80:3310
11. Ni WY, Cheng YT, Grummon DS (2003) Appl. Phys. Lett. 82:2811
12. Shaw GA, Stone DS, Johnson AD, Ellis AB, Crone WC (2003) Appl. Phys. Lett. 83:257
13. Ma XG, Komvopoulos K (2003) Appl. Phys. Lett. 83:3773
14. Ma XG, Komvopoulos K (2004) Appl. Phys. Lett. 84:4274
15. Qian LM, Xiao XD, Sun QP, Yu TX (2004) Appl. Phys. Lett. 84:1076
16. Ni WY, Cheng YT, Grummon DS (2004) Surf. Coat. Technol. 177:512
17. Liu C, Zhao YP, Sun QP, Yu TX, Cao ZX (2005) J. Mater. Sci. 40:1501
18. Liu C, Zhao YP, Yu TX (2005) Mater. Des. 26:465
19. Shaw GA, Trethewey JS, Johnson AD, Drugan WJ, Crone WC (2005) Adv. Mater. 17:1123
20. Zhang YJ, Cheng YT, Grummon DS (2005) J. Appl. Phys. 98:033505
21. Frick CP, Ortega AM, Tyber J, Maksound AEM, Maier HJ, Liu YN, Gall K (2005) Mater. Sci. Eng. A 405:34
22. Huang WM, Su JF, Hong MH, Yang B (2005) Scr. Mater. 53:1055
23. Komvopoulos K, Ma XG (2005) Appl. Phys. Lett. 87:263108
24. Zhang YJ, Cheng YT, Grummon DS (2006) Appl. Phys. Lett. 88:131904
25. Zhang YJ, Cheng YT, Grummon DS (2006) Appl. Phys. Lett. 89:041912
26. Frick CP, Lang TW, Spark K, Gall K (2006) Acta Mater. 54:2223
27. Zhang HS, Komvopoulos K (2006) J. Mater. Sci. 41:5021
28. Wood AJM, Clyne TW (2006) Acta Mater. 54:5607
29. Su JF, Huang WM, Hong MH (2007) Smart Mater. Struct. 16:S137
30. Crone WC, Brock H, Creuziger A (2007) Exp. Mech. 47:133
31. Zhang YJ, Cheng YT, Grummon DS (2007) J. Appl. Phys. 101:053507
32. Zhang YJ, Cheng YT, Grummon DS (2007) J. Mater. Res. 22:2851
33. Johnson KL (1987) Contact Mechanics. Cambridge University Press, Cambridge
34. Fischer-Cripps AC (2004) Nanoindentation, 2nd edn. Springer, New York
35. Cheng YT, Cheng CM (2004) Mater. Sci. Eng. R Rep. 44:91
36. Ni WY, Cheng YT, Grummon DS (2006) Surf. Coat. Technol. 201:1053
37. Zhang YJ, Cheng YT, Grummon DS (2006) Mater. Sci. Eng. A 438:710
38. Ni WY, Cheng YT, Lukitsch MJ, Weiner AM, Lev LC, Grummon DS (2005) Wear 259:842
39. Ni WY, Cheng YT, Lukitsch MJ, Weiner AM, Lev LC, Grummon DS (2004) Appl. Phys. Lett. 85:4028
40. Cheng FT, Shi P, Man HC (2004) Mater. Charact. 52:129
41. Yan WY, Sun QP, Feng XQ, Qian LM (2007) Int. J. Solids Struct. 44:1

Chapter 4
Adhesive Contact of Solid Surfaces

Fuqian Yang

4.1 Introduction

The rapid progress in micro- and nanofabrication techniques over the last several decades has led to the development of various miniature electromechanical systems, such as microvibromotors [1], MEMS RF switches [2, 3], MEMS optical tweezers [4, 5], and MEMS mirrors [6, 7]. These have attracted great interest in using movable micromechanical structures in various MEMS and NEMS devices. The performance and lifetime of MEMS and NEMS devices are limited by the material confinement in submicron structures with large surface area, high electric field, large stress gradient, and other factors normally nonexistent in macroscale structures. On the microscale, material behavior is more controlled by surface-driving effects than by bulk effects. For example, frictional effects play a much more important role than inertia in the control of rotary or linear micromotors; i.e., the start-up conditions strongly depend on static friction and moving/rotational speed on kinetic friction. If there exists electric field, field-assisted contact is present.

One of major concerns for long-term applications of MEMS and NEMS structures is mechanical failure, including surface contact, fracture and fatigue. Surface contact of micromechanical structures can contribute to stiction (adhesive contact), damage, pitting, and surface hardening over local contact area. The presence of stiction is due to high compliance and large surface-to-volume ratio of micromechanical structures, which do not have enough restoring force to overcome the surface interaction after mechanical contact. Stiction can be classified into two categories: release-related stiction and in-service stiction [8]. Release-related stiction appears during the removal of sacrificial layers for the release of micromechanical structures, which basically is due to liquid bridges and is controlled by the Laplace pressure and the surface tension of liquid etchants. Practical solutions to the problem

F. Yang
Department of Chemical and Materials Engineering, University of Kentucky, Lexington, KY 40506
fyang0@engr.uky.edu

F. Yang and J.C.M. Li (eds.), *Micro and Nano Mechanical Testing of Materials and Devices*,
doi: 10.1007/978-0-387-78701-5, © Springer Science+Business Media, LLC, 2008

of release-related stiction include use of dimples and cavity to reduce the contact area [9–11], change of meniscus shape [12], avoidance of liquid–vapor interfaces through supercritical fluid [13], freeze-sublimation drying [14], and drying release methods. However, all of these methods could not prevent the stiction from occurring during the operation of microdevices. When the separation between two solid surfaces reaches interatomic equilibrium distance, chemical bonding (ionic, covalent, or metallic) occurs and attractive force increases profoundly [15]. As attractive force is larger than restoring force, solid surfaces permanently stick to each other causing failure of microdevices – in-service stiction.

Several models have been developed on the adhesive contact between spherical particles or between a spherical particle and a semi-infinite substrate, including Bradley model [16], JKR (Johnson–Kendall–Roberts) theory [17], and DMT (Derjaguin–Müller–Toporov) theory [18]. In the analysis, Bradley [16] calculated the interaction force between two rigid particles using a simple integration of molecular interaction without justification of the force additivity. Assuming that the surface interaction creates a uniform displacement over the contact zone and there is no surface interaction outside the contact zone, Johnson et al. [17] used the Hertz contact theory [19] to obtain the pull-off force required to separate two elastic spheres in frictionless contact. Recently, Li [20] also obtained the same relation by using the Moutier theorem. Derjaguin et al. [18] considered the effect of surface interaction outside the contact zone on the contact deformation of two elastic spheres. They obtained the pull-off force the same as the interaction force given by Bradley [16].

The pull-off force F_a to separate two elastic spherical particles in frictionless contact can be expressed as

$$F_a = \chi \pi R_{\text{eff}} \gamma, \tag{4.1}$$

where $\chi = 3/2$ for the JKR theory and $\chi = 2$ for the DMT theory. Here, $R_{\text{eff}} = R_1 R_2/(R_1 + R_2)$ with R_1 and R_2 being, respectively, the radius of the two spheres. γ is the work of adhesion, which is determined by the difference between the energy per unit area of the contacting surfaces, γ_1 and γ_2, before contact and that of the interface, γ_{12}, after contact ($\gamma = \gamma_1 + \gamma_2 - \gamma_{12}$).

The difference between the JKR theory and the DMT theory was later addressed by Tabor [21], who introduced the Tabor number of μ: the ratio of the maximum neck height in contact zone to intermolecular spacing. The DMT theory provides a good description of adhesive contact for small μ, while the JKR theory prevails for large μ. Such a behavior was also revealed by Müller et al. [22] using numerical calculation and by Maugis [23] using the theory of the Dugdale crack [24].

The development of microfabrication techniques in the last several decades has made use of compliant structures and submicron surface coatings in MEMS and NEMS devices. As the ratio of the contact size to the film thickness increases, the assumption of semi-infinite space becomes invalid and the substrate effect needs to be taken into account. This has limited the use of the JKR theory and the DMT theory in probing the adhesive contact of micromechanical structures and has imposed a tremendous challenge in understanding submicron adhesive behavior and in improving the mechanical design with a least possibility of adhesion.

This chapter covers some basic physics related to adhesive contact of thin films, MEMS structures, and adhesion of thin films to substrates. It summarizes the recent development in studying the mechanical and interfacial behavior of materials.

4.2 Adhesion of a Rigid Axisymmetric Particle to an Elastic Film

A condition for the separation of two elastic solids in contact can be derived from thermodynamics. The total energy U_T of a system consists of the interfacial energy U_S over the contact zone, the stored strain energy U_E in the deformed system, and the potential energy U_P of the system during the contact/separation process.

The interfacial energy is

$$U_S = -A\gamma, \tag{4.2}$$

where A is the contact area. The stored strain energy can be calculated as [17,25–27]

$$U_E = \int_0^{\delta_A} F\,\mathrm{d}\delta - \int_{\delta_A}^{\delta} F|_A\,\mathrm{d}\delta, \tag{4.3}$$

where F is the normal contact force and δ is the relative displacement between the solids. The first term on the right-hand side of (4.3) represents the strain energy stored in the system for the contact area increasing to A without surface interaction, and the second term is the strain energy released due to the effect of the surface interaction when maintaining the same contact area of A. The potential energy of the system is

$$U_P = -F\delta_A. \tag{4.4}$$

The total energy of the system is thus

$$U_T = U_S + U_E + U_P. \tag{4.5}$$

Based on reversible thermodynamics, the stable equilibrium contact between a particle and a film requires [28]

$$\left.\frac{\partial U_T}{\partial \delta_A}\right|_F \geqslant 0 \quad \text{or} \quad \left.\frac{\partial U_E}{\partial \delta_A}\right|_F + \left.\frac{\partial U_P}{\partial \delta_A}\right|_F \geqslant \gamma \left.\frac{\partial A}{\partial \delta_A}\right|_F \equiv -F_a. \tag{4.6}$$

In (4.6), A is considered as a variable, indicating that the contact area is similar during the separation. If this is not the case, the second part of (4.6) is invalid. Equation (4.6) provides the rational to evaluate the adhesion force (the pull-off force) for the separation of a rigid particle from an elastic film. It is worth mentioning that the above discussion is based on elastic deformation, i.e., the energy dissipation is negligible during the contact and the separation. If there is dissipation of finite energy, one needs to incorporate the energy dissipation in calculating the total system energy and use irreversible thermodynamics, which will not be discussed in this work.

4.2.1 Adhesive Contact of an Incompressible Elastic Thin Film

Consider the contact between an axisymmetric particle and an incompressible elastic film deposited on a rigid substrate. The contact between the particle and the film is frictionless. For the contact radius of a much larger than the film thickness of h, the relation between the contact radius and the normal contact force without surface interaction is [29]

$$F = \frac{8\pi\mu}{h} \int_0^a f(r)r \, dr \tag{4.7}$$

for the frictionless contact between the film and the substrate, and

$$F = \frac{6\pi\mu a^3}{h^3} \int_0^a y(r)r \, dr \tag{4.8}$$

for the nonslip contact between the film and the substrate. Here, $f(r)$ is the surface profile of the indenter and μ is the shear modulus of the elastic film. The function $y(r)$ satisfies the following equation

$$\left(\frac{d^2}{dr^2} + \frac{1}{r}\frac{d}{dr} \right) y(r) + \frac{f(r)}{a^3} = 0 \tag{4.9}$$

subjected to the boundary condition: $y(a) = 0$. Using (4.6)–(4.8), one can determine the effect of surface interaction on the relationship between the normal contact load and the contact radius and the pull-off force to separate the particle from the film. For a rigid spherical particle of R in radius, there are [25]

$$F = \begin{cases} \dfrac{\pi\mu a^4}{hR} - 2\pi a^2 \sqrt{\dfrac{2\mu\gamma}{h}} & \text{frictionless contact} \\[3mm] \dfrac{\pi\mu a^6}{32Rh^3} - 2\pi a^3 \left(\dfrac{3\gamma\mu}{32h^3} \right)^{1/2} & \text{nonslip contact} \end{cases} \tag{4.10}$$

and

$$F_a = \begin{cases} 2\pi R\gamma & \text{frictionless contact} \\ 3\pi R\gamma & \text{nonslip contact.} \end{cases} \tag{4.11}$$

The pull-off force is the same as the result given by Bradley [16] and the DMT theory [20] for the frictionless contact. For the nonslip contact, the pull-off force is 3/2 time of that with the frictionless contact. This is likely due to the confinement effect on the deformation of the elastic film.

For a rigid conical particle of a semi-included angle ϑ, there are [25]

$$F = \begin{cases} \dfrac{4\pi\mu a^3 \cot\vartheta}{3h} - 2\pi a^2 \left(\dfrac{2\mu\gamma}{h} \right)^{1/2} & \text{frictionless contact} \\[3mm] \dfrac{\pi\mu a^5 \cot\vartheta}{20h^3} - \dfrac{\pi a^3}{2} \left(\dfrac{3\gamma\mu}{2h^3} \right)^{1/2} & \text{nonslip contact} \end{cases} \tag{4.12}$$

and

$$F_a = \begin{cases} \dfrac{4\pi\gamma \tan^2 \vartheta}{3} \sqrt{\dfrac{2\gamma h}{\mu}} & \text{frictionless contact} \\[2ex] \dfrac{9}{5}\pi h\gamma \tan^{3/2} \vartheta \left(\dfrac{54\gamma}{\mu h}\right)^{1/4} & \text{nonslip contact.} \end{cases} \quad (4.13)$$

In contrast to the adhesive contact between a rigid spherical particle and an elastic thin film, the pull-off force is proportional to the square root of the film thickness for the frictionless contact, while it is proportional to the 3/4 power of the film thickness for the nonslip contact.

For a flat-ended cylindrical particle of a in radius, there are [28]

$$F = \begin{cases} \dfrac{4\pi\mu a^2\delta}{h} & \text{frictionless contact} \\[2ex] \dfrac{3\pi\mu a^4\delta}{8h^3} & \text{nonslip contact} \end{cases} \quad (4.14)$$

and

$$F_a = \begin{cases} \pi a^2 \sqrt{\dfrac{8\mu\gamma}{h}} & \text{frictionless contact} \\[2ex] \dfrac{\pi a^3}{2h} \sqrt{\dfrac{3\mu\gamma}{2h}} & \text{nonslip contact.} \end{cases} \quad (4.15)$$

The pull-off force is inversely proportional to the square root of the film thickness for the frictionless contact, while it is inversely proportional to the 3/2 power of the film thickness for the nonslip contact. Higher pull-off force is required for the nonslip contact between the film and the rigid substrate than that for the frictionless contact.

It is worth pointing out that the above results are based on the frictionless contact between the rigid particle and the elastic film. It is expected that high pull-off force is needed for the nonslip contact, even though there is no closed-form solution for the nonslip contact between a rigid spherical particle and an elastic film and for the nonslip contact between a rigid conical particle and an elastic film. However, the pull-off force is independent of the contact condition for the contact between a rigid spherical particle/flat-ended cylindrical particle and a semi-infinite incompressible elastic material [28, 30].

Consider self-adhesion of poly(dimethylsiloxane) (PDMS). The PDMS samples were made from the commercial product SYLGARD 170 A&B by DOWCORNING. The SYLGARD 170 A&B were first mixed, then left to degas in a weak vacuum system (80 kPa) for about 1 h. After degassing, a small portion of the mixture was coated onto a flat-ended cylindrical indenter. This layer was less than 10-μm thick and its deformation was negligible compared with that of the sample as verified in a finite element simulation. Both the indenter and the sample were put in the oven for curing overnight at room temperature. Stainless steel indenters of five different radii (0.79, 1.19, 1.59, 2.38, and 3.18 mm) were used. The indenter radius is less than the sample thickness, which is about 5 mm.

This suggests that one can treat the PDMS layer as a semi-infinite material. The self-adhesion energy of the PDMS then can be calculated as [28, 31, 32]

$$F_a = 4\sqrt{2\pi\mu\gamma a^3}. \tag{4.16}$$

Figure 4.1 shows a typical loading–unloading curve for the contact of the PDMS at ambient temperature. In contrast to the adhesive contact of PDMS using a rigid spherical indenter [33], there is no hysteresis between loading and unloading. Both the loading and unloading curves display the same linear relationship between the load applied to the indenter and the indenter displacement. During loading, the indenter experienced a jump-in to make adhesive contact. During unloading, the indenter underwent a pull-off force before separation.

By using (4.16) and the pull-off force of the unloading curve as well as the shear modulus of 0.73 MPa [32], the self-adhesion energy for a contact time of 4 mins was calculated to be 28 mJ m^{-2} for the indenter 1.59-mm radius. Figure 4.2 shows the linear relationship between the pull-off force and the 3/2 power of the indenter radius, consistent with the result given by (4.16). The self-adhesion energy can be obtained from the slope of the line. This self-adhesion energy is 29 mJ m^{-2}, which is comparable with 28 mJ m^{-2}, the value obtained previously using the pull-off force alone for the indenter 1.59-mm radius.

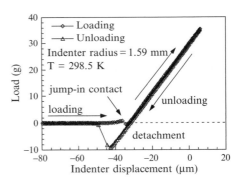

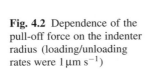

Fig. 4.1 Typical loading–unloading curve for PDMS (both loading and unloading rates were 1 μm s^{-1})

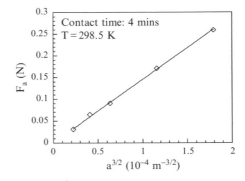

Fig. 4.2 Dependence of the pull-off force on the indenter radius (loading/unloading rates were 1 μm s^{-1})

Fig. 4.3 Time dependence of the adhesion energy for PDMS

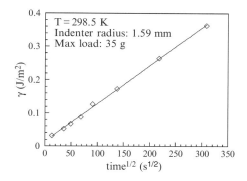

Figure 4.3 shows the linear relationship between the self-adhesion energy and the square root of the contact time, suggesting that diffusion is the dominant mechanism for the development of the self-adhesion between two PDMS surfaces. It is seen that when the contact time is long enough, the self-adhesion energy can be very large, much larger than the surface energy of PDMS, which is only about $20 \, mJ \, m^{-2}$ [34]. In fact, the measurement is the fracture energy, which includes the contribution of diffusion work of plastic deformation accompanying the separation fracture process. As can be seen, Fig. 4.3 has not indicated a saturation after 30 h of contact.

It is seen that a cylindrical indenter with a flat end in contact with a planar surface can avoid the loading–unloading hysteresis encountered in the JKR method using spherical indenters. Without any hysteresis, the elastic constants can be obtained independent of the adhesion at the contact interface. Since the contact area in the case of the cylindrical indenter is unchanged between the time of contact and the time of separation, it is ideal to use the cylindrical indenter to study the development of adhesion energy with contact time.

4.2.2 Adhesive Contact of a Compressible Elastic Thin Film

In contrast to the adhesive contact between an axisymmetric rigid particle and an incompressible elastic thin film, one needs to account for the effect of the compressibility, i.e., the Poisson's ratio, on the contact deformation of compressible elastic films. Similar to Sect. 4.2.1, focus is on frictionless contact between a particle and a film on a rigid substrate under the condition of the contact size much larger than the film thickness.

For frictionless contact between the film and the substrate, the relation between the normal contact force and the contact radius without the surface interaction is [35]

$$F = \frac{2\pi a^2 \mu}{(1-\nu)} \int_0^1 g(t) \mathrm{d}t, \qquad (4.17)$$

where v is the Poisson's ratio. The function $g(t)$ satisfies the Fredholm integral equation

$$g(t) + \frac{1}{\sqrt{2\pi}} \int_0^1 g(u)[K_c(|t-u|) + K_c(|t+u|)]du = -\frac{2\delta}{\pi a}, \tag{4.18}$$

with

$$K_c(\xi) = \sqrt{\frac{2}{\pi}} \int_0^\infty \tilde{f}(u) \cos(\xi k)dk. \tag{4.19}$$

Here, $\tilde{f}(r) = f(r/a)$.

Using (4.6) and (4.17), one obtains the pull-off force and the relationship between the normal contact load and the contact radius as [35]

$$F = \frac{\pi\mu a^4}{2(1-v)hR} - 2\pi a^2 \sqrt{\frac{\mu\gamma}{(1-v)h}} \quad \text{and} \quad F_a = 2\pi\gamma R \tag{4.20}$$

for a rigid spherical particle of R in radius,

$$F = \frac{2\pi\mu a^3 \cot\vartheta}{3(1-v)h} - 2\pi a^2 \sqrt{\frac{\mu\gamma}{(1-v)h}} \quad \text{and} \quad F_a = \frac{8\pi\gamma \tan^2\theta}{3} \sqrt{\frac{(1-v)\gamma h}{\mu}} \tag{4.21}$$

for a rigid conical particle of a semi-included angle ϑ, and

$$F = \frac{2\mu\pi a^2\delta}{(1-v)h} \quad \text{and} \quad F_a = 2\pi a^2 \sqrt{\frac{\mu\gamma}{h(1-v)}} \tag{4.22}$$

for a flat-ended cylindrical particle of a in radius. Equations (4.20)–(4.22) at $v = 0.5$ reduce to the results of (4.10)–(4.15) for the adhesive contact of an incompressible elastic film with frictionless contact between the film and the rigid substrate.

For nonslip contact between an elastic film and a rigid substrate without surface interface, Yang [36] used the perturbation theory to obtain the relationship between the normal contact force and the contact radius to the first order of h/a ($h \ll a$) as

$$F = \frac{2\pi(1-v)E}{(1+v)(1-2v)h} \int_0^a [\delta - f(r)]rdr \tag{4.23}$$

with the relationship between the particle displacement and the contact radius as $\delta = f(a)$. Here, E is the Young's modulus of the elastic film. For adhesive contact between the film and an axisymmetric rigid particle, there are

$$F = \frac{a^4}{4\kappa^2 R} - \frac{a^2}{\kappa}\sqrt{2\pi\gamma} \quad \text{and} \quad F_a = 2\pi\gamma R \tag{4.24}$$

for a rigid spherical particle of R in radius,

$$F = \frac{a^3 \cot\vartheta}{3\kappa^2} - a^2 \frac{\sqrt{2\pi\gamma}}{\kappa} \quad \text{and} \quad F_a = \frac{8\pi\gamma\sqrt{2\pi\gamma}\kappa\tan^2\vartheta}{3} \tag{4.25}$$

for a rigid conical particle of a semi-included angle ϑ, and

$$F = \frac{\delta a^2}{\kappa^2} \quad \text{and} \quad F_a = \frac{\sqrt{2\pi\gamma}a^2}{\kappa} \tag{4.26}$$

for a flat-ended cylindrical particle of a in radius. The parameter of κ is defined as

$$\kappa = \sqrt{(1+\nu)(1-2\nu)h/\pi(1-\nu)E}. \tag{4.27}$$

Equations (4.24)–(4.26) are similar to (4.20)–(4.22) with the same dependence on the contact radius, while the constant $(1-2\nu)$ is in the denominator. Also, they have different dependence on the film thickness as comparing with incompressible elastic films. This suggests that one needs to use (4.10)–(4.15) in evaluating adhesive contact of incompressible elastic films with nonslip contact between film and rigid substrate instead of (4.24)–(4.26).

Considering the effect of the substrate elasticity in the analysis of the adhesive contact between a spherical indenter and a compliant elastic layer, Johnson and his coworkers [37, 38] proposed a semiempirical relation between the contact load and the contact radius as

$$\frac{FR}{E_1'h^3} = \frac{4}{3}\left(\frac{a}{h}\right)^3 \Im\left(\frac{a}{h}, k\right) - \sqrt{\frac{8\pi\gamma R^2}{E_1'h^3}}\left(\frac{a}{h}\right)^{3/2} \Phi\left(\frac{a}{h}, k\right), \tag{4.28}$$

where $E_i' = E_i/(1-\nu_i^2)$ with $i = 1$ and 2, respectively, representing the elastic coating and the substrate and $k = E_2'/E_1'$. The functions $\Im(x, k)$ and $\Phi(x, k)$ are determined by numerical simulation for a variety of combinations in $(a/h, E_2'/E_1')$. In the analysis, the elastic layer was assumed to be perfectly bonded to the elastic half-space. In general, it is very difficult if not impossible to obtain a general expression of the functions $\Im(x, k)$ and $\Phi(x, k)$ over a wide range of a/h and E_2'/E_1'. Sridhar et al. [38] used curve fitting to obtain semianalytical functions $\Im(x, k)$ and $\Phi(x, k)$ for certain range of a/h and E_2'/E_1'. They also considered the effect of the indenter elasticity and introduced an effective plane-strain modulus of the indentation system, in which the indenter can be approximated as a rigid indenter and the bilayer structure as a homogeneous elastic half-space.

Following Maugis' approach [39], Sergici et al. [40] used a Dugdale model to evaluate the JKR–DMT transition in an adhesive contact of a spherical indenter with a layered elastic half-space. Using numerical simulation and varying the Maugis' adhesion parameter, they studied the effects of the elastic moduli ratio of the layer to the substrate and the layer thickness on the pull-off force and the relation between the contact load and the contact radius. They observed a transition from DMT to JKR adhesion. However, they did not give any semiempirical expression to correlate the contact load with the contact radius and the indentation depth, which makes it difficult to use their results in measuring adhesion energy between two solids.

4.3 Indentation-Induced Delamination of Thin Films: Characterization of Interfacial Strength

Reliable measurement of the interfacial strength between a thin film and a substrate plays an important role in the quality assurance of MEMS and NEMS structures and in the understanding of structural degradation occurring in multilayer structures [41, 42]. Chiang et al. [43] were the first to suggest the use of indentation-induced delamination of thin films in measuring the interfacial strength between a film and a substrate. The indentation-induced delamination is due to the development of compression stress during the indentation, resulting in the initiation and propagation of interfacial crack and causing the buckling of the film.

Using the theory of linear fracture mechanics, Evans and Hutchinson [44] and Marshall and Evans [45] developed a semianalytical relationship for measuring the interfacial fracture resistance. They assumed that the interfacial cracks were significantly larger than the film thickness and modeled the delamination region as a clamped circular plate over a rigid substrate. The crack driving force G for the indentation-induced delamination of the film is [44–47]

$$G = \frac{h\sigma_c^2(1-v_f^2)}{2E_f} + [\sigma_r^2 - (\sigma_c - \sigma_b)^2]\frac{(1-\alpha)h(1-v_f)}{E_f}, \qquad (4.29)$$

where E_f and v_f are, respectively, Young's modulus and Poisson's ratio of the film, α is a constant depending on the deformation mode ($\alpha = 1$ for nonbuckling mode), σ_r is the residual stress in the film, and σ_b is the critical buckling stress of the clamped plate. The compression stress σ_c created by indentation is a function of the indentation size in the film.

Considering the substrate effect on the indentation hardness of a bilayer structure, Huang et al. [47] used the composite indentation hardness [48, 49] in the calculation of the indentation size. From the result given by Bhattacharya and Nix [49], they expressed the crack driving force for the Vickers indentation of a compliant film on a hard substrate under nonbuckling mode as

$$G = \frac{(1+v_f)E_f\cot^2\theta}{144(1-v_f)\pi^2\{1+(H_f/H_s-1)\exp[-\beta(\delta/h)^2]\}^3}\frac{a_c^2}{h}\left(\frac{F}{H_s a_c^2}\right)^3, \qquad (4.30)$$

where H_f and H_s are, respectively, the indentation hardness of the film and the substrate, a_c is the radius of the delamination region, θ is the half-angle between opposite edges of a pyramid, and β is a fitting parameter. It is worth pointing out that, to recover (4.30) to (4.29), it requires $\beta = 0$. Huang et al. [47] also formulated the other expression for the crack driving force, using the composite hardness given by Korsunsky et al. [48]. They found that the calculated crack driving force for the ZnO film/Si substrate systems was same whether they used the Bhattacharya and Nix model or the Korsunsky et al.'s model.

Realizing that it may be difficult to measure the delamination size in some conditions, Dehm et al. [50] used a shear-lag model [51] to analyze the initiation of delamination during indentation. Assuming frictionless contact between the film and the substrate over the delamination zone, they established a relationship between the indentation load and the indentation depth for the measurement of the interfacial strength between a thin film and a substrate as

$$F_c = \frac{8\tau_i}{v_f \tan^2 \vartheta} \delta_c^2 \qquad (4.31)$$

for the indentation depth less than the film thickness and

$$F_c - F_s = (2\delta_c - h)\frac{8\tau_i h}{v_f \tan^2 \vartheta} \qquad (4.32)$$

for the indentation penetrating into the substrate. Here, F_c is the indentation load corresponding to the indentation depth of δ_c for the occurrence of the delamination and τ_i is the interfacial shear strength between the film and the substrate. They obtained self-consistent values of the interfacial shear strength for the coating of epitaxial films of (111) copper on (0001) sapphire. They also evaluated the effect of titanium interlayer on the interfacial bonding by varying the interlayer thickness from 0.7 to 110 nm and observed about 40% increase in the interfacial strength for the interlayer of 0.7 nm. Thicker interlayer only caused a marginal further improvement in the interfacial strength. This approach avoids the determination of the delamination size and enables us to measure the interfacial shear strength directly from the indentation loading curve. It is unclear how one can compare the interfacial shear strength with the critical crack driving force.

Rossington et al. [52] evaluated the indentation-induced delamination of ZnO thin films over Si substrates. Delamination occurred either along the interface or in the film parallel to the interface. They suggested that the change in the crack path from the interface to the film was likely due to the indentation-induced buckling of the films. They calculated the interfacial fracture energy using (4.29) as well as the bulking load for the delamination. The calculated buckling loads were not consistent with the buckling theory, which could be due to the indentation-induced deformation in the film. Further study needs to be conducted to evaluate the effect of out-of-plane deformation on the indentation-induced delamination.

Fan et al. [53] used a Knoop indenter to examine the adhesion of diamond coating over polycrystalline copper foil of 1 mm in thickness. They observed that the indentation at small indentation loads caused circular delamination of the coating similar to those by Vickers indentation and Rockwell C indentation. This suggested that the far-field stress was independent of the indenter morphology, as stated by Saint-Venant's principle. They established a relation between the indentation load and the delamination size by considering the sink-in deformation of the diamond coating and observed the critical indentation load of 1.09 GPa. However, they did not address the dependence of the critical indentation load on the interfacial fracture energy or the interfacial shear strength.

Kriese et al. [54] utilized a displacement-controlled indentation to induce initiation and propagation of interfacial delamination of sputtered copper thin films from SiO_2/Si substrates. The effects of heat treatment and interlayer on the interfacial fracture strength were examined. For the film systems without the interlayer, annealing caused the increase in interfacial fracture strength, which likely was due to the relaxation of residual stresses and the formation of intermetallic compounds at the interface. The introduction of the chromium or titanium interlayer resulted in the increase of interfacial strength due to the reduction of residual stresses. The interfacial strength also increased with the film thickness. The results of the effect of heat treatment on the interfacial strength suggest that the dependence of the interfacial strength on microstructure and interfacial reaction can be assessed using the indentation-induced delamination technique.

Lee et al. [55] used indentation-induced delamination to investigate the adhesion of Pt films to Si substrates with a native oxide film. They pointed out the difficulty in using this method for ultrathin films due to relatively small stored strain energy in the film for the creation of delamination. They adopted an overlayer technique to increase the driving force for delamination by sputtering a thick molybdenum layer over the Pt film and used (4.29) to calculate the interfacial fracture strength. They obtained unusually high values, which likely was due to the formation of subsurface cracks in the Si substrates as confirmed by the FIB images. This resulted in complicated stress state in the film, and the assumption used by Marshall and Evans [45] became invalid. The experiments of the indentation-induced delamination became inconclusive. Further study and modeling are needed for using the approach to evaluate local interfacial fracture strength.

Li et al. [56] studied the indentation-induced delamination of thin polystyrene films on glass substrates by using a 90° conical diamond indenter with a tip radius of 1 μm. There was no delamination presented in the 2-μm film for the indentation load less than 200 mN, while the indentation-induced delamination occurred for the indentation load of 15 mN or higher. The indentation-induced subsurface damage initiated in the glass substrate, which led to the formation and propagation of interfacial crack. They used three approaches to calculate the interfacial fracture toughness, including (1) annular-plate model [56], (2) Marshall–Evans model [45], and (3) process-zone model. From the analysis of using the Marshall–Evans model, they suggested that there was no buckling in the film during the indentation and the residual stresses and buckling stress had little role in the calculation of the interfacial fracture strength. They found that the interfacial fracture strengths calculated from all three approaches were in good agreement, while the Marshall–Evans model gave slightly low values. It is unclear which approach is more suitable for the analysis of the indentation-induced delamination in compliant coating systems.

Lu and Shinozaki [58] used a flat-ended cylindrical indenter to measure the interfacial shear strength of polymeric coatings over relatively rigid substrates. They observed the load drop in load–displacement curves, which was due to the initiation of a circular decohesion zone at the interface. Following the approach by Dehm [50] and using a modified shear-lag model, they obtained the critical shear stress at which the delamination initiated as a function of the indentation load and the indentation depth. The increase of rubber content in polystyrene films caused the decrease in the critical interfacial shear stress.

Fig. 4.4 Variation of the
loading stiffness with the
indentation depth for 3-μm
coating at the peak load of
5 mN (reprinted with permission from [59]; Copyright
2006, Elsevier Ltd)

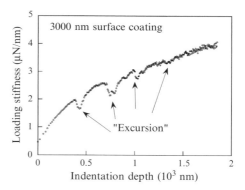

Geng et al. [59] studied the thickness effect on the indentation-induced delamination of SR399 ultrathin polymeric coatings over Si substrate. For the coatings of 47,
125, and 220 nm in thickness, there was only an excursion in the loading curve, representing local interfacial delamination. In contrast to the submicron thin films, the
loading curve for the indentation of 3-μm polymeric coatings had four excursions
as shown in Fig. 4.4. This indicated that the stress state in the thick film was different from the submicron films. Assuming plane-stress state in the submicron films
and following the approach by Dehm [50], Geng et al. related the interfacial shear
strength to the indentation load and the indentation depth. They did not analyze the
delamination of the thick film.

The indentation-induced delamination is complex. It involves triaxial stress state
and interfacial crack, which can cause local buckling from initially curved plate.
This will change the propagation behavior of interfacial crack and require the consideration of the effect of surface contact. In addition, there may involve the formation and propagation of radial cracks in indentation of brittle films over compliant
substrates, which can change the indentation deformation and alter the driving force
for interfacial delamination. New model needs to be developed to account for the
interaction between radial cracks and interfacial crack in the measurement of interfacial fracture strength.

4.4 Adhesion in Micromechanical Structure

To study the adhesion of micromechanical structures, Mastrangelo and Hsu [60]
used an array of cantilever beams of different lengths to measure the adhesion of
polysilicon beam to polysilicon substrate. The cantilever beams were anchored to
the substrate at one end and were suspended at the other end. The principle is that
the cantilever beams longer than the detachment length cannot separate from the
substrate due to less restoring force than the adhesive force. The detachment length
corresponds to the critical length of the cantilever beam above which any cantilever
beam will stick to the substrate. The contact of a cantilever beam with a substrate
can be controlled by mechanical probe or electrostatic force.

For the contact between a cantilever beam and an underlying substrate, the work of adhesion can be calculated as [60, 61]

$$\gamma = \frac{3}{2}\frac{Eg^2t^3}{(l-d)^4} \tag{4.33}$$

for the S-shaped contact and

$$\gamma = \frac{3}{8}\frac{Eg^2t^3}{l^4} \tag{4.34}$$

for only the tip-substrate contact between the beam and the substrate. Here, g is the gap between the beam and the substrate, t is the beam thickness, l is the beam length, and d is the contact length.

Considering that both ends of a compliant plate-like structure are supported by a rigid substrate, Yang [62] studied the adhesive contact between the structure and a rigid substrate. Under the action of uniform pressure of p and surface force, the relation between the work of adhesion and the contact size is

$$(L-d/2)^4 = \frac{16E\gamma w^2 t^3}{3p^2(1-v^2)}\left(\sqrt{1+\left(\frac{3pt}{8w\gamma}\right)^2}-1\right) \tag{4.35}$$

for a simply supported plate and

$$(L-d/2)^4 = \frac{12E\gamma w^2 t^3}{p^2(1-v^2)}\left(\sqrt{1+\left(\frac{pt}{2w\gamma}\right)^2}-1\right) \tag{4.36}$$

for a doubly clamped plate. Here, w is the width of the structure. Equations (4.35) and (4.36) give the work of adhesion at $p = 0$, as

$$\gamma = \frac{3Eg^2t^3}{8(1-v^2)(L-d/2)^4}\begin{cases}1 & \text{simple support}\\4 & \text{clamp at edges}\end{cases} \tag{4.37}$$

which is independent of the width of the structure. The results suggest that a doubly clamped plate-like structure is more difficult to stick to the substrate than a simply supported structure.

It is known that fluctuation can cause deflection of micromechanical structure. This can lead to the occurrence of the jump-in instability for a micromechanical structure in a metastable state and result in the transition from a noncontact deflection to an adhered state. If there is no energy dissipation during the transition for an initial deflection-free plate-like structure under load-free condition, the critical gap for the presence of the jump-in instability is [62]

$$g_c = \frac{3L^2}{16t^2}\sqrt{\frac{6(1-v^2)t\gamma}{E}}\begin{cases}2 & \text{simple support}\\1 & \text{clamp at edges.}\end{cases} \tag{4.38}$$

Equation (4.38) gives a lower bound of the critical gap between a compliant structure and a rigid substrate for the occurrence of adhered contact.

It is worth mentioning that all above results are based on smooth contact between a smooth compliant structure and a rigid smooth substrate. In general, it would be very difficult if not impossible to fabricate molecular smooth surface. One needs to analyze the effect of surface roughness on adhesive contact of micromechanical structure for the design of MEMS and NEMS devices. Analyses using Gaussian distribution and fractal geometry have been summarized by Zhao et al. [61], in which the resultant adhesive force was derived from the integration of multiple asperity contacts. This has imposed a challenge in evaluating the evolution of local stresses. Numerical methods need to be developed for describing the multiscale feature of surface topology and analyzing the multiscale effect on the adhesive contact between solid surfaces.

4.5 Summary

This chapter presents an introduction to the adhesive contact of solid surfaces occurring in surface coatings, MEMS and NEMS structures. The contact deformation itself is complex, and techniques using contact deformation have been developed to evaluate mechanical properties of materials such as nanoindentation, impression testing, and atomic force microscopy. The contact deformation creates triaxial stress and strain states in materials; and local plastic deformation as well as surface interaction can occur simultaneously.

As the characteristic size of mechanical structure continues to shrink, surface interaction becomes a dominant factor in controlling mechanical behavior and mechanical motion of structure and physical object. It can cause malfunction of MEMS and NEMS devices. This has attracted great interest in designing micromechanical structure of a least possibility of adhesion and in controlling or eliminating surface interaction through physicochemical processes. One of the challenges is to understand the fundamental of surface interaction and its effect on micro- and nanomechanical behavior due to the limit of available analytical technology. With the complexity of multiscale physics, one needs to develop multiscale numerical technique to simulate micro- and nanoadhesive contact and compare the simulation results with experimental observation for the characterization of surface interaction and the intrinsic work of adhesion. This will provide us a mechanism in improving mechanical design with a least possibility of adhesion for applications in military and bioengineering.

Acknowledgment The work is supported by the National Science Foundation under Grant No. CMS-0508989.

References

1. Mehregany M, Gabriel KJ, Trimmer WSN (1988) IEEE Trans. Electron Devices 35:719
2. Muldavin JB, Rebeiz GM (2001) IEEE Microw. Wireless Comp. Lett. 11:334
3. Jensen BD, Huang KW, Chow LL-W, Kurabayashi K (2005) Appl. Phys. Lett. 97:103535
4. Ashkin A (1997) Proc. Natl. Acad. Sci. USA 94:4853
5. Bechhoefer J, Wilson S (2002) Am. J. Phys. 70:393
6. Zeng HJ, Wan ZL, Feinerman AD (2006) J. Microelectromech. Syst. 15:1568
7. Joudrey K, Adams GG, McGruer NE (2006) J. Micromech. Microeng. 16:2618
8. Maboudian R, Carraro C (2004) Phys. Chem. 54:35
9. Fan LS (1990) Integrated Micromachinery – Moving Structures in Silicon Chips, Ph.D. Thesis. University of California, Berkeley
10. Fan LS, Tai YC, Muller RS (1989) Sens. Actuators A 20:41
11. Sandeejas FSA, Apte RB, Banyai WC, Bloom DM (1993) In: Proc. 7th Int. Conf. on Solid-State Sensors and Actuators, Transducer 93, Yokohama, Japan. p. 6
12. Abe T, Messner WC, Reed L (1995) J. Microelectromech. Syst. 4:66
13. Mulhern GT, Soane DS, Howe RT (1993) In: Proc. 7th Int. Conf. on Solid-State Sensors and Actuators, Transducer 93, Yokohama Japan. p. 296
14. Guckel H, Sniegowski JJ, Christenson TR, Raissi F (1990) Sens. Actuators A 346:21
15. Komvopoulos K, Yan W (1997) J. Tribol. 119:391
16. Bradley RS (1932) Philos. Mag. 13:853
17. Johnson KL, Kendall K, Roberts AD (1971) Proc. R. Soc. Lond. A 324:301
18. Derjaguin BV, Müller VM, Toporov YP (1975) J. Colloid Interface Sci. 53:314
19. Hertz H (1896) Miscellaneous Papers. Macmillan, London, p. 146
20. Li JCM (2001) Mater. Sci. Eng. A 317:197
21. Tabor D (1977) J. Colloid Interface Sci. 58:1
22. Müller VM, Yushchenko VS, Derjaguin BV (1980) J. Colloid Interface Sci. 77:91
23. Maugis D (1992) J. Colloid Interface Sci. 150:243
24. Dugdale DS (1960) J. Mech. Phys. Solids 8:100
25. Yang FQ (2002) J. Phys. D Appl. Phys 35:2614
26. Yang FQ (2006) Thin Solid Films 515:2274
27. Lin YY, Chen HY (2006) J. Polym. Sci. B Polym. Phys. 44:2912
28. Yang FQ, Li JCM (2001) Langmuir 17:6524
29. Yang FQ (2003) J. Phys. D Appl. Phys. 36:50
30. Yang FQ, Zhang XZ, Li JCM (2001) Langmuir 17:716
31. Kendall K (1971) J. Phys. D Appl. Phys. 4:1186
32. Yang FQ, Zhang XZ, Li JCM (2001) Mater. Res. Soc. Symp. Proc. 649:Q6.5.1
33. Chaudhury MK, Whiteside GM (1991) Langmuir 7:1013
34. Van Krevelen DW (1976) Properties of Polymers: Their Estimation and Correlation with Chemical Structure. Elsevier, Amsterdam, p. 166
35. Yang FQ (2003) Mater. Sci. Eng. A 358:226
36. Yang FQ (2006) Thin Solid Films 515:2274
37. Johnson KL, Sridhar I (2001) J. Phys. D Appl. Phys. 34:683
38. Sridhar I, Zheng ZW, Johnson KL (2004) J. Phys. D Appl. Phys. 37:2886
39. Maugis D (1992) J. Colloid Interface Sci. 150:243
40. Sergici AO, Adams GG, Müftü S (2006) J. Mech. Phys. Solids 54:1843
41. Bahr DF, Robach JS, Wright JS, Francis LF, Gerberich WW (1999) Mater. Sci. Eng. A 259:126
42. Diao DF, Kato K, Hokkirigawa K (1994) J. Tribol. 116:860
43. Chiang SS, Marshall DB, Evans AG (1981) In: Pask JA, Evans AG (eds) Surface and Interfaces in Ceramic and Ceramic/Metal Systems. Plenum, New York, p. 603
44. Evans AG, Hutchinson JW (1984) Int. J. Solids Struct. 20:455
45. Marshall DB, Evans AG (1984) J. Appl. Phys. 56:2632
46. Kriese MD, Gerberich WW, Moody NR (1999) J. Mater. Res. 14:3007

47. Huang B, Zhao MH, Zhang T-Y (2004) Philos. Mag. 84:1233
48. Korsunsky AM, McGurk MR, Bull SJ, Page TF (1998) Surf. Coat. Technol. 99:171
49. Bhattacharya AK, Nix WD (1988) Int. J. Solids Struct. 24:1287
50. Dehm G, Rühle M, Conway HD, Raj R (1997) Acta Mater. 45:489
51. Agrawal DC, Raj R (1989) Acta Metall. 37:1265
52. Rossington C, Evans AG, Marshall DB, Khuri-Yakub BT (1984) J. Appl. Phys. 56:2639
53. Fan QH, Fernandes A, Pereira E, Grácio J (1999) Vacuum 52:163
54. Kriese MD, Moody NR, Gerberich WW (1998) Acta Mater. 46:6623
55. Lee A, Clemens BM, Nix WD (2004) Acta Mater. 52:2081
56. Li M, Palacio ML, Carter CB, Gerberich WW (2002) Thin Solid Films 416:174
57. Rosenfeld LG, Ritter JE, Lardner TJ, Lin MR (1990) J. Appl. Phys. 67:3291
58. Lu Y, Shinozaki DM (2002) J. Mater. Sci. 37:1283
59. Geng K, Yang FQ, Druffel T, Grulke EA (2007) Polymer 48:841
60. Mastrangelo CH, Hsu CH (1992) In: Tech. Digest of 1992 Solid State Sensor and Actuator Workshop, Hilton Head, SC, USA. p. 241
61. Zhao Y-P, Wang LS, Yu TX (2003) J. Adhes. Sci. Technol. 17:519
62. Yang FQ (2004) J. Micromech. Microeng. 14:263

Chapter 5
Nanomechanical Characterization of One-Dimensional Nanostructures

Yousheng Zhang, Eunice Phay Shing Tan, Chorng Haur Sow, and Chwee Teck Lim

5.1 Introduction

One-dimensional (1D) nanostructures belong to an important class of nanomaterials. Since the discovery of carbon nanotubes by Iijima in 1991 [1], 1D nanostructures such as nanotubes [2], nanowires [3], nanorods [4,5], nanobelts [6], nanoribbons [7], and nanofibers [8] have been synthesized and widely investigated. Studies have shown that these nanostructures possess not only unique electrical, thermal, and optical properties, but also outstanding mechanical properties [9,10]. Nanomechanical characterization of 1D nanostructures is so important that it will determine their applications in nanotechnology such as nanoresonators [11], biological sensors [12], nanocantilever [13], piezoelectric nanogenerators [14], nanoelectromechanical systems (NEMS), and tissue engineered scaffolds [15, 16]. It is of great importance to measure the mechanical properties directly from these nanostructures since the properties might have size-dependent behavior at the nanometer length scale. This

Y. Zhang
Nanoscience and Nanotechnology Initiative, National University of Singapore, 2 Science Dr. 3, Singapore 117542
nnizys@nus.edu.sg

E.P.S. Tan
Division of Bioengineering, National University of Singapore, 9 Engineering Dr. 1, Singapore 117576
tanphayshing@yahoo.com

C.H. Sow
Nanoscience and Nanotechnology Initiative, Department of Physics, Blk S12, Faculty of Science, National University of Singapore, 2 Science Drive 3, Singapore 117542
physowch@nus.edu.sg

C.T. Lim
Nanoscience and Nanotechnology Initiative, Division of Bioengineering, Department of Mechanical Engineering, National University of Singapore, 9 Engineering Dr. 1, Singapore 117576
ctlim@nus.edu.sg

F. Yang and J.C.M. Li (eds.), *Micro and Nano Mechanical Testing of Materials and Devices*,   105
doi: 10.1007/978-0-387-78701-5, © Springer Science+Business Media, LLC, 2008

Table 5.1 Elastic modulus of ZnO nanowires

Method	Modulus (GPa)	Ref.
Postbuckling by MEMS test-bed in scanning electron microscope	21	[17]
Tensile test	Within 30% of bulk ZnO	[18]
Mechanical resonance	210 (small diameter) 140 (large diameter)	[19]
Nanoindentation	232 (small diameter, assume fixed–fixed column) 454 (small diameter, assume fixed–pinned column) 117 (large diameter, assume fixed–fixed column) 229 (large diameter, assume fixed–pinned column)	[20]

implies that the mechanical properties of nanostructures cannot be accurately extrapolated from that of the bulk materials.

In this chapter, experimental techniques used in the nanomechanical characterization of 1D nanostructures will be presented. Approaches used to measure the mechanical properties of 1D nanostructures will be the main focus of this chapter. Sample preparation and the challenges in conducting such tests are also discussed.

Experimental study on the mechanical properties of 1D nanostructures is extremely challenging due to their lateral dimensions being a few tens of nanometers and longitudinal dimensions of a few microns. Taking the elastic modulus of ZnO nanowires as an example, different methods yield different results as shown in Table 5.1.

The main challenges for quantitative measurement of the mechanical properties include (i) designing an appropriate test configuration such as fabrication of substrate, scattering/picking/placing/clamping of specimens; (ii) applying and measurement of forces in the nano-Newton level; and (iii) measuring the mechanical deformation at the nanometer length scale. By utilizing the advantages of various types of high-resolution microscopes, nanoscale testing techniques have been developed. These high-resolution microscopes include scanning electron microscope (SEM), transmission electron microscope (TEM), and atomic force microscope (AFM) [21]. In order to manipulate and position small 1D nanostructures under electron microscopes, nanomanipulator with probes [22, 23] or MEMS-based nanoscale material testing system (n-MTS) [24] must be installed in these microscopes. The probes can be AFM tips or tungsten tips with radius of tens to a few hundred nanometers. AFM tips serve as force sensors and tungsten tips are used as picking/placing tools, counterpart electrode, or sample substrate. The microscope will provide the function for measuring the mechanical deformation and specimen dimensions at the nanometer length scale. The n-MTS incorporates a capacitive sensor to measure load electronically at a high resolution. As for the AFM, nano-Newton force can be determined from the deflection of the AFM cantilever multiplied by the cantilever spring constant. The AFM cantilever spring constants range from the order of 0.01 to 100 N m^{-1}, which can be accurately calibrated. High-resolution mapping of surface morphology on almost any type of conductive or nonconductive material can be achieved with AFM. This also provides the information of sample dimensions.

Methods developed and used for measuring the mechanical properties of individual 1D nanostructures can be classified into two different types based on the fundamental instruments previously mentioned: AFM-based and *in situ* electron-microscopy-based testing [25]. Three-point bend test, lateral force microscopy (LFM), and nanoindentation test are classified under the first AFM-based technique; and mechanical resonance, static axial tensile stretching, and compression/buckling tests are classified under the second. For generality, mechanical tests on nanowires are used here for the demonstration of Mechanical Character gates of 1D nanostructures.

5.2 Bend Tests

By using AFM-based technique, different types of bend tests have been developed such as three-point bend test and LFM.

5.2.1 Three-Point Bend Test

Among all the different bend tests, three-point bend test is most commonly used [26–31]. The process and principle for conducting a typical nanoscale three-point bend test can be described as follows. Individual nanowires are deposited onto a substrate with holes or trenches. An AFM cantilever tip is used to apply a small deflection midway along its suspended length using the force mode in AFM contact or tapping mode. Figure 5.1a shows a schematic diagram of a nanowire with mid-span deflected by an AFM tip. Force plot with loading and unloading curves is obtained from the force mode. An example of such a force plot and the method of obtaining the deflection at mid-span of the nanowire (v) are shown in Fig. 5.1b. A reference curve is first obtained by measuring the cantilever deflection over the Z-piezo displacement on a hard substrate, e.g., silicon wafer. The deflection of the nanowire (v) is the difference between the loading and the reference curve.

From the AFM image, the length and diameter of the suspended nanowire can be measured. The Young's modulus (E) of the nanowire is calculated using beam bending theory [32] for three-point bending of a clamped–clamped beam loaded by a midpoint force and is given by

$$E = \frac{PL^3}{192vI},$$

(5.1)

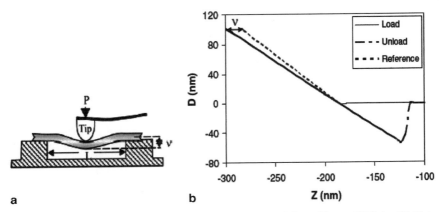

Fig. 5.1 (**a**) Schematic diagram of a nanowire with mid-span deflected by an AFM tip; (**b**) Plot of cantilever deflection (D) vs. vertical displacement of the Z-piezo (Z). (Reprinted with permission from [27]. Copyright [2004], American Institute of Physics)

where P is the maximum force applied, L is the suspended length, v is the deflection of the beam at mid-span, and I is the second moment of area of the beam. The maximum force can be calculated as $P = k_l D_{max}$, where k_l is the AFM cantilever spring constant and D_{max} is the maximum deflection of the cantilever.

The uncertainties involved in the typical three-point bend technique are the undetermined boundary conditions in some cases, the positioning of AFM tip over the exact mid-span of the nanowire, and the accuracy of the length measurement. For instance, the boundaries are always assumed to be clamped at the edge of the trench regardless of whether the two ends of the suspended nanowires are fixed or not. When the fixed ends are not secured well, this assumption may greatly influence the results. Steps to overcome these uncertainties have been published in recent papers [29–31].

Mai [30] presented a method of fitting the force–deflection curves along the entire length of a suspended nanowire under different loading forces, which determine if the measured data are best fitted by the clamped–clamped beam model with the following deflection equation

$$v = \frac{Fa^3(L-a)^3}{3EIL^3} \tag{5.2a}$$

or the free–free beam model with the following deflection equation

$$v = \frac{Fa^2(L-a)^2}{3EIL}, \tag{5.2b}$$

where F is a concentrated load applied at point a away from one of the ends and L is the suspended length. This method improves the accuracy of locating the midpoint of the suspended beam, and thus greatly increases the precision and reliability of the measurements.

Xiong *et al.* [31] proposed a similar approach to the one above. It is worth noting that they have developed a fabrication protocol involving photolithography and subsequent XeF_2 etching to produce trenches on the portion of the Si substrate located underneath the nanofilaments. The merits of this method are the well-clamped ends of the suspended nanofilaments as well as certainty of the length of the nanofilaments.

5.2.2 Lateral Force Microscopy

Wong et al. [33] first reported the utilization of LFM to determine the Young's modulus (E) of SiC nanorods and multiwalled CNTs. Song et al. [34] also used LFM to measure the elastic modulus of vertically aligned ZnO nanowires.

In the study by Wong et al. [33], the SiC nanorods were dispersed randomly on a flat surface and then pinned to a single-crystal MoS_2 substrate by the deposition of a regular array of square pads. The AFM was used to directly measure lateral force and deflection characteristics at varying distances from the pinning point (Fig. 5.2a).

In the LFM measurement of vertically aligned ZnO nanowires, the specimen is in the form of nanowire array with relatively low density or shorter length so that the AFM tip can exclusively reach one nanowire and the growth substrate without touching another nanowire [34]. In AFM contact mode, when the tip comes into contact with the nanowire, the cantilever is twisted and a rapid change in lateral signal is detected by the photodetector (Fig. 5.2b). At the largest bending position, the scanner retracts to the highest position and the lateral signal reaches the maximum value. The lateral force is measured from the torsion of the cantilever. From the experimental point of view, the maximum lateral force and the corresponding deflection of the nanowire can be obtained from the lateral force image and the

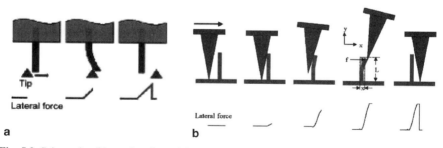

Fig. 5.2 Schematic of beam bending with AFM tip, (**a**) SiC nanorod (From [33]. Reprinted with permission from AAAS) (**b**) ZnO nanowires (Reprinted with permission from [34]. Copyright (2005) American Chemical Society). Before the AFM tip contacts the nanowire, the lateral force, as indicated by the trace at the bottom, is constant. When the tip bends the nanowire, there is a corresponding linear increase in the lateral force with deflection. After the tip has passed over the beam, the lateral force drops to its initial value, and the beam snaps back to its undeflected equilibrium position

topography image, respectively. The length of the nanowire can be derived from the vertical and lateral displacements, h and d, of the AFM image at which the lateral force reaches its maximum. The bending displacement of the nanowire (x_m) and the length of a nanowire (L) are given by [34]

$$x_m = d - h \tan\theta \qquad (5.3a)$$

and

$$L \approx \sqrt{h^2 + x_m^2}, \qquad (5.3b)$$

respectively, where 2θ is the apex angle of the AFM tip.

The elastic modulus of individual nanowire is derived from the relationship between maximum lateral force and maximum bending distance of cantilever beam, which is

$$f = 3EI\frac{y}{x^3}, \qquad (5.4)$$

where f is the bending force and y is the deflection at x distance away from the clamped end.

Wu *et al.* [35] and Heidelberg et al. [36] reported a LFM-based method that unambiguously measures the full spectrum of mechanical properties ranging from Young's modulus, E, to yield strength, plastic deformation and failure. The nanowire under investigation is suspended over a trench in the substrate and fixed by Pt deposits at the trench edges using electron-beam-induced deposition (EBID) method (refer to Fig. 5.3). During the manipulation, the lateral and normal cantilever deflection signals are simultaneously recorded and the corresponding force–displacement $(F–d)$ trace was recorded. From the linear-elastic part of the $F–d$ curves, the Young's modulus E can be determined by the elastic beam-bending theory (refer to (5.2a)).

Heidelberg et al. [36] provided a complete description of the elastic properties in a double-clamped beam configuration over the entire deformation behavior for a

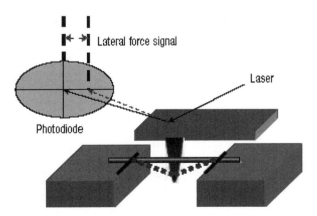

Fig. 5.3 Schematic of a fixed wire in a lateral bending test with an AFM tip. (Reprinted by permission from Macmillan Publishers Ltd: Nature Materials [35], copyright (2005))

variety of wires. They presented a simple closed form solution by using the governing beam equation including the effects of axial forces which is given by

$$EI\frac{d^4w}{dx^4} - T\frac{d^2w}{dx^2} = F_{center}\delta(x - L/2)$$ (5.5)

where w is the deflection of the beam, x is the spatial coordinate along the beam length, T is the tension along the beam, and δ is the Dirac delta function. With this model, it is possible to perform a comprehensive analysis of linear or nonlinear F–d curves. This enables the derivation of linear material constants such as the E as well as the description of the entire deformation behavior from the elastic range to fracture. Hence the yield points for distinctively different systems such as Si and Au nanowires can be identified. As this model allows the yield point to be identified in nanowire systems, it facilitates studies of the evolution of nanowire mechanical properties during work hardening and annealing.

5.3 Nanoindentation

Nanoindentation is a promising approach that does not require clamping of nanowires but instead involves direct indentation of nanomaterials [37–39]. Tan and Lim proposed nanoindentation route to study the mechanical properties of polymeric nanofibers by an AFM tip [40]. One shortcoming of this technique is the AFM tip cannot be perpendicular to the sample surface, thereby causing slip and friction between the AFM tip and the sample surface during indentation. This is due to the angle of approximately 10° that the cantilever makes with respect to the horizontal. The slip friction force makes it impossible for the AFM to measure the indentation load and displacement accurately at the nanoscale. In addition, high indentation loads cannot be reached because of the limitations of AFM cantilever stiffness. The combination of an AFM system and a nanoindenter was used to image a single silver nanowire [37] and then indent the wire with precise placement of the indenter tip on the wire. After the indentation, the same indenter tip was used to image the indentation impression on the wire. The nanoindenter monitors and records the load and displacement of the three-sided pyramidal diamond (Berkovich) indenter during indentation with a force resolution of about 1 nN and a displacement resolution of about 0.2 nm. Figure 5.4 shows the AFM images of the indents on a silver wire and a representative load–displacement curve. The hardness and elastic modulus were calculated from the load–displacement data obtained by nanoindentation.

Nanoindentation hardness is defined as the indentation load divided by the projected contact area of the indentation. It is the mean pressure that a material will support under load. From the load–displacement curve, hardness (H) can be obtained at the peak load and is given by

$$H = \frac{P_{max}}{A},$$ (5.6)

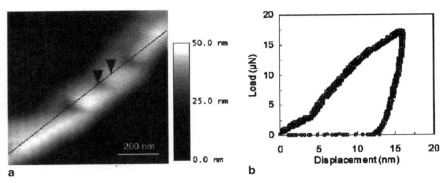

Fig. 5.4 (**a**) AFM images of indents on a silver nanowire; (**b**) a representative nanoindentation load–displacement curve. (Reprinted with permission from [37]. Copyright (2003) American Chemical Society)

where P_{max} is the peak load and A is the projected contact area. The nanoindentation elastic modulus was calculated using the Oliver–Pharr data analysis procedure [41] by fitting the unloading curve with a power-law relation. The unloading stiffness can be obtained from the slope of the initial portion of the unloading curve $S = dP/dh$, where h is the displacement. On the basis of relationships developed by Sneddon [42] for the indentation of an elastic half space by any punch that can be described as a solid of revolution of a smooth function, a geometry-independent relation involving contact stiffness, contact area, and elastic modulus can be derived as follows

$$S = 2\beta\sqrt{\frac{A}{\pi}}E_r, \qquad (5.7)$$

where β is a constant that depends on the geometry of the indenter ($\beta = 1.034$ for a Berkovich indenter) [43] and E_r is the reduced elastic modulus that accounts for the fact that elastic deformation occurs in both the sample and the indenter. E_r is given by

$$\frac{1}{E_r} = \frac{1 - \nu^2}{E} + \frac{1 - \nu_i^2}{E_i}, \qquad (5.8)$$

where E and ν are the elastic modulus and Poisson's ratio for the sample and E_i and ν_i are the same quantities for the indenter, respectively. For diamond, $E_i = 1,141$ GPa and $\nu_i = 0.07$ [41].

The above analysis is based on the standard Oliver–Pharr method. However, the Oliver–Pharr method is based on the indentation of a half-space and it might not be applicable to the indentation of nanowires. Feng et al. [39] presented a model considering contact between the indenter and the nanowire and contact between the nanowire and the substrate. The model has been used to analyze the nanoindentation data for GaN and ZnO nanowires. They found that the Oliver–Pharr hardness may be the rough lower bound of the hardness.

5.4 Mechanical Resonance

Mechanical resonance method was first used to obtain the Young's modulus of CNT in a TEM [44]. Subsequently, the mechanical resonance method was developed to test the Young's modulus of nanowires inside a SEM [45, 46] or under an optical microscope by mechanical or electrical field excitation [47]. Mechanical resonance can be induced in a nanowire when the frequency of the applied periodic force approaches the resonance frequency of the nanowire [48]. This can be done by applying a DC and AC voltage between a sharp tip and the nanowire (refer to Fig. 5.5a) or directly on a piezo bimorph actuator (refer to Fig. 5.5b) and tuning the frequency of the AC voltage.

The relationship between the frequencies and the Young's modulus is derived from the vibration equation of a cantilever

$$f_n = \frac{\beta_n^2}{2\pi L^2}\sqrt{\frac{EI}{\rho A}}, \tag{5.9}$$

where the subscript n represents the nth resonant vibration of a cantilever, f is the resonant frequency, ρ, A, and L are density, cross-sectional area, and length of the nanowire, respectively. β_n is the eigenvalue from the characteristic equation: $\cos\beta_n\cosh\beta_n + 1 = 0$. $\beta_0 = 1.875$, $\beta_1 = 4.694$, $\beta_2 = 7.855$, $\beta_3 = 10.996$ correspond to the first four resonance modes for any cantilever beam. Equation (5.9) can be simply expressed as

$$f_n = \frac{\beta_n^2 D}{8\pi L^2}\sqrt{\frac{E}{\rho}}, \quad f_n = \frac{\beta_n^2 a}{4\pi L^2}\sqrt{\frac{E}{3\rho}}, \quad f_n = \frac{\beta_n^2 a}{4\pi L^2}\sqrt{\frac{5E}{6\rho}}$$

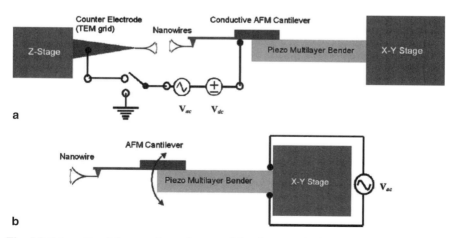

Fig. 5.5 Schematic of the experimental setup of the electromechanical resonance method with excitation coming from (**a**) an electrical field and (**b**) a piezobimorph actuator. (Reprinted with permission from [50]. Copyright [2006], Elsevier)

for a circular cross section with diameter D, a square or rectangular sections with width a, and a hexagon section with side width a, respectively. If the geometrical parameters are known, the Young's modulus can be calculated from the above equations.

5.5 Tensile and Compression/Buckling Tests

Uniaxial tensile loading has been used to measure the fracture strength of nanotubes [50] and nanofibers [51, 52]. In typical tensile tests, a specimen is stretched continuously until it fractures. The elongation of the specimen and the applied load are recorded during the tensile loading process. With the knowledge of the specimen geometry, the Young's modulus, tensile strength, and other tensile properties such as fracture strength and failure strain of the material can be determined.

Ding et al. [49] performed tensile test on boron nanowires by clamping a boron nanowire between two AFM cantilever tips under a tensile load inside a SEM. The stress–strain relationship of a boron nanowire under tension, Young's modulus and fracture strength are obtained. The fracture strength of the Boron nanowire as a function of nanowire length and nanowire diameter was also given. During a typical tensile test, an individual B nanowire was picked up from the source and clamped to two opposing AFM tips. The clamps were formed using the EBID method [53]. The soft cantilever was then gradually moved away from the rigid cantilever by actuating the piezoelectric bender with a DC voltage. A continuously increasing tensile load was applied to the nanowire until it fractured. Through image analysis of a series of SEM images taken during the tensile loading process, the tensile load or strain during the test is obtained and analyzed.

By reversing the loading direction, postbuckling experiments on WS_2 nanotubes were performed in a SEM equipped with a nanomanipulation system [54]. In this experiment a WS_2 nanotube was attached to an AFM cantilever (force constant of $0.05\,N\,m^{-1}$) within the environmental SEM. A mirror-polished silicon wafer was pushed against the mounted nanotube. The nanotube was imaged after each movement and subsequently digitized and analyzed. The postbuckling behavior of the WS_2 nanotubes was studied by applying the elastic theory, which describes large deflections of buckled bars [55]. According to the theory, the large displacement coordinates can be described at any point (x, y) and length (s) along the nanotube for a thin bar with stiffness EI (where E is the Young's modulus and I is the geometrical moment of inertia) under axial force P at the free end (refer to Fig. 5.6).

In the analysis, only the expression for the y coordinate was used: [54]

$$y = \sqrt{\frac{EI}{4P}} \int_0^\alpha \frac{\sin\theta d\theta}{\sqrt{\sin^2\alpha/2 - \sin^2\theta/2}}. \tag{5.10}$$

Young's modulus was calculated according to (5.10) at each y point of the nanotube's curve.

Fig. 5.6 Schematic of the postbuckling shape of a beam that is used for the Elastica analysis

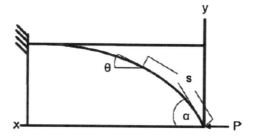

Kaplan-Ashiri et al. [56] measured the Young's modulus of WS$_2$ nanotubes experimentally using the buckling force as observed in the AFM. First, an individual nanotube was attached to a commercial AFM cantilever. A micromanipulator system within the high-resolution SEM chamber was used to attach a single nanotube to the tip of an AFM probe using an adhesive. The mounted nanotubes were limited to 1–3 μm in length to maintain their stiffness and reduce unwanted bending. The cantilever was then placed in an AFM. The force constant of the cantilever was estimated from the plane-view dimensions and free resonant frequency. Subsequently, the WS$_2$ nanotube tip was pushed against the surface, thus applying a controlled and measurable force. The buckling is seen as a discontinuity in the force curve. The force F at the buckling point can be calculated from the force constant of the cantilever and the cantilever deflection at which this took place. The Young's modulus was determined from Euler's force equation and is given by

$$F = \pi^2 \frac{EI}{L^2},\qquad(5.11)$$

where E is the Young's modulus, I is the geometric moment of inertia of the cross section of the nanotube, and L is the length of the nanotube.

5.6 Resonant Contact Atomic Force Microscopy

Resonant contact atomic force microscopy has been used to quantitatively measure the elastic modulus of polymer nanotubes [57] and silver and lead nanowires [58]. The samples were prepared by filtering suspensions of nanotubes or nanowires through poly(ethylene terephthalate) (PET) membranes. These membranes served as supports for the AFM measurements. The pore diameter of the PET membrane was chosen in order to minimize shear deformations.

An oscillating electric field is applied between the sample holder and the microscope head to excite the oscillation of the AFM cantilever in contact with suspended nanotubes/nanowires. The measured resonant frequency of this system is shifted to higher values with respect to the resonant frequency of the free cantilever (see Fig. 5.7a). It is experimentally demonstrated that the system can simply be modeled by a cantilever with the tip in contact with two springs (see Fig. 5.7b). These two

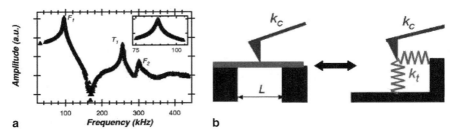

Fig. 5.7 (**a**) Typical resonance spectrum measured for a cantilever in contact with a nanotube (logarithmic ordinate axis). (**b**) Schematic representation of a cantilever resting on a suspended nanotube and of the equivalent mechanical model. (Reprinted with permission from [57]. Copyright [2003], American Institute of Physics)

springs represent the vertical and lateral stiffness associated with the deformation of the nanotube/nanowire in both directions.

The first two flexural vibration modes of the triangular Si_3N_4 cantilever in contact with two springs at the tip position are calculated using the Rayleigh–Ritz approximation. By fitting the experimental result to the calculated frequency shift, the spring stiffness can be determined. The tensile elastic modulus is then simply determined by using the classical theory of beam deflection.

5.7 Sample Preparation

From the experimental point of view, proper sample preparation is vital for successful nanomechanical characterization of 1D nanostructures. Nanomechanical characterization of single nanowires can only be performed on isolated nanowires of sufficient length for characterization, and with one or both ends fixed to a substrate or loading device. In Sect. 5.2.1, two such methods of single-nanowire sample preparation have been mentioned: photolithography with subsequent XeF_2 etching and random dispersion method for bend tests. Techniques that promote large-scale production of isolated, nonoverlapping nanowires with minimized direct manipulation will aid in the mechanical characterization process. Such unique sample preparation techniques [25] are briefly introduced in this section.

DC and AC/DC electric fields were used for the alignment of nanowires [59] and nanotubes [60] by utilizing the dielectrophoresis effect. Well-fabricated, electrically isolated electrodes are typically used to produce electric field gradients in the gap between them. Individual nanostructures are then aligned across the electrodes. This facilitates the mechanical characterization process as the resulting nanostructures are isolated and thus easily manipulated.

Self-assembly is a method of fabricating nanostructures where stable bonds are formed between organic or inorganic molecules and substrates. Rao et al. [61] demonstrated such an approach for large-scale assembly of carbon nanotubes.

Instead of manipulating and aligning nanostructures after their synthesis, direct growth is a very promising method to prepare specimens for nanomechanical characterization without direct manipulation. It does not involve any nanowelding steps like EBID [50] or clamped steps by FIB deposition, which introduce foreign materials onto the surface of nanostructures. Kong et al. [62] reported the synthesis of individual single-walled CNTs on patterned silicon wafers and He et al. [63] succeeded in producing direct growth of Si nanowires between two preexisting single-crystal Si microelectrodes.

5.8 Conclusions

In summary, different methods for the nanomechanical characterization of 1D nanostructures have been presented. Among them, three-point bend test and mechanical resonance method present the most convenient way to conduct such tests. Nanoindentation may be a good choice for very short structures where the influence and effects of shear may be significant when using the other methods. Tensile method has the full capacity for the measurement of elastic modulus, tensile strength, and yield strength. A summary of the different methods and types of materials tested is listed in Table 5.2.

Table 5.2 Summary of nanomechanical characterization methods for ID nanostructures

Methods	Mechanical properties	Materials	References
Three-point bend test	Elastic modulus	ZnO nanobelts	[30, 38]
		Silver nanowires	[29, 64]
		SiO_2 nanowires	[28]
		GaN nanowires	[65]
		Silicon nanowire	[66]
Lateral force bend test	Elastic modulus; Bend strength	SiC nanorods	[33]
		ZnO nanowires/nanorods	[34]
		Gold nanowires	[35]
Nanoindentation	Elastic modulus; Hardness	Silver nanowires	[37]
		ZnO nanobelts/nanowires	[38, 39]
		GaN nanowires	[39]
		ZnS nanobelts	[67]
Tensile test	Elastic modulus; Tensile strength; Yield strength	Boron nanowires	[49]
		WS_2 nanotubes	[54]
Compression/buckling	Elastic modulus	WS_2 nanotubes	[54, 56]
Mechanical resonance	Elastic modulus	SiO_2 nanowires	[28, 46]
		β-Ga_2O_3 nanowires	[68]
		ZnO nanowires	[19, 69]
		Boron nanowires	[49, 70]
		WO_3 nanowires	[71]
		SiO_2/SiC nanowires	[72]
Resonant contact AFM test	Elastic modulus	Silver and lead nanowires	[58]

By using mechanical resonance or three-point bend tests, the size effect of nanowires on the mechanical properties can be observed. The Young's modulus of silver [64], ZnO [19], and WO_3 [71] nanowires increases as diameter decreases, whereas that of GaN nanowires [73] decreases as diameter decreases. The above two phenomena were explained by the surface effect (the lower defect density in nanowires with smaller diameter) and the geometry, respectively. Besides the surface effects and the geometry, the boundary conditions [29, 49, 74], thermal effect under SEM, and residual strain due to fabrication may also play an important role in affecting the mechanical properties measured. Overall, it is still a challenge to quantify the modulus with size effect and to standardize the procedures.

The influence of other factors on the mechanical properties such as effects of structural configuration and defects, time, temperature, electromechanical coupling, and loading conditions need to be explored in the near future. The visualization of initial defect formation or nucleation, propagation, and ultimate failure resulting from defects is also a challenge for further studies.

References

1. Iijima S (1991) Nature 354:56
2. Goldberger J, He RR, Zhang YF, Lee SW, Yan HQ, Choi HJ, Yang PD (2003) Nature 422:599
3. Sun YG, Gates B, Mayers B, Xia YN (2002) Nano Lett. 2:165
4. Zhang DF, Sun LD, Yin JL, Yan CH (2003) Adv. Mater. 15:1022
5. Yang PD, Lieber CM (1996) Science 273:1836
6. Pan ZW, Dai ZR, Wang ZL (2001) Science 291:1947
7. Shi WS, Peng HY, Wang N, Li CP, Xu L, Lee CS, Kalish R, Lee ST (2001) J. Am. Chem. Soc. 123:11095
8. Li D, Xia YN (2004) Adv. Mater. 16:1151
9. Xia YN, Yang PD, Sun YG, Wu YY, Mayers B, Gates B, Yin YD, Kim F, Yan YQ (2003) Adv. Mater. 15:353
10. Kuchibhatla S, Karakoti AS, Bera D, Seal S (2007) Prog. Mater. Sci. 52:699
11. Husain A, Hone J, Postma HWC, Huang XMH, Drake T, Barbic M, Scherer A, Roukes ML (2003) Appl. Phys. Lett. 83:1240
12. Cui Y, Wei QQ, Park HK, Lieber CM (2001) Science 293:1289
13. Hughes WL, Wang ZL (2003) Appl. Phys. Lett. 82:2886
14. Wang ZL, Song JH (2006) Science 312:242
15. Zhang YZ, Lim CT, Ramakrishna S, Huang ZM (2005) J. Mater. Sci. Mater. Med. 16:933
16. Tan EPS, Lim CT (2006) J. Biomed. Mater. Res. A 77A:526
17. Desai AV, Haque MA (2007) Sens. Actuators A Phys. 134:169
18. Hoffmann S, Ostlund F, Michler J, Fan HJ, Zacharias M, Christiansen SH, Ballif C (2007) Nanotechnology 18:205503
19. Chen CQ, Shi Y, Zhang YS, Zhu J, Yan YJ (2006) Phys. Rev. Lett. 96:075505
20. Young SJ, Ji LW, Chang SJ, Fang TH, Hsueh TJ, Meen TH, Chen IC (2007) Nanotechnology 18:225603
21. Qian D, Wagner GJ, Liu WK, Yu MF, Ruoff RS (2002) Appl. Mech. Rev. 55:495
22. Williams PA, Papadakis SJ, Falvo MR, Patel AM, Sinclair M, Seeger A, Helser A, Taylor RM, Washburn S, Superfine R (2002) Appl. Phys. Lett. 80:2574
23. Cumings J, Zettl A (2000) Science 289:602

24. Zhu Y, Espinosa HD (2005) Proc. Natl. Acad. Sci. USA 102:14503
25. Zhu Y, Ke C, Espinosa HD (2007) Exp. Mech. 47:7
26. Cuenot S, Demoustier-Champagne S, Nysten B (2000) Phys. Rev. Lett. 85:1690
27. Tan EPS, Lim CT (2006) Compos. Sci. Technol. 66:1102
28. Ni H, Li XD, Gao HS (2006) Appl. Phys. Lett. 88:043108
29. Chen YX, Dorgan BL, McIlroy DN, Aston DE (2006) J. Appl. Phys. 100:104301
30. Mai WJ, Wang ZL (2006) Appl. Phys. Lett. 89:073112
31. Xiong QH, Duarte N, Tadigadapa S, Eklund PC (2006) Nano Lett. 6:1904
32. Ugural A (1993) Mechanics of Materials. McGraw-Hill, New York, p. 152
33. Wong EW, Sheehan PE, Lieber CM (1997) Science 277:1971
34. Song JH, Wang XD, Riedo E, Wang ZL (2005) Nano Lett. 5:1954
35. Wu B, Heidelberg A, Boland JJ (2005) Nat. Mater. 4:525
36. Heidelberg A, Ngo LT, Wu B, Phillips MA, Sharma S, Kamins TI, Sader JE, Boland JJ (2006) Nano Lett. 6:1101
37. Li XD, Hao HS, Murphy CJ, Caswell KK (2003) Nano Lett. 3:1495
38. Ni H, Li XO (2006) Nanotechnology 17:3591
39. Feng G, Nix WD, Yoon Y, Lee CJ (2006) J. Appl. Phys. 99:074304
40. Tan EPS, Lim CT (2005) Appl. Phys. Lett. 87:123106
41. Oliver WC, Pharr GM (1992) J. Mater. Res. 7:1564
42. Sneddon IN (1965) Int. J. Eng. Sci. 3:47
43. Pharr GM (1998) Mater. Sci. Eng. A Struct. Mater. 253:151
44. Poncharal P, Wang ZL, Ugarte D, de Heer WA (1999) Science 283:1513
45. Yu MF, Wagner GJ, Ruoff RS, Dyer MJ (2002) Phys. Rev. B 66:073406
46. Dikin DA, Chen X, Ding W, Wagner G, Ruoff RS (2003) J. Appl. Phys. 93:226
47. Zhou J, Lao CS, Gao PX, Mai WJ, Hughes WL, Deng SZ, Xu NS, Wang ZL (2006) Solid State Commun. 139:222
48. Wang ZL, Gao RP, Pan ZW, Dai ZR (2001) Adv. Eng. Mater. 3:657
49. Ding WQ, Calabri L, Chen XQ, Kohhaas KM, Ruoff RS (2006) Compos. Sci. Technol. 66:1112
50. Yu MF, Lourie O, Dyer MJ, Moloni K, Kelly TF, Ruoff RS (2000) Science 287:637
51. Tan EPS, Lim CT (2004) Rev. Sci. Instrum. 75:2581
52. Tan EPS, Ng SY, Lim CT (2005) Biomaterials 26:1453
53. Ding W, Dikin DA, Chen X, Piner RD, Ruoff RS, Zussman E, Wang X, Li X (2005) J. Appl. Phys. 98:014905
54. Kaplan-Ashiri I, Cohen SR, Gartsman K, Ivanovskaya V, Heine T, Seifert G, Wiesel I, Wagner HD, Tenne R (2006) Proc. Natl. Acad. Sci. USA 103:523
55. Timoshenko SP, Gere JM (1961) Theory of Elastic Stability. McGraw-Hill, New York
56. Kaplan-Ashiri I, Cohen SR, Gartsman K, Rosentsveig R, Seifert G, Tenne R (2004) J. Mater. Res. 19:454
57. Cuenot S, Fretigny C, Demoustier-Champagne S, Nysten B (2003) J. Appl. Phys. 93:5650
58. Cuenot S, Fretigny C, Demoustier-Champagne S, Nysten B (2004) Phys. Rev. B 69:165410
59. Smith PA, Nordquist CD, Jackson TN, Mayer TS, Martin BR, Mbindyo J, Mallouk TE (2000) Appl. Phys. Lett. 77:1399
60. Chen XQ, Saito T, Yamada H, Matsushige K (2001) Appl. Phys. Lett. 78:3714
61. Rao SG, Huang L, Setyawan W, Hong SH (2003) Nature 425:36
62. Kong J, Soh HT, Cassell AM, Quate CF, Dai HJ (1998) Nature 395:878
63. He RR, Gao D, Fan R, Hochbaum AI, Carraro C, Maboudian R, Yang PD (2005) Adv. Mater. 17:2098
64. Jing GY, Duan HL, Sun XM, Zhang ZS, Xu J, Li YD, Wang JX, Yu DP (2006) Phys. Rev. B 73:235409
65. Ni H, Li XD, Cheng GS, Klie R (2006) J. Mater. Res. 21:2882
66. Tabib-Azar M, Nassirou M, Wang R, Sharma S, Kamins TI, Islam MS, Williams RS (2005) Appl. Phys. Lett. 87:113102
67. Li XD, Wang XN, Xiong QH, Eklund PC (2005) Nano Lett. 5:1982

68. Yu MF, Atashbar MZ, Chen XL (2005) IEEE Sens. J. 5:20
69. Huang YH, Bai XD, Zhang Y (2006) J. Phys. Condens. Matter 18:L179
70. Calabri L, Pugno N, Ding W, Ruoff RS (2006) J. Phys. Condens. Matter 18:S2175
71. Liu KH, Wang WL, Xu Z, Liao L, Bai XD, Wang EG (2006) Appl. Phys. Lett. 89:221908
72. Wang ZL, Dai ZR, Gao RP, Gole JL (2002) J. Electron Microsc. 51:S79
73. Nam CY, Jaroenapibal P, Tham D, Luzzi DE, Evoy S, Fischer JE (2006) Nano Lett. 6:153
74. Chen XQ, Zhang SL, Wagner GJ, Ding WQ, Ruoff RS (2004) J. Appl. Phys. 95:4823

Chapter 6
Deformation Behavior of Nanoporous Metals

Juergen Biener, Andrea M. Hodge, and Alex V. Hamza

6.1 Introduction

Nanoporous open-cell foams are a rapidly growing class of high-porosity materials (porosity $\geqslant 70\%$). The research in this field is driven by the desire to create functional materials with unique physical, chemical, and mechanical properties where the material properties emerge from both morphology and the material itself. An example is the development of nanoporous metallic materials for photonic and plasmonic applications which has recently attracted much interest. The general strategy is to take advantage of various size effects to introduce novel properties. These size effects arise from confinement of the material by pores and ligaments, and can range from electromagnetic resonances [1] to length scale effects in plasticity [2, 3].

In this chapter, we focus on the mechanical properties of low-density nanoporous metals and how these properties are affected by length scale effects and bonding characteristics. A thorough understanding of the mechanical behavior will open the door to further improve and fine-tune the mechanical properties of these sometimes very delicate materials, and thus will be crucial for integrating nanoporous metals into products. Cellular solids with pore sizes above $1\,\mu m$ have been the subject of intense research for many years, and various scaling relations describing the mechanical properties have been developed [4]. In general, it has been found that the most important parameter in controlling their mechanical properties is the relative density, i.e., the density of the foam divided by that of solid from which the foam is made. Other factors include the mechanical properties of the solid material and the foam morphology such as ligament shape and connectivity. The characteristic

J. Biener and A.V. Hamza
Nanoscale Synthesis and Characterization Laboratory, Lawrence Livermore National Laboratory, Livermore, CA 94550, US
biener2@llnl.gov, hamza1@llnl.gov

A.M. Hodge
Aerospace and Mechanical Engineering Department, University of Southern California, Los Angeles, CA 90089
ahodge@usc.edu

F. Yang and J.C.M. Li (eds.), *Micro and Nano Mechanical Testing of Materials and Devices*,
doi: 10.1007/978-0-387-78701-5, © Springer Science+Business Media, LLC, 2008

internal length scale of the structure as determined by pores and ligaments, on the other hand, usually has only little effect on the mechanical properties. This changes at the submicron length scale where the surface-to-volume ratio becomes large and the effect of free surfaces can no longer be neglected. As the material becomes more and more constraint by the presence of free surfaces, length scale effects on plasticity become more and more important and bulk properties can no longer be used to describe the material properties. Even, the elastic properties may be affected as the reduced coordination of surface atoms and the concomitant redistribution of electrons may soften or stiffen the material. If, and to what extent, such length scale effects control the mechanical behavior of nanoporous materials depends strongly on the material and the characteristic length scale associated with its plastic deformation. For example, ductile materials such as metals which deform via dislocation-mediated processes can be expected to exhibit pronounced length scale effects in the submicron regime, where free surfaces start to constrain efficient dislocation multiplication.

In this chapter, we limit our discussion to our own area of expertise which is the mechanical behavior of nanoporous open-cell gold foams as a typical example of nanoporous metal foams. Throughout this chapter, we review our current understanding of the mechanical properties of nanoporous open-cell foams including both experimental and theoretical studies.

6.2 Processing Techniques

Nanoporous metal foams have been made in various shapes such as nanowires [5, 6], thin films [7, 8], and macroscopic 3D samples [8, 9]. Their morphology ranges from disordered three-dimensional network structures (sponge-like morphology) to well-defined periodic thin film structures with excellent long-range order, which are mostly used for photonic applications. This section provides a short review of the synthesis of nonperiodic nanoporous metal foams, which can be prepared in the form of very uniform millimeter-sized 3D objects. A typical example of the sponge-like open-cell foam morphology of np-Au is shown in Fig. 6.1a.

Synthesis techniques include top-down (dealloying) and bottom-up strategies (filter casting and templating) as well as combinations thereof. Many of these techniques have been developed or further refined at Lawrence Livermore National Laboratory with the ultimate goal to design a new class of three-dimensional nanoporous metals for high-energy-density laser experiments. This application requires the fabrication of millimeter-sized, defect-free monolithic samples of nanoporous materials with well-defined pore size distributions (including hierarchical porosities) and adjustable densities down to a few atomic percent. Yet, the material has to be mechanical robust enough to be machined. An example of such a laser target is shown in Fig. 6.1b.

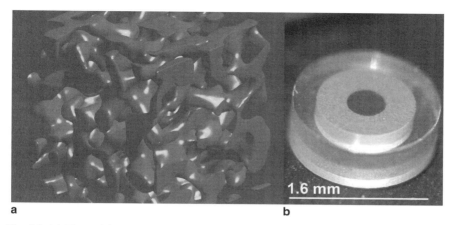

Fig. 6.1 (**a**) Focused ion beam nanotomography (FIB-nt) image showing the 3D structure of nanoporous Au. This state-of-the-art FIB-nt image was collected by L. Holzer, Ph. Gasser, and B. Münch from EMPA, Switzerland. The resolution of $2 \times 3 \times 6$ nm presents the current record of the FIB-nt method and allows one to determine microstructural parameters responsible for mechanical size effects of nanoporous materials. (**b**) Machined ultralow-density Au foam target for high-energy-density laser experiments. The picture was provided by courtesy of C. Akaba, LLNL

6.2.1 Dealloying

Dealloying is by far the most often applied method to generate macroscopic 3D as well as 2D (thin film) and 1D (nanowire) samples of low-density nanoporous metal foams for nanomechanical testing. In metallurgy, dealloying is defined as selective corrosion (removal) of the less noble constituent from an alloy, usually via dissolving this component in a corrosive environment [10,11]. In the case of binary solid-solution alloys with a narrow compositional range around $A_{0.7}B_{0.3}$ (where A is the less noble alloy constituent), this process can lead to spontaneous pattern formation, i.e., development of a three-dimensional bicontinuous nanoporous structure while maintaining the original shape of the alloy sample. The material can be very uniform, even on a macroscopic length scale, and typically exhibits a specific surface area in the order of a few m^2 g^{-1} [12]. The best-studied example is the formation of nanoporous gold (np-Au) via selective removal of Ag from a Ag–Au alloy [7,8,11], but other alloys such as $AuAl_2$ (np-Au) [13], MnCu (np-Cu) [14], and CuPt (np-Pt) [15] have also been successfully dealloyed. The process can be easily extended to two-dimensional thin film samples by using commercially available white-gold leaves with a thickness of a few hundred nanometers [7] or thin sputter-deposited alloy films [16,17]. Even, one-dimensional structures such as nanoporous Au nanowires can be fabricated by using a combination of electrochemical templating and dealloying [5]. The characteristic sponge-like open-cell foam morphology of np-Au is illustrated by the focused ion beam nanotomography image shown in Fig. 6.1a.

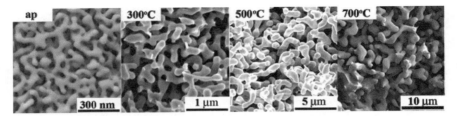

Fig. 6.2 Cross-sectional SEM micrographs of as-prepared (*ap*) and annealed np-Au: 300, 500, and 700°C. Note the self-similarity of the structure while increasing the feature size by more than a factor of 30. The relative density of the materials remains constant at approximately 30%

Nanoporous Au is ideally suited to study length scale effects in nanoporous metal foams due to its unique annealing behavior: the length scale of both pores and ligaments can be easily adjusted by a simple thermal treatment in ambient over a wide range from 3 nm to the micron length scale [18–20]. Most notably, the material rearrangement during annealing does not affect the relative density or relative geometry of the material (ligament connectivity or ligament, pore, and sample shape) [18, 21]. An example of such an annealing experiment is shown in Fig. 6.2.

The dealloying technique can also be used to introduce even more complicated morphologies such as hierarchical porosities. For example, bimodal pore size distributions can be realized by using a three-step dealloying/annealing/dealloying strategy where a micron-scale ligament structure is first generated by dealloying followed by annealing, and nanoscale porosity is finally reintroduced into the ligaments by a second dealloying step. This approach either requires the use of a ternary alloy as starting material or relies on metal deposition after the first dealloying/annealing step [22]. For example, we have used ternary Cu–Ag–Au alloys to prepare low-density (∼10 at.% relative density) nanoporous Au samples with a bimodal pore size distribution. However, preparation of the appropriate homogeneous single phase ternary alloy is challenging and limits the applicability of this approach.

6.2.2 Bottom-Up Approaches

Although the bottom-up approach to nanostructured metal foams is very versatile regarding the range of accessible densities and morphologies, very little is known about the mechanical properties of the resulting materials. The technique usually involves the use of sacrificial organic materials such as polystyrene (PS) microbeads as templates to generate a nanostructured porous solid [23, 24]. The dimensions of the template directly determine the length scale of the porosity in the final structure. The technique also provides a powerful approach to create materials with complex hierarchical porosities, specifically in combination with dealloying.

Fig. 6.3 Ultralow-density metal foam $(1\,\mathrm{gcc^{-1}})$ prepared from hollow 1-μm diameter $Ag_{0.7}Au_{0.3}$ shells. The SEM micrograph was provided by courtesy of G.W. Nyce, LLNL

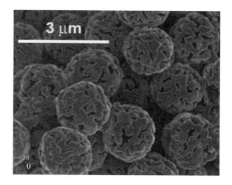

In the case of Au foams, the synthesis starts with the preparation of large quantities of Au or Ag–Au coated core–shell particles. These human-made building blocks are then assembled (casted) into a monolithic porous structure using a procedure analogous to the slip-casting process of ceramic [25] and metal [24] particles. Finally, the pure Au foam sample is obtained by removing the PS template by a simple heat treatment (Fig. 6.3). If Ag–Au coated core–shell particles were used as a starting material, one can easily introduce hierarchical porosities by adding a dealloying step. The technique outlined above has been successfully used to fabricate millimeter-sized, defect-free monolithic samples of nanoporous Au with well-defined pore size distributions and densities down to a few atomic percent (see for example the laser target shown in Fig. 6.1b).

6.3 Deformation Behavior

6.3.1 Macroporous Foams

Before going into further details on the deformation behavior of nanoporous foams, it is beneficial to cover the basics of foam mechanics. To keep this discussion short, we will only address the deformation behavior of open-cell foams. A complete discussion can be found in "Cellular Solids" by Gibson and Ashby [4].

Our knowledge on foam mechanics comes almost exclusively from the study of macroporous foams with cell sizes exceeding $1\,\mu m$. As stated in the introduction, the behavior of such materials is governed by the properties of the base material and the porosity, or more specifically, the relative density (ρ^*/ρ_s), which is defined as the density of the foam (ρ^*) divided by the density of the material it is made of (ρ_s). A variety of scaling relations have been developed to predict and describe the properties of cellular solids using these two parameters. For example, by compiling experimental data collected from a wide variety of open-cell foams, it can be shown that the yield strength (σ) and the Young's modulus (E) of open-cell foams can be described by the following scaling relations:

$$\sigma^* = C_2\sigma_s(\rho^*/\rho_s)^{3/2} \qquad (6.1)$$

$$E^* = C_1 E_s(\rho^*/\rho_s)^2, \qquad (6.2)$$

where * refers to the foam properties and s refers to the bulk properties, and $C_1 = 1$ and $C_2 = 0.3$ are fitting constants. It is important to note that these scaling relations contain no explicit length scale dependence and assume that the material properties of the ligaments such as the yield strength and the Young's modulus are size independent and equal to the bulk value. As will be discussed below, this assumption is no longer valid for metal foams with submicron features.

Experimentally, the mechanical behavior of foams is typically studied in compression; however, tensile testing has been used as well [26, 27]. In the case of a uniaxial compression test, the typical stress–strain deformation curve for an elastic–plastic foam will be composed of a linear elastic regime followed by a collapse plateau followed by a densification regime (Fig. 6.4).

Indentation is another frequently used technique to assess hardness and yield strength data of cellular solids [28–32]. In the case of low-density foams ($\rho^*/\rho_s \leqslant 0.3$), it is generally assumed that the indentation hardness (H) is approximately equal to the yield strength, $H \sim \sigma_y$, rather than following the $H \sim 3\sigma_y$ relationship typically observed for fully dense materials [4]. The reason for this is that the material under the indenter is not constrained by the surrounding material as it densifies with very little lateral spread (which translates into an effective plastic Poisson's ratio ~ 0). The $H \sim \sigma_y$ relationship has been validified by a number of studies and is used throughout this review to assess the yield strength of low-density metal foams from nanoindentation hardness data. However, it has to be pointed out that this relationship is only valid for foams with at least 60% porosity and when

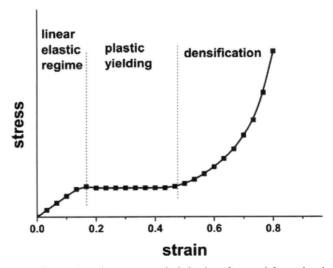

Fig. 6.4 Representative compressive stress–strain behavior of a metal foam showing regimes of elastic and plastic deformation until densifications dominate

issues such as indentation size effects and densification are properly accounted for. A detailed discussion on this topic can be found in a recent publication [21].

6.3.2 Nanoporous Foams

In the case of nanoporous materials, the majority of studies have relied on nanoindentation as a tool to measure mechanical properties such as yield strength and Young's modulus [33–35]. The technique has the advantage of being experimentally simple and, compared with other techniques, of having relatively low requirements regarding sample size or defect concentration. Furthermore, nanoindentation can be easily applied to thin films and irregularly shaped samples as long as the surface roughness is low. Other experimentally more complex methods are pillar micro-compression tests [36, 37] and film or beam-bending tests [16, 35, 38, 39]. Examples of nanoindentation and beam deflection tests are shown in Fig. 6.5. In the case of

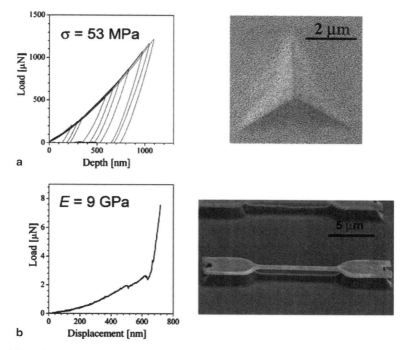

Fig. 6.5 Typical examples of nanomechanical tests performed on nanoporous Au. (**a**) Series of nanoindentation load–displacement curves using a Berkovich tip (*left*) and a SEM micrograph showing a typical residual impression of such a test (*right*). This specific test was performed on nanoporous Au with a relative density of 0.25 and an average ligament width of 30 nm, and reveals a yield strength of 53 MPa. (**b**) Load–displacement curve of a deflective tensile test performed on microfabricated dog-bone samples (*left*). Curve fitting to nonlinear beam theory reveals an elastic modulus of ∼9 GPa (relative density ∼0.35, ligament width 20–40 nm). SEM micrograph of a dog-bone sample for deflective tensile tests (*right*). The tensile test data were provided by courtesy of J.W. Kysar, Columbia University, New York

nanoindentation, the yield strength is typically obtained by assuming that $\sigma_y \sim H$ as discussed above. Given that more studies have addressed the yield strength, we will start with a discussion of the yield strength data, followed by a short review on the elastic properties of nanoporous Au.

6.3.2.1 Strength and Modulus

Nanoporous Au has been the preferred material in the majority of studies in this field [21,33–36,40]. As outlined above, the material is ideally suited to study length scale effects on plasticity in nanoporous metal foams as the pore size is experimentally simple to control, for example by annealing. Another material system studied is nanoporous Cu [14]. The two most often reported properties are yield strength and elastic modulus, and the reported values for these properties range from 15 to 240 MPa and 7 to 40 GPa, respectively (Fig. 6.6). Surprisingly, and despite its high porosity, nanoporous Au (240 MPa) can be stronger than bulk Au (200 MPa for severely worked Au single crystals). It is instructive to compare these values with the predictions made by the Gibson and Ashby scaling relations. One can calculate the yield strength and the Young's modulus of nanoporous Au as a function of relative density using (6.1) and (6.2), and the bulk properties of Au as input. The yield strength of bulk Au is very dependent on the sample history, and values as low as 2 MPa for well annealed and as high as $\sim$200 MPa for severely worked single crystals have been reported (see for example [41] and references wherein). To be conservative, we will use the high-end value of $\sim$200 MPa and the average value of the elastic modulus $\sim$80 GPa [42]. Assuming that the relative densities range from 25 to 42%, (6.1) and (6.2) then predict yield strength values from 7.5 to 16 MPa and elastic modulus values from 5 to 14 GPa (see solid lines in Fig. 6.6). Clearly,

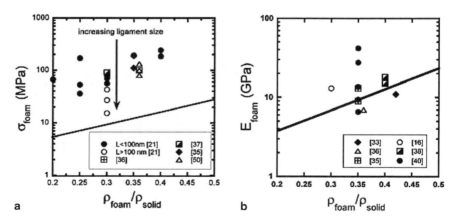

Fig. 6.6 Calculated (*lines*) and experimental (*symbols*) yield strength (**a**) and Young's modulus (**b**) data of nanoporous Au, plotted as a function of the relative density of the material. For comparison, the yield strength and the Young's modulus of fully dense Au are $\sim$200 MPa and 80 GPa, respectively. Note that np-Au can be stronger than bulk Au despite its high porosity

the experimental data do not follow the prediction made by the scaling relations: specifically, the yield strength data are higher than predicted, sometimes by more than an order of magnitude. The Young's modulus seems to be better described by the scaling equations, but the data are still very scattered. Furthermore, one can observe length scale effects not predicted by the Gibson and Ashby scaling relations. For example, the yield strength of nanoporous Au with relative density of 30% increases from $\sim$15 to $\sim$100 MPa as the ligament size decreases from 900 to 40 nm [21]. In conclusion, only for the largest ligaments, a reasonable agreement between experiment and theory is observed.

To compile data obtained from samples with different relative densities and check for length scale effects, it is instructive to use (6.1) and (6.2) to calculate the mechanical properties of the ligament material and plot the result as a function of the ligament size. The result of such a data compilation is shown in Fig. 6.7. Figure 6.7a reveals that the calculated strength of the ligament material increases with decreasing feature size, independent of the test method. This is in contrast to previous studies on macrocellular foams, which found that for a given porosity the cell size had a minimal effect in the mechanical behavior [43, 44]. Clearly, this is not true for nanoporous foams. In contrast, as revealed by Fig. 6.7a, the majority of studies on nanoporous Au observed that the yield strength of the ligaments in nanoporous Au approaches the theoretical shear stress $G/2\pi$ of Au ($\sim$4.3 GPa) as the ligament size decreases to the 10-nm length scale.

The high yield strength of the ligaments of np-Au is consistent with recent nanomechanical measurements performed on submicron Au columns [41, 45, 46] and nanowires [47]. Both of these test specimens closely resemble the ligaments in foams and mechanical tests reveal the general trend that "smaller is stronger." For example, Greer and Nix [45, 46] reported that the yield strength of submicron Au

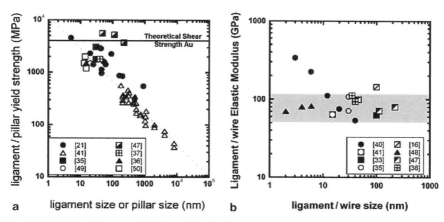

Fig. 6.7 Calculated ligament yield strength (**a**) and modulus (**b**) as a function of ligament diameter. Included are data obtained by nanoindentation [21,33,35,37], microcompression [36,37,41], beam bending [16, 35, 38], thin film buckling [40], nanowire bending [47], and MD [48, 49] tests. For comparison, the theoretical shear stress of Au is $\sim$4.3 GPa and the Young's modulus ranges from 42 to 116 GPa depending on crystal orientation

columns fabricated by focused ion beam micromachining approaches the theoretical yield strength of Au as the column diameter decreases to a few hundred nanometer. Specifically, the yield strength data reported by Volkert and Lilleodden [41] are in excellent agreement with the size dependence observed for nanoporous Au: the yield strength of submicron Au columns follows a power law d^{-n}, where d is the column diameter and n is 0.6. Figure 6.7a suggests that this size dependence continues to be valid down to 10 nm, the smallest ligament size which has been tested so far.

The size dependence of the yield strength observed in column microcompression tests has been explained by "dislocation starvation," i.e., the absence of efficient, easy-to-activate dislocation sources in submicron samples due to the presence of free surfaces. The argument goes along the following line: the smaller the sample volume (column/ligament), the smaller the dislocation source which can still be accommodated, the higher the stresses which are required to initiate yield. In this context, np-Au can be envisioned as a three-dimensional network of defect-free, ultrahigh-strength Au nanowires [37]. Length scale effects in plasticity can also be studied theoretically using molecular dynamics (MD) simulations [48, 49]. Preliminary MD results show indeed that it is difficult to incorporate stable dislocation sources (such as a Frank-Read source) in nanometer-sized ligaments. These simulations also reveal that plastic deformation of nanometer-sized Au columns is still dislocation mediated and that these dislocations nucleate in the vicinity of step edges on free surfaces at stress levels close to the theoretical strength of the material (Fig. 6.8) [49]. Dislocation-mediated plasticity of nanometer-sized ligaments in nanoporous Au has recently also been observed experimentally using an in situ transmission electron microscope nanoindentation technique [50].

Additional factors which potentially could contribute to the higher-than-predicted strength are testing method, densification, and grain size. For example, nanoindentation hardness and strength data obtained from fully dense materials show frequently higher values than those obtained by macroscopic tests, which is commonly attributed to the so-called "indentation size effect." We discard this possibility as the strength of nanoporous Au has been tested by a variety of different testing methods and generally good agreement has been found. Specifically, the microcompression tests on nanoporous Au performed by Volkert et al. [36] have shown that the measured strength is independent of the test volume, consistent with the idea that the size effects in nanoporous Au are dominated by the ligament/pore dimensions and not by the sample size. Similarly, densification effects can be ruled out as the increase of strength with decreasing ligament size was observed for a fixed initial foam density. Finally, the grain structure of the ligaments or the lack thereof is still a subject of debate and is beyond the scope of this discussion [9]. However, even if we assume that the ligaments develop a nanocrystalline grain structure during dealloying, this would still not account for the high strength found experimentally since the yield strength of nanocrystalline Au [51, 52] is only four times higher than that of coarse-grained material.

The elastic properties of nanoporous Au have attracted less interest, and consequently there are fewer data available which, in addition, appear to be more scattered (Figs. 6.6b and 6.7b). Most studies report Young's modulus values in the range of

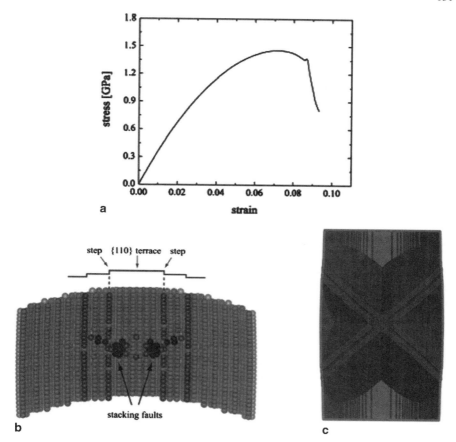

Fig. 6.8 Mechanical response of nanoscale Au pillars under compressive loading studied by MD. The displayed data were obtained from a 20 × 40-nm Au pillar compressed at 0 K: (**a**) compressive stress–strain curve demonstrating the high yield strength of nanoscale Au ligaments, (**b**) nucleation of partial dislocations near the step edges revealing dislocation-mediated plasticity, and (**c**) final state after 10% compression. The MD results were provided by courtesy of L.A. Zepeda-Ruiz and B. Sadigh, LLNL

5–13 GPa consistent with the predictions made by (6.2) if one uses the bulk modulus of Au (∼80 GPa) and relative densities ranging from 0.25 to 0.4. An exception to this is the recent study by Mathur and Erlebacher [40] who reported a fourfold increase in the Young's modulus, from ∼10 to ∼40 GPa, as the ligament diameter decreases, from >12 to 3 nm. The narrow range of most of the available data seems to indicate that there are little or no size effects in the elastic properties. In principle, the high surface-to-volume ratio of the nanoscale ligaments in nanoporous Au could give rise to a size effect in the elastic properties: One can think of the ligaments as core–shell structures, where the core has bulk-like properties and the shell exhibits elastic properties reflecting the reduced coordination number of surface atoms [53].

Depending on the actual electron redistribution caused by reduced coordination, the surface can become either softer or stiffer. Recent theoretical studies on Cu [54] showed that the effect of free surfaces on elasticity depends also on crystal orientation. However, even in the extreme case of a two-layer system, the Young's modulus did not change by more than 50%.

Mathur and Erlebacher [40] attributed the increased stiffness of ultrafine nanoporous Au to a combination of surface stress, density increase during dealloying due to shrinkage, and a higher bending stiffness of smaller ligaments. However, the "smaller-is-stiffer" trend could also be caused by an increased level of residual Ag in sub-10-nm material. The preparation of this ultrafine material requires dealloying conditions which strongly suppress surface diffusion. Experimentally, this can be achieved either by using dealloying potentials close to the Au oxidation potential [55] or by performing the dealloying process at low temperatures [20]. Both methods have been shown to increase the amount of residual Ag which affects both the relative density and the elastic properties of the material. Unfortunately, many studies do not report the residual silver content, thus making the comparison from study to study more difficult. The effect of residual Ag on the elastic properties of nanoporous Au has recently been studied by Doucette et al. [56], who found that Ag doping generally increases the Young's modulus of nanoporous Au.

As mentioned above, one can think of nanoporous Au as a 3D network of Au nanowires. Thus, it is instructive to compare the data discussed above with the literature on Au nanowires. Wu et al. [47] studied mechanical properties of Au nanowires ranging from 40 to 250 nm in diameter and reported a Young's modulus of $\sim$70 GPa, independent of the nanowire diameter and close to the value of bulk gold. Similar observations, namely a bulk-like behavior, have been reported for submicron Au columns [41]. The mechanical properties of sub-10-nm Au nanowires and pillars have been the subject of several molecular dynamics studies. Generally, it was observed that the Young's modulus is only weakly dependent on the wire/pillar diameter (for wires which do not undergo a phase transformation [53]) and close to the value of bulk gold [48, 57]. Furthermore, it has been shown that the surface stress in these ultrahigh surface-to-volume ratio systems mainly affects the strength by creating a compressive stress state in the wire/pillar core. In conclusion, most data presently available do not support the idea of a pronounced size effect in the elastic properties of nanoporous Au.

6.3.2.2 Fracture Behavior

As discussed above, nanoporous Au has unique mechanical properties under compressive loading conditions where it combines high strength with high porosity. On the other hand, the material is notoriously brittle which severely limits its usefulness in terms of applications [18, 58]. Actually, the brittleness of nanoporous Au is the reason why scientists started to study the mechanical behavior of nanoporous Au in the first place: Ag–Au alloys were used to study stress corrosion cracking, i.e., brittle failure of otherwise ductile materials in a corrosive environment [59–61].

These studies demonstrated that the formation of a thin surface layer of nanoporous gold is responsible for crack initiation and brittle failure of the uncorroded portion of the sample. Given the fact that Au is the most malleable metal, this brittle behavior is surprising. The questions which immediately arise in this context are: What causes the macroscopic brittleness of np-Au? Is the normal dislocation-mediated plasticity suppressed in nanoscale Au ligaments, or is the brittleness simply a consequence of the macroscopic morphology? These are the questions which will be addressed in the following.

The brittleness generally seems to increase with increasing Ag content of the Ag–Au master alloy, at least if dealloying is performed under free corrosion conditions: for example, samples made from $Ag_{70}Au_{30}$ alloys tend to be fairly robust, whereas samples made from $Ag_{80}Au_{20}$ alloys are generally more fragile. This trend seems to correlate with shrinkage and the buildup of tensile stress during dealloying which can induce crack formation [8]. In this context, it is interesting to note that thin films and nanowires can accommodate this tensile stress better than macroscopic 3D samples as they are less constraint with regard to shrinkage: for example, nanoporous Au nanowires have been successfully prepared from $Ag_{82}Au_{18}$ alloys [5], whereas a 3D sample prepared from such Ag-rich alloys would typically disintegrate during dealloying due to extensive cracking. The key to crack-free material is to avoiding dealloying conditions which lead to extensive buildup of tensile stress. Indeed, fracture-resistant nanoporous Au can be prepared by using dealloying conditions which promote stress relaxation, i.e., at elevated temperatures and lower dealloying potentials [55].

As described above, brittle failure of nanoporous Au is observed under tensile loading conditions. Microstructural characterization of the fracture surfaces reveals that nanoporous Au fails by a combination of intra- and intergranular fracture modes [58]: intergranular fracture, where the crack follows the grain structure of the original Ag–Au master alloy, reveals itself by the characteristic Rock candy morphology of the fracture surface, whereas intragranular fracture gives rise to smooth and featureless regions. On a microscopic level, however, characteristic necking features reveal ductile failure due to overloading of individual ligaments. In particular, regions of intragranular fracture exhibit a high density of disrupted ligaments (Fig. 6.9). Intergranular fracture surfaces, on the other hand, exhibit a very smooth morphology with only a few disrupted ligaments. This difference can be explained by the presence of extended two-dimensional void-like defects, which can be observed along the original grain boundaries of the master alloy. The formation of these defects can be explained by Ag enrichment at the grain boundaries, which leads to the development of a reduced density material during dealloying. Indeed, Ag surface segregation during annealing has been reported for the Ag–Au system [62]. The fracture mode described here – brittle on a macroscopic length scale, but microscopically ductile – is not a unique feature of nanoporous gold, but has recently even been observed for glass which is the common example for brittle failure [63].

The two-dimensional void-like defects described in the previous paragraph presumably act as crack nucleation sites due to local stress enhancement (Fig. 6.10). Ligaments connecting the regions on opposite sides of a defect experience the

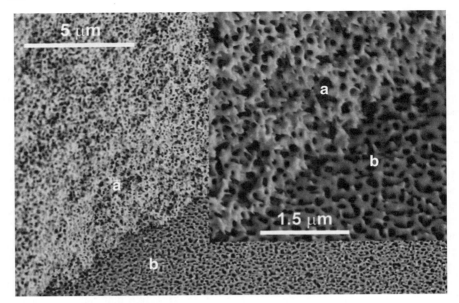

Fig. 6.9 Fracture appearance of nanoporous Au of both intragranular (**a**) and intergranular regions (**b**). The *inset* shows the same region at higher magnification and reveals the ductile nature of the fracture

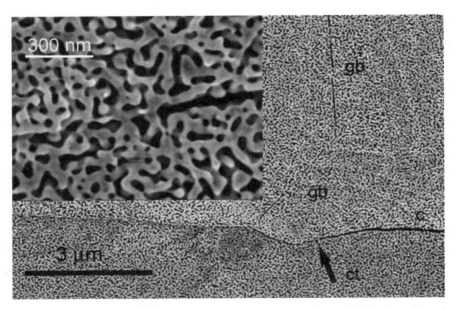

Fig. 6.10 Crack propagation along 2D defects, which are formed during dealloying along the grain boundaries of the original Ag–Au alloy. A crack (*c*) propagates from the right to the left along a defect. The remnant of a second grain boundary (*gb*) can be seen in the *upper right*. *Inset*: a higher magnification image of the crack tip (*ct*) region reveals the void-like character of the 2D defects. The SEM results were provided by courtesy of A. Wittstock, LLNL

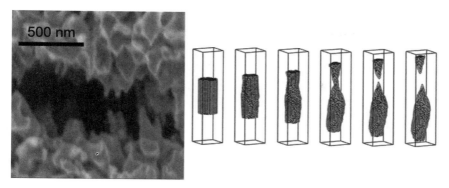

Fig. 6.11 Both SEM micrographs (*left*) and MD simulations (*right*) reveal the ductile failure behavior of nanoscale Au ligaments under tensile loading. The Au pillar used in the simulation is 5×10 nm and snapshots of the configuration were taken at 0, 25, 50, 100, 125, and 150% strain. The MD results were provided by courtesy of L.A. Zepeda-Ruiz, LLNL

highest stress fields and are the first to fail. Once an unstable crack is formed, the crack propagates along the 2D defects until intersecting with another 2D defect at an oblique angle, where the fracture may or may not switch from intergranular to intragranular (see Fig. 6.9).

The microscopic ductility of nanoporous Au is particularly obvious in the vicinity of crack tips, where highly strained ligaments still bridging the crack can be found (Fig. 6.11). These ligaments show pronounced necking and their elongation suggests strain values exceeding 100%. This microscopic ductility is also observed in molecular dynamics simulations on defect-free nanoscale Au columns under tensile loading (Fig. 6.11). Here, the elongation to failure is in the order of 100%, consistent with the finding of highly strained ligaments near crack tips. The microscopic ductility is a remarkable result in the context of the macroscopic brittleness of np-Au, but is consistent with the fact that Au is the most malleable metal. This immediately raises the question of what causes the macroscopic brittleness of np-Au. A qualitative answer can be found by applying a random fuse network model [64, 65]: according to this model, "brittle" failure can be expected for a sufficiently narrow ligament-strength distribution, regardless if the ligaments fail microscopically in a ductile or brittle manner. In the limit of a narrow ligament-strength distribution, rupture of the weakest ligament initiates the catastrophic failure of the network structure by overloading adjacent ligaments. The unstable crack then propagates quickly through the bulk of the material following the path of least resistance. This explanation is consistent with the narrow pore size/ligament width distribution of np-Au observed experimentally, which implies uniform failure strength. The overall strength of a randomly fused network is determined by the largest "critical" defect; i.e., the defect that causes the highest stress concentration at its edge. In the case of nanoporous Au, the two-dimensional void-like defects formed during dealloying along the grain boundaries of the Ag–Au master alloy serve as crack nucleation sites by concentrating the stress on adjacent ligaments. Thus, instead of plastic deformation of the whole sample, the failure of a few ligaments triggers the brittle fracture of the network. This conclusion may be used to improve the mechanical properties of nanoporous Au by introducing a broader ligament-strength distribution and by reducing the number and size of defects.

6.4 Summary

In this chapter, we have reviewed the mechanical behavior for nanoporous Au. In contrast to macroporous foams, the yield strength of nanoporous Au increases dramatically with decreasing feature size. Consequently, a nanoporous metal foam can be envisioned as a three-dimensional network of ultrahigh-strength nanowires, thus bringing together two seemingly conflicting properties: high strength and high porosity. This highly unusual combination of material properties opens a new door for engineering applications. The elastic properties, on the other hand, seem to be much less affected by the nanoscale morphology of the material and can be reasonably well described by the scaling relations derived from macroporous foams. Currently, the macroscopic brittleness of np-Au presents a major obstacle to applications. This brittleness seems to arise from the network structure rather than reflecting a microscopic brittleness, and thus could potentially be overcome, for example by introducing a broader ligament-strength distribution and by eliminating two-dimensional defects. As progress is achieved in understanding the mechanics of nanoporous materials, further research must include different types of nanoporous metals to develop more general trends.

Acknowledgments This work was performed under the auspices of the U.S. Department of Energy by Lawrence Livermore National Laboratory under Contract DE-AC52-07NA27344. The authors would like to acknowledge Beat Münch, Philippe Gasser, Lorenz Holzer, Craig Akaba, Greg Nyce, Jeffrey Kysar, Luis Zepeda-Ruiz, Babak Sadigh, and Arne Wittstock for providing original figures.

References

1. Maier SA (2007) Plasmonics: Fundamentals and Applications. Springer, New York
2. Arzt E (1998) Acta Mater. 46:5611
3. Sieradzki K, Rinaldi A, Friesen C, Peralta P (2006) Acta Mater. 54:4533
4. Gibson LJ, Ashby MF (1997) Cellular Solids: Structure and Properties. Cambridge University Press, Cambridge, UK
5. Liu Z, Searson PC (2006) J. Phys. Chem. B 110:4318
6. Yoo SH, Park S (2007) Adv. Mater. 19:1612
7. Ding Y, Kim YJ, Erlebacher J (2004) Adv. Mater. 16:1897
8. Parida S, Kramer D, Volkert CA, Rosner H, Erlebacher J, Weissmuller J (2006) Phys. Rev. Lett. 97:035504
9. Hodge AM, Hayes JR, Caro JA, Biener J, Hamza AV (2006) Adv. Eng. Mater. 8:853

10. Newman RC, Corcoran SG, Erlebacher J, Aziz MJ, Sieradzki K (1999) MRS. Bull. 24:24
11. Erlebacher J, Aziz MJ, Karma A, Dimitrov N, Sieradzki K (2001) Nature 410:450
12. Tulimieri DJ, Yoon J, Chan MHW (1999) Phys. Rev. Lett. 82:121
13. Cortie MB, Maaroof AI, Stokes N, Mortari A (2007) Aust. J. Chem. 60:524
14. Hayes JR, Hodge AM, Biener J, Hamza AV, Sieradzki K (2006) J. Mater. Res. 21:2611
15. Pugh DV, Dursun A, Corcoran SG (2003) J. Mater. Res. 18:216
16. Zhu JZ, Seker E, Bart-Smith H, Begley MR, Kelly RG, Zangari G, Lye WK, Reed ML (2006) Appl. Phys. Lett. 89:133104
17. Dixon MC, Daniel TA, Hieda M, Smilgies DM, Chan MHW, Allara DL (2007) Langmuir 23:2414
18. Li R, Sieradzki K (1992) Phys. Rev. Lett. 68:1168
19. Kucheyev SO, Hayes JR, Biener J, Huser T, Talley CE, Hamza AV (2006) Appl. Phys. Lett. 89:053102
20. Qian LH, Chen MW (2007) Appl. Phys. Lett. 91:083105
21. Hodge AM, Biener J, Hayes JR, Bythrow PM, Volkert CA, Hamza AV (2007) Acta Mater. 55:1343
22. Ding Y, Erlebacher J (2003) J. Am. Chem. Soc. 125:7772
23. Nyce GW, Hayes JR, Hamza AV, Satcher JH (2007) Chem. Mater. 19:344
24. Hayes JR, Nyce GW, Kuntz JD, Satcher JH, Hamza AV (2007) Nanotechnology 18:275602
25. Callister WD (2003) Materials Science and Engineering: An Introduction. Wiley, New York
26. Motz C, Pippan R (2001) Acta Mater. 49:2463
27. Olurin OB, Fleck NA, Ashby MF (2000) Mater. Sci. Eng. A 291:136
28. Andrews EW, Gioux G, Onck P, Gibson LJ (2001) Int. J. Mech. Sci. 43:701
29. Liu Z, Chuah CSL, Scanlon MG (2003) Acta Mater. 51:365
30. Wilsea M, Johnson KL, Ashby MF (1975) Int. J. Mech. Sci. 17:457
31. Toivola Y, Stein A, Cook RF (2004) J. Mater. Res. 19:260
32. Ramamurty U, Kumaran MC (2004) Acta Mater. 52:181
33. Biener J, Hodge AM, Hamza AV, Hsiung LM, Satcher JH (2005) J. Appl. Phys. 97:024301
34. Hakamada M, Mabuchi M (2007) Scr. Mater. 56:1003
35. Lee D, Wei X, Chen X, Zhao M, Jun SC, Hone J, Herbert EG, Oliver WC, Kysar JW (2007) Scr. Mater. 56:437
36. Volkert CA, Lilleodden ET, Kramer D, Weissmuller J (2006) Appl. Phys. Lett. 89:061920
37. Biener J, Hodge AM, Hayes JR, Volkert CA, Zepeda-Ruiz LA, Hamza AV, Abraham FF (2006) Nano Lett. 6:2379
38. Seker E, Gaskins JT, Bart-Smith H, Zhu J, Reed ML, Zangari G, Kelly R, Begley MR (2007) Acta Mater. 55:4593
39. Lee D, Wei XD, Zhao MH, Chen X, Jun SC, Hone J, Kysar JW (2007) Model. Simul. Mater. Sci. Eng. 15:S181
40. Mathur A, Erlebacher J (2007) Appl. Phys. Lett. 90:061910
41. Volkert CA, Lilleodden ET (2006) Philos. Mag. 86:5567
42. Davis JR (1998) Metals Handbook. ASM International, Materials Park
43. Yamada Y, Wen C, Shimojima K, Mabuchi M, Nakamura M, Asahina T, Aizawa T, Higashi K (2000) Mater. Trans. JIM 41:1136
44. Nieh TG, Higashi K, Wadsworth J (2000) Mater. Sci. Eng. A 283:105
45. Greer JR, Oliver WC, Nix WD (2005) Acta Mater. 53:1821
46. Greer JR, Nix WD (2005) Appl. Phys. A Mater. Sci. Process. 80:1625
47. Wu B, Heidelberg A, Boland JJ (2005) Nat. Mater. 4:525
48. Koh SJA, Lee HP (2006) Nanotechnology 17:3451
49. Zepeda-Ruiz LA, Sadigh B, Biener J, Hodge AM, Hamza AV (2007) Appl. Phys. Lett. 91:101907
50. Sun Y, Ye J, Shan Z, Minor AM, Balk TJ (2007) JOM 59:54
51. Hodge AM, Biener J, Hsiung LL, Wang YM, Hamza AV, Satcher Jr JH (2005) J. Mater. Res. 20:554

52. Tanimoto H, Fujita H, Mizubayashi H, Sasaki Y, Kita E, Okuda S (1996) Mater. Sci. Eng. A 217:108
53. Diao JK, Gall K, Dunn ML (2004) J. Mech. Phys. Sol. 52:1935
54. Zhou LG, Huang HC (2004) Appl. Phys. Lett. 84:1940
55. Senior NA, Newman RC (2006) Nanotechnology 17:2311
56. Doucette R, Biener M, Biener J, Cervantes O, Hamza AV, Hodge AM (2008) Scr. Mater. (to be submitted)
57. Wu HA (2006) Mech. Res. Commun. 33:9
58. Biener J, Hodge AM, Hamza AV (2005) Appl. Phys. Lett. 87:121908
59. Friedersdorf F, Sieradzki K (1996) Corrosion 52:331
60. Sieradzki K, Newman RC (1985) Philos. Mag. A 51:95
61. Kelly RG, Frost AJ, Shahrabi T, Newman RC (1991) Metall. Trans. A 22:531
62. Meinel K, Klaua M, Bethge H (1988) Phys. Stat. Sol. A 106:133
63. Celarie F, Prades S, Bonamy D, Ferrero L, Bouchaud E, Guillot C, Marliere C (2003) Phys. Rev. Lett. 90:075504
64. Duxbury PM, Leath PL, Beale PD (1987) Phys. Rev. B 36:367
65. Kahng B, Batrouni GG, Redner S, de Arcangelis L, Herrmann HJ (1988) Phys. Rev. B 37:7625

Chapter 7
Residual Stress Determination Using Nanoindentation Technique

Zhi-Hui Xu and Xiaodong Li

7.1 Introduction

Residual stress is defined as the stress that remains in a material without an external load being applied to the material. It originates from the misfits between regions and exists widely in the devices and components for various engineering applications, such as thin films and lines deposited on substrates in microelectronic and optoelectronic devices and components, surface coatings designed to influence optical, magnetic, thermal, environmental, and tribological performance, and joined components [49, 67]. The existing residual stress field in these components has a strong effect on their performance and plays a vital role in their failure behavior. Thus, determination of residual stress field is very important in both scientific and technological perspectives.

Various methods have been developed to measure the residual stresses in materials, such as neutron and X-ray tilt [16], beam bending, hole drilling [12], layer removal [35], indentation crack [31] and methods using nanoindentation technique. Among all these methods, the methodologies based on nanoindentation technique have recently attracted extensive research attentions [1–3, 8, 19, 25–27, 39, 40, 42–44, 47, 50, 56, 57, 62, 64–66]. Unlike the other methods, nanoindentation technique has the capability of deforming materials at a very small scale and allows the determination of residual stresses to be performed at micro/nanoscale.

The purpose of this chapter is to summarize the recent research efforts on the determination of residual stress using nanoindentation technique. Emphasis is placed on the influence of residual stress on nanoindentation behavior and the various methodologies developed on the basis of nanoindentation technique. Discussion on the limitation of each method has been also given where it is appropriate. Future research directions have also been pointed out after the summary of state-of-art nanoindentation technique for residual stress measurement.

Z.-H. Xu and X.D. Li
Department of Mechanical Engineering, University of South Carolina, 300 Main Street, Columbia, SC 29208, USA
xuzhihui@engr.sc.edu, LIXIAO@engr.sc.edu

F. Yang and J.C.M. Li (eds.), *Micro and Nano Mechanical Testing of Materials and Devices*,
doi: 10.1007/978-0-387-78701-5, © Springer Science+Business Media, LLC, 2008

7.2 Nanoindentation Method

Nanoindentation is a very powerful method for characterizing the mechanical behavior of materials at very small length scale [7,30,33,34]. This technique is based on high-resolution instruments that continuously monitor the load and displacement of a diamond indenter as it is pushing into and then withdrawing from a material. The load-displacement data obtained from the indentation process, which is often referred to as P–h curve of indentation, contain abundant information of material deformation and can be used to determine different mechanical properties of materials, such as hardness, elastic modulus, fracture toughness, yield stress, and work hardening rate [6,9,32,51,52,58,59,63]. A big advantage of nanoindentation test over conventional hardness test is that the contact area of nanoindentation can be directly determined from the P–h curve with the known geometry of the indenter. This feature makes it particularly suitable for measuring the mechanical properties of materials at small scales where accurate determination of the contact area would be an extremely difficult task for conventional hardness test.

A typical load-penetration depth curve of nanoindentation usually consists of two parts, loading and unloading, as shown in Fig. 7.1. The loading part normally includes the elastic-plastic deformation of the material and can be expressed as

$$P = Ah^2, \tag{7.1}$$

where P is the indentation load, h is the penetration depth measured from surface, and A is a constant that is dependent on the geometry of the indenter and the mechanical properties of the material. The relationship in (7.1) has been confirmed by the experiments [14,37], the finite element analyses [10,21], and the dimensional analysis [4,60] of indentation on elastic-plastic materials.

For a pure elastic indentation, the P–h curve can be described by

$$P = Bh_{\text{el}}^m, \tag{7.2}$$

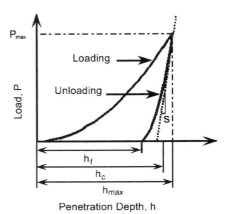

Fig. 7.1 A typical load-penetration depth curve of a depth-sensing indentation

where h_{el} is the elastic depth of the indentation, B is a constant that is related to the elastic properties of materials and the geometry of indenter, and m is a constant, which equals 1, 1.5, and 2 for a flat cylindrical punch, sphere or parabola of rotation, and cone, respectively [38]. For indentation on elastic–plastic materials, the unloading part of the P–h curve is mainly elastic. Experiments conducted on a variety of materials have revealed that the unloading curve is still well-described by (7.2) with a modification of the surface perturbation [33], that is,

$$P = B(h - h_{\mathrm{f}})^m, \tag{7.3}$$

where h_f is the final penetration depth after complete unloading. The value of m obtained from the indentations of different materials ranges from about 1.2 to 1.5 for Berkovich indentation [33] and equals 2 for Vickers indentation [36].

A crucial step to determine the hardness and elastic modulus of materials from analysing the load-penetration depth curve of an indentation is the calculation of the projected contact area of the indentation under peak load. Two things must be known to calculate the projected contact area. One is the geometry of the indenter, i.e., the area function, $A = f(h)$, that relates the cross-sectional area of the indenter to the distance from its tip. This area function can be determined using either direct methods, such as directly measuring the indents made on soft material with transmission electron microscope [34] or the indenter itself with a scanning force microscope [17], or indirect methods, such as the calibration method suggested by [33]. The other parameter needed for the calculation of the projected contact area is the contact depth at the peak load. As shown in Fig. 7.2, the contact depth at peak load is given by

$$h_{\mathrm{c}} = h_{\max} - h_{\mathrm{s}}, \tag{7.4}$$

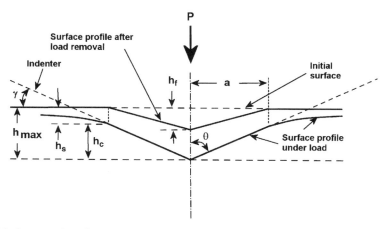

Fig. 7.2 Cross-section of an indentation and the parameters used in the analysis of nanoindentation [33]

where h_c is the contact depth, h_{max} is the maximum penetration depth, which can be directly determined from the load-penetration depth curve, and h_s is the deflection of the surface at the perimeter of the contact, which is given by Sneddon's equation [38]

$$h_s = \varepsilon \frac{P_{max}}{S}, \qquad (7.5)$$

where ε is a geometric constant and $\varepsilon = 0.72$ for cone, $\varepsilon = 0.75$ for paraboloid of revolution, and $\varepsilon = 1$ for flat punch. S is the contact stiffness at the initial unloading and $S = dP/dh$, which can be directly obtained by either linear curve fitting of the top one-third of the unloading data [7] or differentiating (7.3) at h_{max} [33]. The contact depth is then given by

$$h_c = h_{max} - \varepsilon \frac{P}{S}, \qquad (7.6)$$

and the projected contact area can be calculated from the relation

$$A_c = f(h_c). \qquad (7.7)$$

With the projected contact area so determined, the hardness of materials can be calculated using (7.1) and the elastic modulus is determined by

$$E_r = \frac{\sqrt{\pi}}{2} \frac{S}{\sqrt{A_c}} \qquad (7.8)$$

with

$$\frac{1}{E_r} = \frac{1 - v_i^2}{E_i} + \frac{1 - v_s^2}{E_s}, \qquad (7.9)$$

where E_r is the so-called reduced modulus because of a nonrigid indenter, E_i is the elastic modulus of the indenter, E_s is the elastic modulus of the specimen, v_i and v_s are the Poisson's ratio of the indenter and the specimen, respectively. It should be noted that (7.8) is based on the solution of Sneddon [38] for the elastic deformation of an isotropic elastic materials with a flat-ended cylindrical punch and should be modified with a certain factor when it is applied to analyze the indentation data of different indenter shapes [15, 20, 46] and different materials [15, 46].

7.3 Determination of Residual Stress with Nanoindentation Technique

Generally, there are two ways that nanoindentation technique can be used to determine the surface residual stresses. One is by making an indentation in a residual stress field that generates radial cracks at the corner of the indent. The lengths of these cracks are sensitive to the magnitude and sign of the residual stress state where the indent is made. By measuring the crack lengths of indentations on stressed surfaces and comparing them with the crack length of indentations on unstressed surface, the residual stress can be estimated [5, 11, 13, 18, 22–24, 31, 41, 44, 45, 48, 53–56]. Obviously the methods developed by this way can only be applied to brittle materials like ceramics where cracking usually occurs when the indentation load is larger than certain threshold value. The other way is based on the influence of residual stress on the P–h curve of nanoindentation. Residual stresses are found to have significant effects on the contact area, the loading curve and unloading curve of nanoindentation, which may be used for the determination of residual stress [1–3, 8, 19, 25–27, 39, 40, 42, 43, 47, 50, 57, 62, 64–66]. All the methods developed in this way often rely on extensive finite element modeling and can be applied to various materials and stress states.

7.3.1 Estimation of Residual Stress Using Indentation Fracture Technique

The indentation fracture technique for measuring residual stress is based on the classical fracture mechanics [22]. When indentation is made on the surface of a brittle material with a moderate force, a permanent impression is often formed with radially oriented cracks at the corner of the indent as shown in Fig. 7.3.

For sharp indenters such as Vickers or Knoop for macro/micro-indentation and Berkovich or cube corner for nanoindentation, a radial-median and lateral crack system is often formed during indentation. For the purpose of creating crack, sharper cube corner is normally preferred as it can easily generate cracks at small

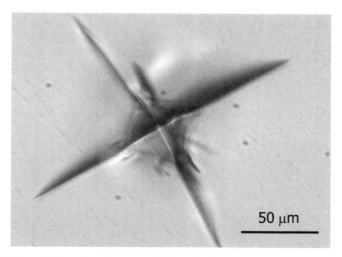

Fig. 7.3 Radial cracks formed at the corner of Vickers indentation

Fig. 7.4 Radial-median and
lateral crack systems of Vickers indentation. (**a**) *Top view*
showing Vickers indent and
surface traces of the radial-
median cracks; (**b**) *Cross-
sectional view* showing the
plastic deformation zone, the
lateral cracks, and the radial-
median cracks

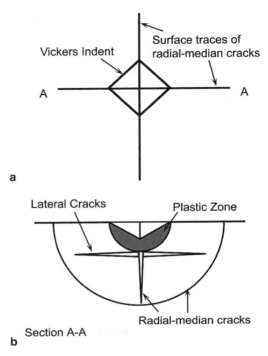

indentation force [28, 29, 61]. Schematic of the radial-median and lateral crack system of Vickers indentation is shown in Fig. 7.4.

When indentation load is low, the radial cracks normally form at the corners of
indent during unloading and maybe not extend deep into materials. These cracks
are also called Palmqvist cracks [13, 22]. When indentation load is high enough,
the median cracks nucleate and grow just underneath the plastic zone during loading and are circular in shape at the initial stages. In its fully developed form, the
median crack may extend to the surface of the sample with a final semicircular
shape [24]. In most cases, both the radial and median cracks, which are indistin-
guishable from each other, occur during the indentation process. Lateral cracks also
form underneath the plastic zone and propagate in plane that is parallel with the
indented surface and have a tendency to grow toward the free surface [5, 22, 41, 48].

Using the classic indentation fracture mechanics, the fracture toughness of
elastic–plastic brittle materials can be directly related to the length of radial-median
cracks at the surface. For residual stress free materials at equilibrium, the fracture
toughness K_c is given by [22, 23]

$$K_c = \chi P / c_0^{3/2}, \tag{7.10}$$

where c_0 is the crack length without residual stress and χ is a dimensionless constant, which is given by

Fig. 7.5 Schematic showing the composite stress intensity at the crack tip [53]

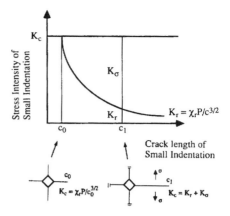

$$\chi = \xi_0(\cot\theta)^{2/3}(E/H)^{1/2}, \tag{7.11}$$

with ξ_0 is dimensionless constant, which depends on the nature of the deformation, and θ is the indenter half-angle. For materials with prevailing residual stresses, composite stress intensity will act at the crack tip (Fig. 7.5) and at equilibrium the fracture toughness is given by [53, 54].

$$K_c = \chi P \big/ c^{3/2} \pm \psi\sigma_r c^{1/2}, \tag{7.12}$$

where ψ is dimensionless constant and σ_r is the residual stress. The first term on the right hand of (7.12) represents the stress intensity due to indentation load, while the second term corresponds to the contribution of the residual stress, which is added to the first term for tensile stresses and subtracted for compressive stresses. Combining (7.10) and (7.12) and noting that the same peak load is involved in both expressions, the unknown residual stress can be solved as

$$\sigma_r = \pm K_c \left[\frac{1 - (c_0/c)^{3/2}}{\psi\sqrt{c}} \right]. \tag{7.13}$$

The state of the residual stresses can be defined by comparing c and c_0. A tensile stress will act to extend the residual stress-free crack $(c > c_0)$, while a compressive stress will shorten it $(c < c_0)$.

Although the indentation fracture technique has been developed for indentations on bulk materials at macro/micro-scale, this technique can also be modified for measuring residuals stress at nanoscale [18, 44, 45] and extended to apply to determine residual stress in thin films and coatings [11, 56]. However, the difficulty in measuring the crack length at nanoscale may severely limit the accuracy of determination of residual stress using the indentation fracture technique.

7.3.2 *Determination of Residual Stress from Indentation P – h curve*

The indentation *P–h* curve contains a wealth of information about the deformation behavior of the test materials. This curve can be used for not only the analysis of mechanical properties such as elastic modulus, hardness, work-hardening exponent, and fracture toughness but also the determination of the prevailing residual stress in materials. Both theoretical and experimental investigations show that residual stresses have significant effect on the *P–h* curve and the mechanical properties measured by nanoindentation [1, 2, 8, 19, 25, 39, 40, 43, 47, 50, 57, 62, 64–66]. Several methodologies have been developed for the determination of the residual stresses from the indentation *P–h* curves.

Suresh and Giannakopoulos [39] have developed a general methodology for the determination of the surface residual stresses by sharp instrumented indentation. Their method is based on the indentation contact area difference between a residual stress-free material and the same material with residual stress. Through theoretical analysis, Suresh and Giannakopoulos have shown that the real contact area of indentation on materials with residual stress is larger than that of a virgin one at the existence of compression residual stress while smaller for tensile residual stress. This difference is reflected in the loading curves of indentation on materials with and without residual stresses (Fig. 7.6). As expected, material with compressive residual stress requires larger force to be penetrated to the same depth as the one without residual stress while material with tensile residual stress requires smaller force. Invoking the invariance of contact pressure in the presence of residual stress, the relation between the residual stress and the ratio of the real contact area, A_c, of material with residual stress to that of the same material without residual stress, A_0, indented to the same depth, is derived as

$$\frac{A_c}{A_0} = \left(1 + \frac{\sigma_r}{H}\right)^{-1} \text{ (for tensile residual stress)} \tag{7.14}$$

$$\frac{A_c}{A_0} = \left(1 + \frac{\sigma_r \cdot \text{Sin}\gamma}{H}\right)^{-1} \text{ (for compressive residual stress),} \tag{7.15}$$

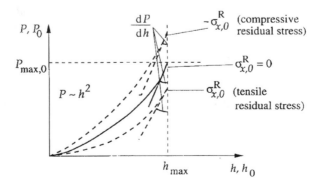

Fig. 7.6 Schematic of indentation load-penetration depth curves for the surfaces with and without residual stress [39]

where γ is the angle between the surface of the conical indenter and the contact surface. A step-by-step analysis procedure has also been recommended for extracting the elastic residual stress [39]. Recently, [25, 65] have shown that the above method can be successfully used to estimate biaxial surface stress in artificial strained steel sample and diamond-like-carbon and gold coatings on silicon substrate. Limitations to the application of this method include the requirement of a reference sample without residual stress, which is often not available, and the difficulty in the application to very soft materials where the indentation deformation is dominated by plasticity and the influence of the elastic residual stress on the contact area is relatively small.

An alternative way to measure the residual stress is using a blunt spherical indenter. Taljat and Pharr [43] has suggested that spherical indentation is more sensitive to residual stress effects than sharp indentation. They have shown that indentation load-penetration depth curve in a transition regime between elastic contact at small loads and fully developed plastic contact at large loads, the so-called elastic–plastic transition, is affected by residual stress in a potentially measurable way as shown in Fig. 7.7. Swandener et al. [40] have performed spherical nanoindentation test on aluminum alloys stressed to prescribed levels of biaxial tension and compression. They found that nanoindentation load-displacement curves are shifted to larger penetration depth by tensile stress while smaller depth by compression stress. On the basis of these observations, Swadener et al. have developed two analysis methods for measurement of residual stress using spherical indenter. The first method based on the fact that the measured depth or contact radius at the onset of yielding is affected by the stress in a way that can be analyzed by Hertzian contact mechanics. For spherical indentation, the residual stress can be related to the contact radius by

$$\frac{\sigma_r}{\sigma_y} = 1 - \frac{3.72}{3\pi}\left(\frac{E_r a}{R\sigma_y}\right)_0, \qquad (7.16)$$

where R is the indenter radius, a is the contact radius, and σ_y is the yield stress. Apparently, if an independent estimate of yield stress is available, residual stress

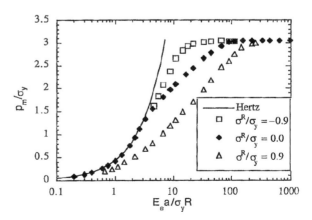

Fig. 7.7 Finite element prediction of the effect of residual stress on mean pressure [40, 43]

can be determined using (7.16) with experimental measurement of $(E_r a/R\sigma_y)_0$, which can be done by extrapolating the experimental data of h_f/h_{max} and $E_r a/R\sigma_y$ to $h_f/h_{max} = 0$.

The second method proposed by [40] is based the empirical Tabor relationship between hardness and yield stress [42], that is,

$$H = \kappa\sigma_y, \qquad (7.17)$$

where κ is the constrain factor. For materials with residual stress, experimental observations indicate that (7.17) should be modified as:

$$H + \sigma_r = \kappa\sigma_y. \qquad (7.18)$$

If the variation of $\kappa\sigma_y$ with $E_r a/R\sigma_y$ can be established by experiments in a reference material in a known state of stress, (7.18) may be used to determine residual stress since hardness can be measured by indentation test.

Limitations in the application of the above two methods are that the first method requires that the yield stress of materials is independently measured, while the second method requires a reference specimen in a known stress state. For a test material, these information may not be readily available and extra tests are needed to determine either the yield stress of material or the stress state of reference material. An attempt to use those methods in measuring residual stress in thin films has also shown that their practical application are almost impossible because of the difficulty arising from the factors such as surface roughness and substrate effects [27].

Another method for measuring the residual stress is based on the effect of the prevailing residual stress on the unloading curve of nanoindentation. It has been observed from nanoindentation tests on coatings [25, 26] and bulk materials [19, 62] that residual stress has a clear effect on the elastic recovery of indentation. Through extensive finite element simulations, Xu and Li [50, 62] have systematically investigated the influence of residual stress on the elastic recovery of nanoindentation. They have found that elastic recovery parameter h_e/h_{max} has a linear relationship with the ratio of σ_r/σ_y as shown in Fig. 7.8. A simple equation has been derived [62]

$$\frac{h_e}{h_{max}} = -\alpha\frac{\sigma_r}{\sigma_y} + \beta, \qquad (7.19)$$

where α and β are constants, β is the h_e/h_{max} ratio at $\sigma_r = 0$, and α is dependent only on the E/σ_y ratio while independent of strain hardening behavior of materials, that is,

$$\alpha = 10.53\left(\frac{E}{\sigma_y}\right)^{-1.25}. \qquad (7.20)$$

On the basis of (7.19) and (7.20), a novel method for estimation of residual stress from elastic recovery of nanoindentation has been proposed. Combining the standard three-point bending technique, nanoindentations have been carried out to successfully determine the surface residual stress in a mechanically polished fused

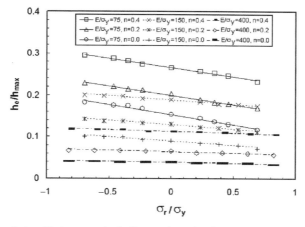

Fig. 7.8 Linear relationship between the h_e/h_{max} ratio and σ_r/σ_y ratio for materials with different E/σ_y ratios and strain-hardening exponents, n [62]

quartz beam [62]. An advantage of this method is that there is no prerequisite of the knowledge of the stress state of a reference sample or any particular mechanical properties of the test materials. However, as the method rely on an accurate determination of the h_e/h_{max} ratio, experimental factors such as surface roughness may result in large error in the experimental determination of the h_e/h_{max} ratio from the unloading curve of indentation. Moreover, as shown in Fig. 7.8, the sensitivity of the h_e/h_{max} ratio to the residual stress is inversely proportional to the E/σ_y ratio of materials. This also limits the application of the current method to very soft materials with high E/σ_y ratios.

Recently dimensional analysis has been used to find the basic relationships between indentation parameters, such as hardness, stiffness, and piling up of materials, and unknown parameters, such as residual stress, yield stress, and elastic modulus [3, 57]. Extensive finite element simulations have been carried out to build these basic relations. Using the so determined relationship, residual stress can then be estimated by reverse analysis method. Apparently these methods based on dimensional and reverse analyses require no reference samples for comparison. Although there is no prerequisite for the application of these methods, the basic relationships developed by finite element simulations with certain specific boundary conditions will limit the application of these methods, and no direct experimental validation of these methods is available.

7.4 Summary and Future Outlook

Nanoindentation technique has been widely used for mechanical characterization of materials at micro/nanoscale. The indentation load-penetration depth curve contains a wealth of information about the deformation behavior and can be used to determine various mechanical properties. This chapter has summarized the current research work on the development and application of methodologies for measuring residual stresses using nanoindentation technique.

The existence of residual stresses affects the cracking and indentation behavior of materials. Fundamental relationships developed on the basis of indentation fracture mechanics or theoretical and finite element analysis of nanoindentation provide the bases for developing different methods for determination of residual stresses using nanoindentation technique. On the one hand, method based on indentation fracture is only applicable to brittle materials, where fracture of materials occurs for indentation force beyond certain threshold value. In the application of indentation fracture method to nanoscale measurement of residual stress, the difficulty in accurate determination of the crack length may limit the accuracy of residual stress measurement. On the other hands, methods based on residual stress effects on nanoindentation behavior may be applied to any elastic plastic materials. Different methods have been developed to estimate residual stresses from the loading curve, elastic recovery of the unloading curve, contact area difference, hardness difference, and relationship of dimensional analysis and inverse analysis. Application of these methods may be limited by the complexity of the real stress state, the prerequisite of reference sample or known material properties, or the accuracy of the primary parameter determined by nanoindentation technique.

Clearly, one of the future research directions may be to further validate the current available methods for residual stress measurement by carrying out nanoindenations on materials with known stress states. Knowledge of the limitations in the application of these methods is very important for the successful determination of the residual stresses. Another future research direction may be to develop a novel close-form method for mapping the residual stress field at small scale. The information should be essential to the successful design and manufacture and the reliability prediction of micro/nanostructures, such as micro/nano-electromechanical systems.

Acknowledgement Financial support for this work was provided by the National Science Foundation (Grant No. EPS-0296165), the South Carolina Space Grant Consortium-NASA, the South Carolina EPSCoR grant, and the University of South Carolina NanoCenter Seed Grant. The content of this information does not necessary reflect the position or policy of the government and no official endorsement should be inferred.

Notation

A Constant
A_0 Contact area without residual stress
A_c Projected contact area
a Contact radius
B Constant
c Crack length

c_0 Crack length without residual stress
E Elastic modulus
E_i Elastic modulus for the indenter
E_r Reduced elastic modulus
E_s Elastic modulus for the sample
H Hardness
h Indentation depth
h_c Contact depth
h_{el} Elastic depth of indentation
h_f Final depth after complete unloading
h_{max} Penetration depth at the peak load
h_s Deflection of indented surface
K_c Fracture toughness
m Constant
P Indentation load
R Indenter radius
S Contact stiffness
α Constant
β Constant
χ Dimensionless constant
ε Geometric constant
γ Angle between the indenter surface and the contact surface
κ Constrain factor
v_i Poisson's ratio for the indenter
v_s Poisson's ratio for the sample
θ Indenter half-angle
σ_r Residual stress
σ_y Yield stress
ξ_0 Dimensionless constant
ψ Dimensionless constant

References

1. Bolshakov A, Oliver WC, Pharr GM (1996) J. Mater. Res. 11:760
2. Carlsson S, Larsson PL (2001) Acta Mater. 49:2179
3. Chen X, Yan J, Karlsson AM (2006) Mater. Sci. Eng. A 416:139
4. Cheng YT, Cheng CM (1998) J. Appl. Phys. 84:1284
5. Cook RF, Roach DH (1986) J. Mater. Res. 1:589
6. Dao M, Chollacoop N, Van Vliet KJ, Venkatesh TA, Suresh S (2001) Acta Mater. 49:3899
7. Doerner MF, Nix WD (1986) J. Mater. Res. 1:601
8. Giannakopoulos AE (2003) J. Appl. Mech. Trans. ASME 70:638
9. Giannakopoulos AE, Suresh S (1999) Scr. Mater. 40:1191
10. Giannakopoulos AE, Larsson PL, Vestergaard R (1994) Int. J. Solids Struct. 31:2679
11. Gruninger MF, Lawn BR, Farabaugh EN, Wachtman Jr JB (1987) J. Am. Ceram. Soc. 70:344

12. Gupta BP (1973) Exp. Mech. 13:45
13. Hagan JT, Swain MV (1978) J. Phys. D Appl. Phys. 11:2091
14. Hainsworth SV, Chandler TF, Page TF (1996) J. Mater. Res. 11:1987
15. Hay JC, Bolshakov A, Pharr GM (1999) J. Mater. Res. 14:2296
16. Hehn L, Zheng C, Mecholsky JJ, Hubbard CR (1995) J. Mater. Sci. 30:1277
17. Herrmann K, Jennett NM, Wegener W, Meneve J, Hasche K, Seemann R (2000) Thin Solid Films 377–378:394
18. Kese K, Rowcliffe DJ (2003) J. Am. Ceram. Soc. 86:811
19. Kese KO, Li ZC, Bergman B (2004) J. Mater. Res. 19:3109
20. King RB (1987) Int. J. Solids Struct. 23:1657
21. Larsson PL, Giannakopoulos AE, Söderlund E, Rowcliffe DJ, Vestergaard R (1996) Int. J. Solids Struct. 33:221
22. Lawn BR (1993) Fracture of Brittle Solids, 2nd edn. Cambridge University Press, London
23. Lawn BR, Fuller ER (1975) J. Mater. Sci. 10:2016
24. Lawn BR, Evans AG, Marshall DB (1980) J. Am. Ceram. Soc. 63:574
25. Lee YH, Kwon D (2002) J. Mater. Res. 17:901
26. Lee YH, Kwon D, Jang JI (2003) Int. J. Mod Phys. B 17:1141
27. Lepienski CM, Pharr GM, Park YJ, Watkins TR, Misra A, Zhang X (2004) Thin Solid Films 447–448:251
28. Li XD, Bhushan B (1998) Thin Solid Films 315:214
29. Li XD, Diao D, Bhushan B (1997) Acta Mater. 45:4453
30. Loubet JL, Georges JM, Marchesini O, Meille G (1984) J. Tribol. 106:43
31. Marshall DB, Lawn BR (1977) J. Am. Ceram. Soc. 60:86
32. Mata M, Alcala J (2003) J. Mater. Res. 18:1705
33. Oliver WC, Pharr GM (1992) J. Mater. Res. 7:1564
34. Pethica JB, Hutchings R, Oliver WC (1983) Philos. Mag. A 48:593
35. Read WT (1951) J. Appl. Phys. 22:415
36. Sakai M (2003) J. Mater. Res. 18:1631
37. Sakai M, Nakano Y (2002) J. Mater. Res. 17:2161
38. Sneddon IN (1965) Int. J. Eng. Sci. 3:47
39. Suresh S, Giannakopoulos AE (1998) Acta Mater. 46:5755
40. Swandener JG, Taljat B, Pharr GM (2001) J. Mater. Res. 16:2091
41. Swain MV (1976) J. Mater. Sci. Lett. 11:2345
42. Tabor D (1951) The Hardness of Metals. Clarendon, Oxford
43. Taljat B, Pharr GM (2000) Mater. Res. Soc. Symp. Proc. 594:519
44. Tandon R (2007) J. Eur. Ceram. Soc. 27:2407
45. Tandon R, Buchheit TE (2007) J. Am. Ceram. Soc. 90:502
46. Troyon M, Huang L (2005) J. Mater. Res. 20:610
47. Tsui TY, Oliver WC, Pharr GM (1996) J. Mater. Res. 11:752
48. Whittle BR, Hand RJ (2001) J. Am. Ceram. Soc. 84:2361
49. Withers PJ, Bhadeshia HKDH (2001) Mater. Sci. Technol. 17:355
50. Xu ZH, Li XD (2005) Acta Mater. 53:1913
51. Xu ZH, Rowcliffe D (2002) Philos. Mag. A 82:1893
52. Zeng K, Chiu CH (2001) Acta Mater. 49:3539
53. Zeng K, Rowcliffe DJ (1995) Acta Metall. Mater. 43:1935
54. Zeng K, Giannakopoulous AE, Rowcliffe DJ (1995) Acta Metall. Mater. 43:1945
55. Zeng K, Giannakopoulous AE, Rowcliffe D, Meier P (1998) J. Am. Ceram. Soc. 81:689
56. Zhang TY, Chen LQ, Fu R (1999) Acta Mater. 47:2869
57. Zhao M, Chen X, Yan J, Karlsson AM (2006) Acta Mater. 54:2823
58. Cheng YT, Cheng CM (2004) Mater. Sci. & Eng. R 44:91
59. Li XD, Bhushan B (2002) Mater. Charact. 48:11
60. Cheng YT, Cheng CM (1999) Int. J. Solids Structures 36:1231
61. Li XD, Bhushan B (1999) Thin Solid Films 355-356:330
62. Xu ZH, Li XD (2006) Phil. Mag. 86:2835

63. Oliver WC, Pharr GM (2004) J. Mater. Res. 19:3
64. Carlsson S, Larsson PL (2001) Acta Mater. 49:2193
65. Lee YH, Kwon D (2004) Acta Mater. 52:1555
66. Lee YH, Kwon D (2004) Exp. Mech. 44:55
67. Withers PJ, Bhadeshia HKDH (2001) Mater. Sci. Technol. 17:366

Chapter 8
Piezoelectric Response in the Contact Deformation of Piezoelectric Materials

Fuqian Yang

8.1 Introduction

Micro- and nanoelectromechanical systems (MEMS and NEMS) using low-dimensional piezoelectric structures (piezoelectric thin films, nanowires, and nanobelts) [1–10] have potential applications in many areas, including biosensors, actuators, and motion-controllers due to intrinsic electromechanical coupling. The electromechanical coupling provides a unique route for sensing mechanical stimuli from the change in electric potential/field, and for controlling structural deformation via electrical loading [11], which determines the performance and lifetime of micro- and nanodevices during the device operation. It becomes of primary interest to experimentally and theoretically investigate the piezoelectric behavior of materials on both the micro- and nanoscales under electrical and mechanical loading for the development, design, and process control of MEMS and NEMS devices.

Piezoelectricity is an interaction between mechanical deformation and electric field. Several techniques have been used to characterize the piezoelectric behavior of piezoelectric structures and materials, which are based on either the direct piezoelectric effect or the inverse piezoelectric effect. The direct piezoelectric effect is that mechanical deformation produces electric polarization, and the inverse piezoelectric effect represents electric field-induced mechanical strain [12]. The techniques using the direct piezoelectric effect include stress-induced charge (Berlincourt method) and indentation, and the techniques using the inverse piezoelectric effect consist of laser interferometers, laser scanning vibrometers, and piezoresponse force microscope. The Berlincourt method, the laser interferometers, and the laser scanning vibrometers can be readily used in determining the piezoelectric behavior of bulk piezoelectric materials, while it is very difficult if not impossible to apply these

F. Yang
Department of Chemical and Materials Engineering, University of Kentucky, Lexington, KY 40506
fyang0@engr.uky.edu

F. Yang and J.C.M. Li (eds.), *Micro and Nano Mechanical Testing of Materials and Devices*,   155
doi: 10.1007/978-0-387-78701-5, © Springer Science+Business Media, LLC, 2008

techniques to evaluate the electromechanical functionality on the nanoscale in low-dimensional piezoelectric structures and materials. This is due to the size constraint and the sensitivity/resolution of the techniques, which limit their applications in characterizing low-dimensional structures and materials.

Advances in the micro- and nanofabrication of nanoelectronics and micro- and nanodevices have resulted in the development of surface force microscopy, including nanoindentation and scanning force microscopy (SFM), which has been used to evaluate the mechanical behavior of low-dimensional structures. The principle of the nanoindentation and SFM techniques is based on the surface–contact interaction through direct contact and/or intermediate contact between the probe-tip and the surface of specimen. The sensitivity and resolution are determined by the stiffness of the plate for the nanoindentation and the cantilever beam for the SFM. The advantages of using the surface force techniques involve the use of a small amount of materials such as low-dimensional structures and possibly the evaluation of the local behavior of heterogeneous materials.

The use of contact mechanics was proposed by Lefki and Dormans [13] in measuring the piezoelectric coefficients of lead zirconate titanate (PZT) thin films through continuous charge integration. A similar technique was later used by Fu et al. [14] to measure the piezoelectric effect of $PbZr_{0.53}Ti_{0.47}O_3$ thin films. Suresh and his co-workers [15–17] extended the indentation technique to evaluate the electric response of 1–3 piezoelectric ceramic–polymer composite, lead zirconate titanate, and barium titanate by monitoring the electric current passing through the conductive indenter and the counter electrode. They did not evaluate the piezoelectric response. Recently, Rar et al. [18] modified nanoindentation to assess the piezoresponse of polycrystalline lead zirconate titanate and $BaTiO_3$ piezoceramics by applying an oscillating voltage between the indenter-tip and the back-electrode, similar to the technique of the piezoresponse force microscopy (PFM), as demonstrated by Birk et al. [19].

In parallel, SFM modulated by electric field, such as electric force microscopy [20] and PFM [19, 21], has been developed to map surface charges and/or examine local electromechanical behavior. The concept of PFM was demonstrated first by Birk et al. [19], using scanning tunneling microscopy to measure local piezoelectric activity of piezoelectric thin films made of vinylidene fluoride-trifluoroethylene copolymer. The principle of the PFM technique is based on monitoring the deflection of a cantilever beam responding to local surface oscillation of a piezoelectric material, which is modulated by an AC electric voltage between the conductive tip of the SFM and the counter electrode underneath the piezoelectric material [19, 21–24]. The development of MEMS and NEMS devices in the last decade has made the PFM as an important technique for evaluating nanopiezoelectric activities of low-dimensional structures and materials. This has imposed a tremendous challenge in understanding the local piezoelectric activities for quantifying the piezoelectric behavior of materials.

This chapter is devoted to the common theme of the contact deformation in piezoelectric materials and its applications in the indentation testing and the PFM for the characterization of the piezoelectric behavior in piezoelectric materials.

It summarizes the dependence of the contact deformation on the electromechanical interaction and the interconnection that correlate local piezoelectric activities with piezoelectric properties and the deformation behavior. It reviews the achievements in evaluating the fracture behavior of piezoelectric materials using the indentation technique. In particular, it emphasizes the need for much-deeper studies of the electromechanical interaction present in the contact deformation of piezoelectric materials.

8.2 Fundamental Equations

Crystals deform when subjected to mechanical loading. In certain classes of crystals such as ZnO and ZnS, the deformation leads to the displacement of ions and results in electrical polarization in a crystal. This produces accumulation of electric charges and measurable difference of electrical potential. Such phenomenon is called the direct piezoelectric effect, as discovered by Pierre and Jacques Currie. On the other hand, an electric field can cause the distortion of the crystal, i.e., the inverse piezoelectric effect, as predicted by Lippmann and verified by the Curies. Figure 8.1 shows the atomic origin of piezoelectricity. In a stress-free state, the center of positive charges is the same as that of negative charges. The resultant charges contributed from positive charges and negative charges are zero, and there is no electrical potential gradient over the crystal. Under the action of mechanical loading, the crystal is deformed to form electrical dipole layers by moving the positive charges closer at one side and the negative charges at the other. This introduces electrical polarization and electrical field in the crystal. The strength of the stress-induced electrical polarization is a function of the bonding strength of the crystal and is represented by piezoelectric coefficients.

The phenomenological theory of linear piezoelectricity is established from the principles of thermodynamics [12]. Parton and Kudryavtsev [25] have summarized

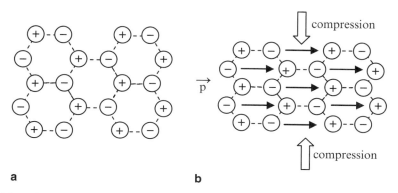

Fig. 8.1 (a) Stress-free state; (b) stress-induced electrical polarization – piezoelectricity

the constitutive relation and piezocoupling coefficients in linear piezoelectric systems. For a piezoelectric material, the equilibrium equations in a Cartesian coordinate system, x_i $(i = 1,2,3)$, are

$$\sigma_{ij,i} = 0 \text{ and } D_{i,i} = 0, \tag{8.1}$$

where σ_{ij} is the stress tensor, D_i is the electric displacement vector, a comma denotes the partial differentiation with respect to the coordinate x_i, and the Einstein summation convention over repeated indices is used. The constitutive relation of an anisotropic, linear piezoelectric material is

$$\sigma_{ij} = c_{ijkl}\varepsilon_{kl} - e_{kij}E_k \text{ and } D_i = e_{ikl}\varepsilon_{kl} + \in_{ik}E_k, \tag{8.2}$$

where ε_{ij} is the strain tensor, E_i is the electric field intensity, c_{ijkl} is the elastic stiffness tensor measured in a constant electric field intensity, e_{ikl} is the piezoelectric tensor measured in the possession of a spontaneous electric field, and $\in_{ik}$ is the dielectric tensor. The interchange symmetry of the tensors gives

$$c_{ijkl} = c_{ijlk} = c_{jikl} = c_{jilk} = c_{klij}, e_{kij} = e_{kji}, \text{ and } \in_{ij} = \in_{ji}. \tag{8.3}$$

The relation between the components of the strain tensor and the components of the displacement vector, u_i, is given by

$$\varepsilon_{ij} = (u_{i,j} + u_{j,i})/2. \tag{8.4}$$

The electric field intensity can be described as

$$E_i = -\phi_{,i}, \tag{8.5}$$

where the electrical potential, ϕ, satisfies the Laplacian equation.

The effect of the crystalline symmetry on the independent components in the physical property tensors has been summarized by Nye [26]. The following is a list of the elastic stiffness matrix, the piezoelectric matrix, and the dielectric matrix for the symmetrical classes of 32, 3m, 6mm, and $\overline{4}3$ m. The dielectric matrix is [27]

$$\begin{pmatrix} \in_{11} & 0 & 0 \\ 0 & \in_{22} & 0 \\ 0 & 0 & \in_{33} \end{pmatrix}$$ with $\in_{11} = \in_{22}$ for the classes of 32, 3 m, and 6 mm, and $\in_{11} = \in_{22} = \in_{33}$ for the class of $\overline{4}3$ m.

The piezoelectric matrix is [27]

$$
\begin{array}{cccc}
\text{Class 32} & \text{Class 3m} & \text{Class 6mm} & \text{Class } \overline{4}3\text{m} \\
\begin{pmatrix} e_{11} & 0 & 0 \\ -e_{11} & 0 & 0 \\ 0 & 0 & 0 \\ e_{41} & 0 & 0 \\ 0 & -e_{41} & 0 \\ 0 & -e_{11} & 0 \end{pmatrix}
&
\begin{pmatrix} 0 & -e_{22} & e_{13} \\ 0 & e_{22} & e_{13} \\ 0 & 0 & e_{33} \\ 0 & e_{51} & 0 \\ e_{51} & 0 & 0 \\ e_{22} & 0 & 0 \end{pmatrix}
&
\begin{pmatrix} 0 & 0 & e_{13} \\ 0 & 0 & e_{13} \\ 0 & 0 & e_{33} \\ 0 & e_{51} & 0 \\ e_{51} & 0 & 0 \\ 0 & 0 & 0 \end{pmatrix}
&
\begin{pmatrix} 0 & 0 & 0 \\ 0 & 0 & 0 \\ 0 & 0 & 0 \\ e_{41} & 0 & 0 \\ 0 & e_{41} & 0 \\ 0 & 0 & e_{41} \end{pmatrix}.
\end{array}
$$

The elastic stiffness matrix is [27]

$$
\begin{array}{ccc}
\text{Classes 32 and 3 m} & \text{Class 6 mm} & \text{Class } \bar{4}3\text{ m} \\[4pt]
\begin{pmatrix}
c_{11} & c_{12} & c_{13} & c_{14} & 0 & 0 \\
c_{12} & c_{11} & c_{13} & -c_{14} & 0 & 0 \\
c_{13} & c_{13} & c_{33} & 0 & 0 & 0 \\
c_{14} & -c_{14} & 0 & c_{44} & 0 & 0 \\
0 & 0 & 0 & 0 & c_{44} & c_{14} \\
0 & 0 & 0 & 0 & c_{14} & c_{66}
\end{pmatrix}
&
\begin{pmatrix}
c_{11} & c_{12} & c_{13} & 0 & 0 & 0 \\
c_{12} & c_{11} & c_{13} & 0 & 0 & 0 \\
c_{13} & c_{13} & c_{33} & 0 & 0 & 0 \\
0 & 0 & 0 & c_{44} & 0 & 0 \\
0 & 0 & 0 & 0 & c_{44} & 0 \\
0 & 0 & 0 & 0 & 0 & c_{66}
\end{pmatrix}
&
\begin{pmatrix}
c_{11} & c_{12} & c_{12} & 0 & 0 & 0 \\
c_{12} & c_{11} & c_{12} & 0 & 0 & 0 \\
c_{12} & c_{12} & c_{11} & 0 & 0 & 0 \\
0 & 0 & 0 & c_{44} & 0 & 0 \\
0 & 0 & 0 & 0 & c_{44} & 0 \\
0 & 0 & 0 & 0 & 0 & c_{44}
\end{pmatrix}
\end{array}
$$

with $c_{66} = (c_{11} - c_{12})/2$ for the symmetrical classes of 32, 3 m, and 6 mm.

8.3 General Solutions of Axisymmetrical Piezoelectric Coupling Problems

Several techniques have been used to derive the analytical solutions of piezoelectric deformation when subjected to mechanical and electrical loading. They include Stroh formalism [28–31], Green's function [32–36], and integral transform [37–42]. In general, it is very difficult to obtain closed-form solutions for anisotropic materials. There are limited analytical solutions available for electromechanical deformation of piezoelectric materials, depending on the geometry and the loading condition. Most analytical solutions have been reported for transversely isotropic piezoelectric materials.

For a semi-infinite transversely isotropic piezoelectric material of the hexagonal crystal class 6 mm with deformation symmetrical about the poled direction, the general solutions of the field variables are [38]

$$
u_r(r,z) = \int_0^\infty U(\zeta,z) J_1(\zeta r)\,d\zeta, \tag{8.6}
$$

$$
[u_z(r,z), \phi(r,z)] = \int_0^\infty [V(\zeta,z), \Phi(\zeta,z)] J_0(\zeta r)\,d\zeta, \tag{8.7}
$$

where u_r and u_z are, respectively, the displacement components along the radial and poled directions, and $J_0(\cdot)$ and $J_1(\cdot)$ are, respectively, the Bessel functions of the first kind of order 0 and 1. The auxiliary functions (U, V, Φ) can be expressed as [38]

$$
(U, V, \Phi) = (\Delta_{21}, \Delta_{22}, \Delta_{23}) f \tag{8.8}
$$

with

$$
\Delta_{21} = (c_{13} + c_{44})\zeta \frac{d}{dz}\left(\in_{11} \zeta^2 - \in_{33} \frac{d^2}{dz^2}\right) - (e_{31} + e_{15})\zeta \frac{d}{dz}\left(-e_{15}\zeta^2 + e_{33}\frac{d^2}{dz^2}\right)
$$

$$\Delta_{22} = \left(\in_{11} \zeta^2 - \in_{33} \frac{d^2}{dz^2} \right) \left(c_{44} \frac{d^2}{dz^2} - c_{11} \zeta^2 \right) + (e_{31} + e_{15})^2 \zeta^2 \frac{d^2}{dz^2}$$

$$\Delta_{23} = - \left(-e_{15} \zeta^2 + e_{33} \frac{d^2}{dz^2} \right) \left(c_{44} \frac{d^2}{dz^2} - c_{11} \zeta^2 \right) - (c_{13} + c_{44})(e_{31} + e_{15}) \zeta^2 \frac{d^2}{dz^2}.$$

Here, the solution of f is a function of $\exp(k\zeta z)$, in which k is the root of the algebraic equation,

$$ak^6 - bk^4 + ck^2 - d = 0 \tag{8.9}$$

with a, b, c, and d are constants, depending on the piezoelectric properties of piezoelectric materials. For detailed information of a, b, c, and d, see [38].

For a semi-infinite transversely isotropic piezoelectric material in the space of $z > 0$, the function f has four possible solutions, depending on the properties of k_i^2:

(a) For $k_1^2 \neq k_2^2 \neq k_3^2 > 0$,

$$f = A_1(\zeta)e^{-k_1\zeta z} + A_2(\zeta)e^{-k_2\zeta z} + A_3(\zeta)e^{-k_3\zeta z}. \tag{8.10}$$

(b) For $k_1^2 \neq k_2^2 = k_3^2 > 0$,

$$f = A_1(\zeta)e^{-k_1\zeta z} + [A_2(\zeta) + A_3(\zeta)\zeta z]e^{-k_2\zeta z} \tag{8.11}$$

(c) For $k_1^2 = k_2^2 = k_3^2 > 0$,

$$f = [A_1(\zeta) + A_2(\zeta)\zeta z + A_3(\zeta)\zeta^2 z^2]e^{-k_1\zeta z}. \tag{8.12}$$

(d) For $k_1^2 > 0$ and k_2^2, $k_3^2 < 0$ or k_2^2 and k_3^2 being a pair of conjugate complex roots, the k_2 and k_3 are a pair of conjugate complexes $-\delta \pm i\tilde{\omega}$. The solution of f is

$$f = A_1(\zeta)e^{-k_1\zeta z} + A_2(\zeta)e^{-\delta\zeta z}\cos \tilde{\omega}\zeta z + A_3(\zeta)e^{-\delta\zeta z}\sin\tilde{\omega}\zeta z. \tag{8.13}$$

Here δ and $\tilde{\omega} > 0$, and A_i $(i = 1,2,3)$ are functions of ζ to be determined by the boundary conditions.

Using (8.6)–(8.13), one can obtain the distribution of the displacement, stresses, electric displacement, and electrical potential.

8.4 Indentation of Piezoelectric Materials

8.4.1 Contact Mechanics of Semi-Infinite Transversely Isotropic Piezoelectric Materials

The contact problem of piezoelectric materials is complex due to anisotropic nature. The closed-form solutions are only available for the contact between a semi-infinite transversely isotropic piezoelectric material and a rigid axisymmetric solid. However, the results are controversial.

Matysiak [37] was the first to address the axisymmetric contact of a linear piezo-electric halfspace by a conductive indenter, using the Hankel transform. He obtained the dependence of the indentation load on the contact radius for the indentation by a flat-ended indenter and a spherical indenter and noted the contribution of piezoelec-tric coefficients to the indentation deformation. Giannakopoulos and Suresh [43] followed Matysiak's approach and summarized the relations between the indenta-tion load and the contact area for the indentations by three different conductive rigid indenters, including flat-ended indenter, spherical indenter, and conical inden-ter. The dependence of the indentation load on the contact area for the spherical indenter and the conical indenter is similar to that for the indentation of an elastic halfspace with coefficients being a function of piezoelectric constants.

Chen and Ding [44] used potential theory and the superposition principle to obtain the relationship between the indentation load and the contact area. They obtained the singular distribution of the electric displacement vector and found that the indentation depth is proportional to the contact area, independent of the elec-trical potential over the indenter. In contrast to the results given by Matysiak [37] and Giannakopoulos and Suresh [43], the relation between the indentation load and the contact area is a function of the applied electrical potential over the conductive indenter. Karapetian et al. [45] noted the incorrect boundary condition used in the work of Giannakopoulos and Suresh [43] for the indentation by a rigid flat-ended indenter, while they did not discuss the difference between the results of Chen and Ding [41] and those of Giannakopoulos and Suresh [43]. They used the correspon-dence principle and the superposition principle to obtain the field distribution in-side crystal and claimed that their results are the same as those given by Chen and Ding [41]. The concern on their approach [44, 45] is that they did not apply the displacement constraint (the normal displacement is zero over the contact area) to the subproblem with applied electrical loading. This likely causes the discrepancy between their results and those given by Matysiak [37] and Giannakopoulos and Suresh [43].

For the indentation of a semi-infinite transversely isotropic piezoelectric material by a rigid conductive flat-ended indenter of radius a with an electric potential of ϕ, the contact area remains same. The dependence of the indentation load of F and the resultant induced electric charge of Q over the contact area on the indentation depth h is [37,43]

$$F = 4a(C_1 h - C_2 \phi), \tag{8.14}$$

$$Q = 4a(C_3 \phi + C_4 h), \tag{8.15}$$

where h is the indentation depth and C_i $(i = 1,2,3,4)$ are constants, depending on the material properties of the piezoelectric material. From (8.14) and (8.15), there are

$$S = \frac{dF}{dh} = 4aC_1, \tag{8.16}$$

$$F = [C_1 Q - 4a\phi \, (C_3 C_1 + C_2 C_4)] C_4^{-1}, \tag{8.17}$$

$$\dot{Q}/\dot{F} = C_4/C_1. \tag{8.18}$$

Here, the dot represents the time derivative. The contact stiffness, S, is proportional to the indenter radius. The indentation load is a linear function of the electric-polarized charge, and the charging rate (i.e., electric current) is proportional to the indentation loading rate.

For the indentation by a rigid conductive spherical indenter of radius R with an electric potential of ϕ, the indentation depth changes with the contact radius of a as

$$h = \frac{a^2}{R} + C_5\phi. \tag{8.19}$$

The dependence of the indentation load and the resultant induced electric charge on the indentation depth is [37, 43]

$$F = \frac{8C_1 a^3}{3R} = \frac{8C_1}{3} R^{1/2}(h - C_5\phi)^{3/2}, \tag{8.20}$$

$$Q = \frac{8C_6 a^3}{3R} + 4a\phi(C_3 + C_4 C_5) = \frac{8C_6}{3} R^{1/2}(h - C_5\phi)^{3/2}$$
$$+ 4\phi(C_3 + C_4 C_5)R^{1/2}(h - C_5\phi)^{1/2}. \tag{8.21}$$

Here C_5 and C_6 are two constants, determined by the material properties of the piezoelectric material. Equations (8.20) and (8.21) give

$$S = 4C_1 R^{1/2}(h - C_5\phi)^{1/2}, \tag{8.22}$$

$$Q = \frac{C_6 F}{C_1} + 2\phi(C_3 + C_4 C_5)\left(\frac{3RF}{C_1}\right)^{1/3}, \tag{8.23}$$

$$\dot{Q}/\dot{F} = \frac{C_6}{C_1} + \frac{2\phi(C_3 + C_4 C_5)F^{-2/3}}{3}\left(\frac{3R}{C_1}\right)^{1/3}. \tag{8.24}$$

In contrast to the indentation by a rigid flat-ended indenter, the contact stiffness is also a function of the applied electrical potential. Both the resultant electrical charge and the charging rate are nonlinear functions of the indentation load. For $\phi = 0$, the resultant electric charge is proportional to the indentation load and the charging rate to the indentation loading rate, similar to the results for the indentation by a rigid flat-ended indenter.

For the indentation by a rigid conductive conical indenter of semi-included angle θ with an electric potential of ϕ, the indentation depth changes with the contact radius as [43]

$$h = \frac{1}{2}\pi a \cot\theta + C_5\phi. \tag{8.25}$$

The dependence of the indentation load and the resultant induced electric charge on the indentation depth is [43]

$$F = \pi a^2 C_1 \cot\theta = \frac{4C_1}{\pi\cot\theta}(h - C_5\phi)^2, \tag{8.26}$$

$$Q = \pi a^2 C_6 \cot\theta + 4a\phi(C_3 + C_4 C_5) = \frac{4C_6}{\pi\cot\theta}(h - C_5\phi)^2$$
$$+ \frac{8\phi(C_3 + C_4 C_5)}{\pi\cot\theta}(h - C_5\phi), \tag{8.27}$$

which give

$$S = \frac{8C_1}{\pi\cot\theta}(h - C_5\phi), \tag{8.28}$$

$$Q = \frac{C_6 F}{C_1} + 4\phi(C_3 + C_4 C_5)\left(\frac{F}{\pi C_1 \cot\theta}\right)^{1/2}, \tag{8.29}$$

$$\dot{Q}/\dot{F} = \frac{C_6}{C_1} + \frac{2\phi(C_3 + C_4 C_5)F^{-1/2}}{(\pi C_1 \cot\theta)^{1/2}}. \tag{8.30}$$

Similar to the spherical indentation, the contact stiffness depends on the electrical potential applied to the conductive indenter. Both the resultant electric charge and the charging rate are a function of the indentation load. For $\phi = 0$, the ratio of the resultant electric charge to the indentation load is the same as that of the charging rate to the indentation loading rate, independent of the indentation load.

Equations (8.14)–(8.30) provide the rational base to evaluate the local piezoelectric interaction in transversely isotropic piezoelectric materials and to characterize the piezoelectric properties of piezoelectric materials by using the techniques of indentation and surface force microscopy. It is worth mentioning that the above results can only be applied to quasistatic indentation deformation of piezoelectric materials for the deformation rate and the indentation velocity much less than the speed of sound waves in materials. For high speed indentation, dynamic effect needs to be included in the analysis.

8.4.2 Indentation Testing of Piezoelectric Materials

Three different methods have been developed to characterize piezoelectric properties, using the indentation technique. They involve different schemes in electrical characterization or electrical modulation for measuring or controlling the indentation behavior of piezoelectric materials.

8.4.2.1 Charge Integration Technique

This method was developed by Lefki and Dormans [13], using continuous charge integration to measure the electrical charge along the loading direction. Figure 8.2 shows the schematic setup for the characterization of the effective piezoelectric coefficients using the direct piezoelectric effect. The total electric charge generated during the indentation is measured as an electrical voltage over a capacitor.

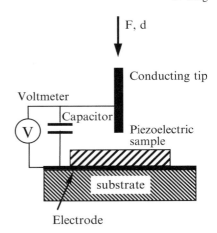

Fig. 8.2 Schematic of the setup for the characterization of the effective piezoelectric coefficients using the direct piezoelectric effect

Assuming uniform distribution of electric charge over the contact area, one can obtain the normal component of the apparent electrical displacement vector over the contact surface by calculating the apparent charge density. The effective piezoelectric coefficient d_{33}^{dp}, called the charge constant, can be determined by

$$d_{33}^{dp} \equiv \left.\frac{\partial D_3}{\partial \sigma_{33}}\right|_{\vec{E}} = \frac{\tilde{D}_3}{\tilde{\sigma}_{33}} = \frac{Q}{F} = \frac{C_C V}{F}, \tag{8.31}$$

where D_3 and σ_{33} are, respectively, the normal component of the electrical displacement vector and the stress tensor along the loading direction; $\tilde{D}_3$ and $\tilde{\sigma}_{33}$ are, respectively, the average value of D_3 and σ_{33} over the contact surface; $\vec{E}$ is the electric field intensity; Q and F are, respectively, the total electric charge and the indentation load; C_C is the capacitance of the capacitor; and V is the electrical voltage measured over the capacitor. Under simple compression (i.e., the contact size is much larger than the thickness of the piezoelectric material), the charge constant is a function of material properties as [13]

$$d_{33}^{dp} = e_{33} - 2e_{31}\left(s_{13}^E + \nu_S/E_S\right)/\left(s_{11}^E + s_{12}^E\right), \tag{8.32}$$

where ν_S and E_S are, respectively, the Poisson ratio and Young's modulus of the substrate, and s_{11}^E, s_{12}^E, and s_{13}^E are the components of the compliance matrix of the piezoelectric material subjected to constant electric field intensity. Obviously, this method cannot independently measure the piezoelectric coefficients, such as e_{33}.

It is interesting to note that electrical potential difference will be created between the conducting tip and the counter electrode during the indentation, which needs to be taken into account in the analysis. It is worth mentioning that (8.32) becomes invalid for the indentation size being compatible to or smaller than the thickness of the piezoelectric material. It would be inappropriate to use (8.17), (8.23), and (8.29) to calculate d_{33}^{dp} since the analyses do not consider the effect of the indentation-induced electric potential difference.

Lefki and Dormans [13] evaluated the dependence of the effective piezoelectric coefficient on the composition of the unpoled OMCVD PZT (lead zirconate titanate) thin films. The maximum piezoelectric effect occurred at the morphotropic phase boundary, and the presence of the piezoelectric effect in the unpoled PZT films was due to the internal stress-induced alignment of ferroelectric domains. The polling of the films dramatically increased their piezoelectric properties, as expected.

Fu et al. [14] measured the effective piezoelectric coefficient, d_{33}^{dp}, of $PbZr_{0.53}Ti_{0.47}O_3$ thin films formed by chemical solution deposition. They observed piezoelectric relaxation in the piezoelectric response during the loading and unloading. They were able to describe the relaxation process using the stretched exponential law [46,47] by introducing the intrinsic and relaxed effective piezoelectric coefficients. The piezoelectric relaxation was due to the motion of domain walls of non-180° domains, which contributed 15% to the total piezoelectric coefficient. Fu et al. [48] also observed higher value of the effective piezoelectric coefficient for c-axis-oriented $PbZr_{0.53}Ti_{0.47}O_3$ thin films, which was likely due to the substrate deformation as suggested by Barzegar et al. [49].

The charge integration technique has successfully measured the effective piezoelectric coefficient of PZT thin films. It is expected that this technique can be extended to the nanoindentation testing for evaluating the size dependence of piezoelectric behavior in low-dimensional piezoelectric structures. This will require the development of new models to take account of the effect of size confinement.

8.4.2.2 Electric Current Technique

Suresh and his co-workers [15–17] used the indentation technique to assess the mechanical and electrical responses of piezoelectric solids by monitoring the pass of electric current through the conducting indenter and the counter electrode. Figure 8.3 shows the schematic setup. The conducting indenter is connected to the corresponding counter electrode through an Amper meter, which records the electric current generated during the indentation. The principle of the electric current technique is similar to the charge integration technique with the measurement of the electrical current as a function of the indentation load/depth instead of the total electrical charge. It should be pointed that Suresh and his co-workers [15–17,50] did not use this technique to evaluate the effective piezoelectric coefficient of piezoelectric materials.

From the Sect. 8.4.1, one obtains the effective piezoelectric coefficient as

$$d_{33}^f = C_4/C_1 \qquad (8.33)$$

for the indentation by a conducting flat-ended cylindrical indenter,

$$d_{33}^s = \frac{C_6}{C_1} + \frac{2\phi(C_3 + C_4C_5)F^{-2/3}}{3}\left(\frac{3R}{C_1}\right)^{1/3} \qquad (8.34)$$

Fig. 8.3 Schematic of the
setup for the characterization
of the mechanical and electrical response of piezoelectric
materials using the electric
current technique

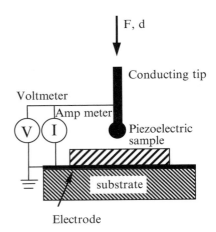

for the indentation by a conducting spherical indenter, and

$$d^c_{33} = \frac{C_6}{C_1} + \frac{2\phi(C_3 + C_4 C_5)F^{-1/2}}{(\pi C_1 \cot\theta)^{1/2}} \tag{8.35}$$

for the indentation by a conducting conical indenter. In contrast to the indentation by
a conducting flat-ended cylindrical indenter, the effective piezoelectric coefficient is
a function of the indentation load for the indentation by a conducting spherical or
conical indenter. It should be mentioned that the parameter ϕ represents the electrical voltage applied between the conducting indenter and the counter electrode and
the self-generated electric potential difference is not included in (8.33)–(8.35).

Sridhar et al. [50] evaluated the effects of poling, poling direction, indentation
speed, and polarization loss on the indentation-induced electric current through a
conducting spherical indenter in the indentation of PZT-4 and $(Ba_{0.917}Ca_{0.083})TiO_2$.
A power-law relationship was observed between the indentation velocity and the
electric current. They focused their work on the dependence of the electric current
on the indentation velocity and the poling direction without addressing the effect
of inelastic deformation. It is unclear if there was piezoelectric relaxation present
during the indentation, which could influence the amount of electric charges passing
through the conducting indenter.

Using the same technique, Saigal et al. [17] studied the dependence of
indentation-induced electric current on the indentation velocity and the indentation
size for a 1–3 piezoelectric ceramic–polymer composite. They found that the electric current increased with the increase in the indentation velocity and the indenter
size and was a nonlinear function of the indentation load. There existed discrepancy
between the experimental results and the proposed model at small indentation load,
and there was no quantitative result on the dependence of the electric current on
the indenter size. This could be due to the dependence of the indentation-induced
electric charge on the indentation load and the dynamic effect, which was neglected
in the analysis.

Sridhar et al. [15] used conical indenter to evaluate the time-dependence of the indentation-induced electric current in PZT-4 and $(Ba_{0.917}Ca_{0.083})TiO_2$. The experimental results were larger than the theoretical prediction, which likely was due to the occurrence of inelastic deformation and the time-dependence of the piezoelectric interaction.

Algueró et al. [51] studied the stress-induced electrical depolarization in lanthanum-modified lead titanate thin film, using the nanoindentation. A conducting spherical indentation was used to locally pole the unpoled $(Pb, La)TiO_3$ films, and the electric current passing through the indenter was measured for the characterization of the depolarization process. There was electric current–load hysteresis in an indentation loading–unloading cycle, suggesting the presence of piezoelectric response in the films and the depolarization in the loading phase. There was the size dependence of the depolarization with more depolarization in thicker films, which likely was due to initially higher electrical polarization. They did not assess the piezoelectric response of thin films.

Koval et al. [52] used nanoindentation to study the stress-electrical relaxation in lead zirconate titanate films with a Zr/Ti ratio of 30/70. In contrast to using the charge integration technique, they measured the electric current passing through the conducting spherical indenter and calculated the electrical charge generated in the indentation by the integration of the electric current curves with respect to time. They found that the electrical relaxation followed the stretched exponential law and the dielectric relaxation followed a single logarithmic-time function. For constant indentation loading rate, one then could use (8.24) to determine the effective piezoelectric coefficient if a constant electrical voltage was applied. The size-dependence of the relaxation time was observed, which increased with the increase in the indentation load and the decrease in the film thickness. It is unclear how they measured the capacitance of the piezoelectric system. They did not evaluate the effective piezoelectric constants and the relaxation behavior at the beginning of the indentation.

From the results given by Lefki and Dormans [13] and the normal load method as standardized in IEEE standards [53], Koval et al. [54] realized that the ratio of the indentation load to the indentation-generated charge represents the effective piezoelectric coefficient consisting of the intrinsic and extrinsic effective piezoelectric coefficients in nanoindentation and suggested that the indentation-induced electric current included the contributions from the stress-induced polarization and the domain motion. They evaluated the piezoelectric response of lead zirconate titanate thin films under nanoindentation. The indentation-induced electric current increased with the indentation load, which resulted in a charge–force hysteresis loop. The effect of domain motion on the indentation-induced electric charge was expressed by the Rayleigh-type behavior. The effective piezoelectric coefficient for unpoled thin films decreased with the indentation load in the loading phase due to the stress-depolarization of the films underneath the indentation, while, in the poled films, the effective piezoelectric coefficient increased with the indentation load. Such a phenomenon could be associated with the domain motion of $90°$ domains.

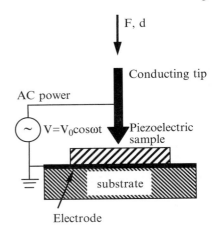

Fig. 8.4 Schematic of the setup for the characterization of the piezoelectric response using an oscillating electric voltage between a conducting indenter and a bottom electrode

8.4.2.3 Electrical Modulation Technique

Figure 8.4 shows the schematic setup of the electrical modulation technique in nanoindentation, as introduced by Algueró et al. [55] for assessing the ferroelectric hysteresis loops in (Pb, La)TiO$_3$ thin films and by Rar et al. [18] for studying electromechanical coupling and pressure-induced behavior in ferroelectric materials. The conducting indenter is used as an electrode, and the piezoelectric material is mounted on an electrode. An oscillating electrical voltage is applied between a conducting indenter and the bottom electrode to actuate the piezoelectric material. Either the electric current through the indenter [55] or the surface deformation [18] is monitored. The dependence of electric current on electric voltage can be used to probe the indentation-induced ferroelectric hysteresis, and the dependence of surface deformation on electric voltage to assess local mechanical response under the action of localized electrical loading.

Assume that the frequency of the applied electrical voltage is much smaller than the resonance frequency of the system and the energy dissipation is negligible. One can neglect the dynamic effect on the indentation behavior and use the results for static indentation. For a displacement–control indentation with a constant indentation depth, (8.20) gives

$$\frac{dF}{dt} = -4C_1C_5R^{1/2}(h - C_5\phi)^{1/2}\frac{d\phi}{dt} \tag{8.36}$$

for a spherical indentation, and (8.26) gives

$$\frac{dF}{dt} = -\frac{8C_1C_5}{\pi\cot\theta}(h - C_5\phi)\frac{d\phi}{dt} \tag{8.37}$$

for a conical indentation. For a load–control indentation with a constant indentation load, there is

$$\frac{dh}{d\phi} = C_5 \tag{8.38}$$

for the indentation by a spherical/conical indenter. Comparing (8.33)–(8.35) with (8.36)–(8.38), one notes that the material behavior examined by using the electrical modulation technique is different from those determined by using the charge integration technique and the electric current technique. It should be emphasized that one needs to consider the dynamic effect if the amplitude of the surface oscillation is compatible with the indentation depth and the indentation-induced electrical potential difference is neglected in deriving (8.36)–(8.38).

Algueró et al. [52] investigated the indentation-induced ferroelectric hysteresis loops in $(Pb, La)TiO_3$ thin films. They demonstrated the effect of the indentation deformation on the current density–electric field hysteresis loops and the ferroelectric hysteresis loops, which could be due to the domain motion. The increase in the indentation load from 100 to 500 mN caused the decrease of the maximum polarization. Their results showed the stress-dependence of ferroelectric behavior.

Rar et al. [18] used nanoindentation to study the electromechanical coupling and pressure-induced dynamic phenomenon in polycrystalline lead zirconate titanate and barium titanate. Surface oscillation occurred under the action of an oscillating electrical voltage due to piezoelectric interaction, and the surface oscillation was a function of the loading–unloading sequence in the barium titanate. It is unclear how they used the results given by Karapetian et al. [45] to relate the amplitude of the surface oscillation to the applied electrical voltage under the action of constant indentation load.

8.5 Piezoresponse Force Microscopy

Atomic force microscope (AFM) has become one of important techniques in the study of local surface behavior and in the modification of surface functionality. It has been used to evaluate local elastoplasticity [56,57] and to engineer surface structure and surface manipulation on the nanoscale [58,59]. Based on the principle of AFM, several new surface force microscopy have been developed, including electric force microscope [60,61], magnetic force microscope [62,63], and piezoresponse force microscope [64,65]. Among them, piezoresponse force microscope (PFM) has become one of important techniques for probing local electromechanical interaction in ferroelectric materials on the nanoscale through local imaging and polarization.

The concept of the PFM is based on the electrical modulation technique, as introduced by Birk et al. [19] using the inverse piezoelectric effect. There are two different approaches in the electrical modulation. The first one [19,66,67] uses surface coatings over the sample surfaces as electrodes, and the surface deformation is monitored by a conducting AFM tip. The advantage is that relatively uniform deformation is created in the sample. The second one [64,65,68,69] uses a conducting AFM tip as one of the electrodes and the sensing element. The surface deformation due to the electrical modulation is superposed on the indentation deformation by the contact between the AFM tip and the sample surface. The localized deformation and polarization are complicated, and there is no closed form solution to relate the piezoresponse to the material properties for anisotropic piezoelectric and ferroelectric materials except the transversely isotropic piezoelectric materials.

For the PFM using surface coatings over piezoelectric structure as electrodes, uniform electric field causes relatively uniform deformation in the structure along the field direction. The change in the film thickness, Δz, can be expressed as [70]

$$\Delta z = \pm z d_{33}^{dp} E_3 = \pm d_{33}^{dp} V, \tag{8.39}$$

where z is the film thickness, E_3 is the component of the electric field intensity along the thickness direction, and the $\pm$ is associated with the effective piezoelectric coefficients of antiparallel domains. It is worth pointing out that the d_{33}^{dp} measured is always less than the intrinsic piezoelectric coefficient, e_{33}, due to the substrate confinement and (8.39) is valid only for quasistatic deformation. When subjected to an oscillation electric voltage of $V_0 \cos \omega t$, there exists inelastic deformation due to mechanical and dielectric losses, and (8.39) becomes

$$\Delta z = \alpha d_{33}^{dp} V_0 \cos(\omega t + \vartheta), \tag{8.40}$$

where α ($0 < \alpha < 1$) is a constant, depending on the energy loss in the dynamic deformation and ϑ is the phase difference between the applied electrical voltage and the surface oscillation.

For the PFM using a conducting tip as an electrode and a sensing element of the surface oscillation of the sample, there is no simple relation between the applied electrical voltage and the surface deformation. The surface deformation consists of (1) the contribution from the contact between the tip and the piezoelectric materials as discussed by Giannakopoulos and Suresh [43], Chen and Ding [44], and Karapetian et al. [45]; (2) the contribution from the electromechanical interaction between the tip and the material [71, 72]; and (3) the contribution from the electromechanical interaction between the cantilever-beam and the surface of the sample [73].

Considering the effect of electric stresses (the Maxwell stress) on the piezoelectric deformation, Kalinin and Bonnel [74, 75] divided the piezoelectric response in the PFM into two regimes (1) strong indentation and (2) weak indentation. In the strong indentation, the indentation deformation dominates with a constant electric potential over the contact area, while, in the weak indentation, the electromechanical interaction is described by the image charge method. They constructed the PFM contrast mechanism maps, which depend on the tip size, indentation load, and electromechanical interaction.

Currently, most work has focused on electromechanical imaging of ferroelectric materials (see Fig. 8.5) and local electromechanical hysteresis (see Fig. 8.6). There are only a few reports on the characterization of the effective piezoelectric coefficients of piezoelectric structure and materials.

Zavala et al. [77] studied the piezoelectric behavior of lead zirconate titanate films using SFM. A silicon tip was in contact with the surface of the thin film, which polarized the material around the contact by applying electrical voltage between the

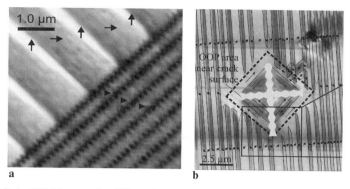

Fig. 8.5 (**a**) An PFM image of a $90°$-*aa*-domain area in a BaTiO$_3$, (**b**) an PFM image of the indented $90°$-*aa*-in-plane domain area. Reprinted with permission from [76]. Copyright [2005], American Institute of Physics

Fig. 8.6 Piezoelectric loop of a lead zirconate titanate film. Reprinted with permission from [76]. Copyright [1997], American Institute of Physics

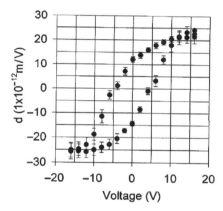

tip and the counter electrode. The effective piezoelectric coefficient decreased with the increase in the indentation load, which could be due to the change in the morphotropic phase boundary composition. The intrinsic mechanisms for this phenomenon are still unclear.

Christman et al. [67] used a conducting diamond AFM tip in contact with a top electrode in characterizing the piezoelectric constants of X-cut single-crystal quartz and ZnO thin films. The electrical interaction between the tip and the electric field could introduce significant error in monitoring the surface oscillation, which depended on the spring constant of the cantilever beam and the contact force as well as the scanner calibration constant. There was a linear relation between the amplitude of the surface oscillation and the applied voltage, which was used to calculate the effective piezoelectric constant according to (8.39).

Zhao et al. [66] evaluated the piezoelectric behavior of ZnO nanobelts. They observed the effective piezoelectric coefficient of ZnO is higher than the corresponding bulk ZnO and decreases with the frequency. The reason for such a behavior is still unclear. It might be due to the perfect crystallinity in the nanobelts and the clamping effect from the substrate. They did not address the size effect.

Zhong et al. [78] studied the thickness dependence of the effective piezoelectric constants in epitaxial (001) $PbZr_{0.2}Ti_{0.8}O_3$ thin films from 5 to 30 nm in thickness. Surface coating of $SrRuO_3$ was used as electrodes. The effective piezoelectric coefficient increased with the film thickness and approached a constant for the film thickness larger than 11 nm. They suggested that the variation of the piezoelectric behavior was due to intrinsic effects including incomplete screening, interfacial discontinuities, and electron-band offset effects. It is unclear if substrate deformation and surface stress had significant effect on the surface oscillation of nanoscale piezoelectric films.

In practice, it is not trivial to quantify the piezoelectric properties directly from the dynamic contact deformation without knowledge of the microstructure and crystal orientation in piezoelectric structures and materials. The dynamic piezoelectric response is complex, which includes the contribution of piezoelectric coupling, electromechanical interaction, local polarization, and domain motion. There is no closed-form solution addressing the scaling-effect of piezoelectric response even though there are some analytical solutions for the contact deformation of transversely isotropic piezoelectric materials. It would be very difficult, if not impossible, to incorporate the domain motion and crystal orientation in these solutions. Numerical technique needs to be developed to establish the dynamic piezoelectric response of piezoelectric structures and materials in PFM.

8.6 Indentation Fracture of Piezoelectric Materials

Indentation at high indentation loads will inevitably create cracks around the indent as shown in Fig. 8.7. They are radial cracks developed around a Berkovich indent on a ZnS nanobelt, in which the cracks emanate at the sharp edges of the indent. The indentation fracture can provide valuable information on the fracture mechanisms in brittle materials and determine material fracture parameters, such as fracture toughness and crack-velocity exponents [80]. A dimensional analysis [81]

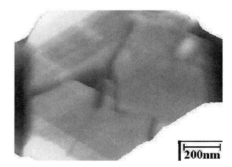

Fig. 8.7 Indentation-induced cracks in a ZnS nanobelt

200nm

suggests that the fracture toughness, K_c, is a function of material properties and the indentation deformation by a Vickers indenter, which can be expressed as

$$K_c = g\left(F, H_v, \frac{c}{a}, \frac{E}{H_v}, \frac{R}{a}\right),$$ (8.41)

where H_v and E are, respectively, the Vickers hardness and Young's modulus of the material, c the size of the radial crack, $\bar{a}$ the size of the indent, and R the size of the plastic deformation zone underneath the indentation. Different empirical relations have been proposed, including

$$K_c = 0.16 H_v \bar{a}^{1/2} \left(\frac{c}{a}\right)^{-3/2} \quad \text{or} \quad K_c = 0.0732 H_v \bar{a}^{1/2} \left(\frac{E}{H_v}\right)^{0.4} \left(\frac{c}{a}\right)$$ (8.42)

as given by Evans and Charles [81],

$$K_c = 0.028 H_v \bar{a}^{1/2} \left(\frac{E}{H_v}\right)^{1/2} \left(\frac{c}{a}\right)^{-3/2}$$ (8.43)

as given by Lawn et al. [82], and

$$K_c = 0.016 \left(\frac{E}{H_v}\right)^{1/2} \frac{F}{c^{3/2}}$$ (8.44)

as given by Anstis et al. [83]. Equations (8.42)–(8.44) provide the rational basis of using the indentation technique to characterize the fracture behavior of brittle materials.

Tobin and Pak [84] studied the indentation crack of PZT-8 ceramics in an electrostatic field. Cracks initiated at the sharp edges of the indent and propagated along the directions parallel and perpendicular to the electric field (the poling direction). The crack size was a function of the poling direction with longer crack length in the direction the same as the poling direction and shorter crack length in the opposite direction to the poling direction. This trend suggested that electric field can enhance or inhibit the growth of cracks, depending on the relative direction of electric field to the poling direction.

Wang and Singh [85] evaluated the crack propagation in a PZT EC-65 material, when indenting the material by a Vickers indenter in an electrostatic field. For unpoled samples, the crack length in the direction perpendicular to the electric field increased with the electric field while the crack length in the direction parallel to the electric field decreased with the electric field intensity. In poled PZT, the cracks propagated anisotropically in an electric field similar to the crack growth without the electric field. When subjected to electric field the same direction as the poling direction, the electric field suppressed the crack growth in both directions. For electric field opposite to the poling direction, the crack length in the perpendicular direction increased with the electric field intensity, while the crack length in the parallel direction was more or less independent of the electric field intensity. This trend is different from the results given by Tobin and Pak [84].

Jiang and Sun [86] studied the crack created by the Vickers indentation as a semicircular surface crack in a semi-infinite space. The plastic zone underneath the indentation was modeled as a one-dimensional piezoelectric rod element, which elongated under the action of an electric field along the poling direction. Such an extension would result in the development of wedging force, causing the change in the crack opening displacement. In contrast, the size of the rod was small when a high electric field was applied in the direction opposite to the polling direction. They introduced effective indentation load in (8.44) by incorporating the field-induced deformation in the plastic zone, which predicted the field-dependence of the crack growth in electric field.

Fu and Zhang [87] investigated the effect of electric field on the indentation-fracture test of a poled lead zirconate titanate by a Vickers indenter and used (8.44) to calculate the fracture toughness. The application of electric field resulted in the decrease in fracture toughness. The electric field opposite to the poling direction had stronger influence on the indentation crack. Large scattering in the fracture toughness was present under the action of electric field, which could be due to the inter-action of structural flaws with the electric field.

Shindo et al. [88] evaluated the indentation fracture of P-7, 5-D, and N-6 poled PZT ceramics in electric field for both closed circuit condition (zero electric field) and open circuit condition (zero surface charge density). The tests were conducted at a Vickers indentation at a load of 9.8 N. The crack size parallel to the poling direction was less than that normal to the poling direction. The electric field assisted the indentation-crack growth in the poling direction, similar to the results given by Tobin and Pak [84]. The energy release rate increased with the electric field in the poling direction for the crack normal to the poling direction, while it was independent of the electric field for the crack parallel to the poling direction.

Popa and Calderon-Moreno [89] examined the propagation of indentation-induced cracks under the action of electric field. The cracks were created in a PZT $PbZrTiO_3$ by a Vickers indenter without applying electric field. An electric field was applied to the material after completely removing the indentation load. The cracks grew more than twice of their initial size at a typical poling field intensity of $1.5\,kV\,mm^{-1}$. High electric field resulted in the rupture of the material. There was no effect of ferroelastic toughening in the crack growth during poling.

Huang et al. [90] studied the effect of electric field on the propagation of the indentation-induced crack in silicon oil during the unloading phase of a PZT-5 ceramics. The residual stresses created in the indentation did not provide enough driving force for the propagation of radial cracks. The electric field $(0.2\,kV\,cm^{-1} < E < 5.25\,kV\,cm^{-1})$ led to the delayed propagation of the indentation cracks, which were arrested after the growth of 10–30 µm in length. Instantaneous crack propagation occurred for electric field of larger than $5.25\,kV\,cm^{-1}$. A field-dependent criterion was proposed for the crack propagation in electric field.

The indentation fracture of piezoelectric materials in electric field is complex. It involves domain switch, domain motion, multiaxial stress state, piezoelectric interaction, and stress evolution during poling. New models need to be developed to consider the scaling effect and the contribution of piezoelectric interaction in the calculation of fracture toughness. The effects of electromechanical loading require more research on understanding the field-controlled fracture behavior for micro- and nanostructural applications of low-dimensional piezoelectric materials.

8.7 Conclusion

This chapter presents an introduction to the applications of the indentation technique and surface force microscopy to the investigation of piezoelectric behavior in piezoelectric materials and ferroelectric materials, especially to low-dimensional structures and materials. In general, the local piezoelectric response of piezoelectric materials can be examined by the electrical-controlled indentation technique and SFM. With the complexity of three-dimensional electroelastic and elastoplastic deformation involving domain switch and domain motion, one needs to combine numerical simulation with electromechanical indentation for the analysis and interpretation of experimental results. In addition, surface imaging using SFM plays an essential role in correlating piezoelectric response with microstructure and piezoelectric properties.

Related to different electrical-controlled indentation techniques, the piezoresponse is quite different. Qualitatively one could examine local deformation-induced electric behavior by using the electrical charge integration technique and the electric current technique and the electric-induced deformation by using electrical modulation technique and PFM. It is a great challenge to quantitatively characterize piezoelectric properties of materials especially low-dimensional structures and materials, since the analyses of quasistatic deformation may be inapplicable to the techniques involving dynamic electromechanical behavior even for the first-order approximation. A microstructure-based electromechanical model is needed that accounts for the evolution of microstructures and dynamic interaction for future application in microscale and nanoscale smart structures.

Acknowledgment The work is supported by Kentucky Science and Engineering Foundation.

References

1. Setter N, Waser R (2000) Acta Mater. 48:151
2. Chu MW, Szafraniak I, Scholz R, Harnagea C, Hesse D, Alexe M, Gosele U (2004) Nat. Mater. 3:87
3. Zhang XY, Zhao X, Lai CW, Wang J, Tang XG, Dai JY (2004) Appl. Phys. Lett. 85:4190
4. Arnold M, Avouris P, Pan ZW, Wang ZL (2003) J. Phys. Chem. B 107:659
5. Wang XD, Summers CJ, Wang ZL (2004) Nano Lett. 4:423
6. Huang MH, Mao S, Feick H, Yan HQ, Wu YY, Kind H, Weber E, Russo R, Yang PD (2001) Science 292:1897

7. Yu C, Hao Q, Saha S, Shi L, Kong XY, Wang ZL (2005) Appl. Phys. Lett. 86:063101
8. Buchine BA, Hughes WL, Degertekin FL, Wang ZL (2006) Nano Lett. 6:1155
9. Wang ZL, Song JH (2006) Science 312:242
10. Wang XD, Zhou J, Song JH, Liu J, Xu NS, Wang ZL (2006) Nano Lett. 6:2768
11. Yang FQ (2003) In: Shindo Y (ed) Mechanics of Electromagnetic Material Systems and Structures. WIT, Boston, p. 171
12. Devonshire AF (1954) Adv. Phys. 3:85
13. Lefki L, Dormans GJM (1994) J. Appl. Phys. 76:1764
14. Fu D, Ishikawa K, Minakata M, Suzuki H (2001) Jpn. J. Appl. Phys. 40:5683
15. Sridhar S, Giannakopoulos AE, Suresh S (2000) J. Appl. Phys. 87:8451
16. Ramamurty U, Sridhar S, Giannakopoulos AE, Suresh S (1999) Acta Mater. 47:2417
17. Saigal A, Giannakopoulos AE, Pettermann HE, Suresh S (1999) J. Appl. Phys. 86:603
18. Rar A, Pharr GM, Oliver WC, Karapetian E, Kalinin SV (2006) J. Mater. Res. 21:552
19. Birk H, Glatz-Reichenbach J, Jie L, Schreck E, Dransfeld K (1991) J. Vac. Sci. Technol. B 9:1162
20. Meyer E, Hug HJ, Benewitz R (2003) Scanning Probe Microscopy – The Lab on a Tip. Springer, Berlin
21. Huey BD, Nath R, Garcia RE, Blendell JE (2005) Microsc. Microanal. 11:6
22. Güthner P, Dransfeld K (1992) Appl. Phys. Lett. 61:1137
23. Franke K, Besold J, Haessler W, Seegenbarth C (1994) Surf. Sci. 302:L283
24. Kolosov O, Gruverman A, Hatano J, Takahashi K, Tokumoto H (1995) Phys. Rev. Lett. 74:4309
25. Parton VZ, Kudryavtsev BA (1988) Electromagnetoelasticity. Gordon and Breach, New York
26. Nye JF (1957) Physical Properties of Crystals. Oxford University Press, Oxford
27. Zhang TY, Zhao MH, Tong P (2001) Adv. Appl. Mech. 38:147
28. Barnett DM, Lothe J (1975) Phys. Stat. Sol. B 67:105
29. Pak YE (1992) Int. J. Fract. 54:79
30. Park SB, Sun CT (1995) Int. J. Fract. 70:203
31. Suo Z, Kuo CM, Barnett DM, Willis JR (1992) J. Mech. Phys. Solids 41:1155
32. Feltan F, Schneider GA, Saldaña J, Kalinin SV (2004) J. Appl. Phys. 94:563
33. Gao CF, Noda N (2004) Acta Mech. 172:169
34. Lu P, Williams FW (1998) Int. J. Solids Struct. 35:651
35. Ding HJ, Chen B, Liang J (1997) Int. J. Solids Struct. 34:3042
36. Bielski W, Matysiak S (1979) Bull. L'Acad. 33:25
37. Matysiak S (1985) Bull. L'Acad. 27:369
38. Yang FQ (2004) Q. J. Appl. Math. Mech. 57:2529
39. Yang FQ (2004) J. Mater. Sci. 39:2811
40. Yang FQ (2004) J. Mater. Sci. 39:2811
41. Yang FQ (2001) Int. J. Solids Struct. 38:3813
42. Wang BL, Han JC (2006) Arch. Appl. Mech. 76:367
43. Giannakopoulos AE, Suresh S (1999) Acta Mater. 47:2153
44. Chen WQ, Ding HJ (1999) Acta Mech. Solida Sin. 12:114
45. Karapetian E, Kachanov M, Kalinin SV (2005) Philos. Mag. 85:1017
46. Angell CA, Ngai KL, Mckenna GB, McMillan PF, Martin SW (2000) J. Appl. Phys. 88:3113
47. Hong JW, Jo W, Kim DC, Cho SM, Nam HJ, Lee HM, Bu JU (1999) Appl. Phys. Lett. 75:3183
48. Fu DS, Sukuzi H, Ogawa T, Ishikawa K (2002) Appl. Phys. Lett. 80:3572
49. Barzegar A, Damjanovic D, Ledermann N, Muralt P (2003) J. Appl. Phys. 93:4756
50. Sridhar S, Giannakopoulos AE, Suresh S, Ramamurty U (1999) J. Appl. Phys. 85:380
51. Algueró M, Bushby AJ, Reece MJ, Poyato R, Ricote J, Calzada ML, Pardo L (2001) Appl. Phys. Lett. 79:3830
52. Koval V, Reece MJ, Bushby AJ (2005) Ferroelectrics 318:55
53. IEEE (1988) IEEE Standard on Piezoelectricity ANSI/IEEE Std. 176–1987. The Institute of Electrical and Electronic Engineers, New York, p. 47
54. Koval V, Reece MJ, Bushby AJ (2005) Appl. Phys. Lett. 97:074301

55. Algueró M, Calzada ML, Bushby AJ, Reece MJ (2004) Appl. Phys. Lett. 85:2023
56. Lee HJ, Kim JH, Cho K, Kang JY, Baek CW, Kim JM, Choa SH (2006) Int. J. Mod. Phys. B 20:3781
57. Weihs TP, Hong S, Bravman JC, Nix WD (1988) J. Mater. Res. 3:931
58. Tomita Y, Hasegawa Y, Kobayashi K (2005) Appl. Surf. Sci. 244:107
59. Shklyaev AA, Shibata M, Ichikawa M (2001) J. Vac. Sci. Technol. B 19:103
60. Belaidi S, Girard P, Leveque G (1997) J. Appl. Phys. 81:1023
61. Bluhm H, Wadas A, Wiesendanger R, Meyer KP, Szczesniak L (1997) Phys. Rev. B 55:4
62. Sueoka K, Okuda K, Matsubara N, Sai F (1991) J. Vac. Sci. Technol. B 9:1313
63. Barnes JR, O'Shea SJ, Welland ME (1994) J. Appl. Phys. 76:418
64. Gruverman A, Auciello O, Tokumoto H (1996) J. Vac. Sci. Technol. B 14:602
65. Lehnen P, Dec J, Kleemann W (2000) J. Phys. D Appl. Phys. 33:1932
66. Zhao MH, Wang ZL, Mao SX (2004) Nano Lett. 4:587
67. Christman JA, Woolcott RR, Kingon AI, Nemanich (1998) Appl. Phys. Lett. 73:3851
68. Güthner P, Dransfeld K (1992) Appl. Phys. Lett. 61:1137
69. Poyato R, Huey BD (2006) J. Mater. Res. 21:547
70. Gruverman A, Kalinin SV (2006) J. Mater. Sci. 41:107
71. Binnig G, Quate CF, Gerber C (1986) Phys. Rev. Lett. 56:930
72. Sarid D (1991) Scanning Force Microscopy. Oxford University Press, New York
73. Hong S, Shin H, Woo J, No K (2002) Appl. Phys. Lett. 80:1453
74. Kalinin SV, Bonnell DA (2002) Phys. Rev. B 65:125408
75. Kalinin SV, Bonnell DA (2004) In: Alexe M, Gruverman A (eds) Nanoscale Characterization of Ferroelectric Materials. Springer, Berlin, p. 1
76. Schneider GA, Scholz T, Muñoz-Saldaña J, Swain MV (2005) Appl. Phys. Lett. 86:192903
77. Zavala G, Fendler JH, Trolier-McKinstry S (1997) J. Appl. Phys. 81:7480
78. Zhong S, Alpay SP, Nagarajan V (2006) J. Mater. Res. 21:1600
79. Yang FQ, Jiang CB, Du WW, Zhang ZQ, Li SX, Mao SX (2005) Nanotechnology 16:1073
80. Lawn B (1993) Fracture of Brittle Solids. Cambridge University Press, New York
81. Evans AG, Charles EA (1976) J. Am. Ceram. Soc. 59:371
82. Lawn BR, Evans AG, Marshall DB (1980) J. Am. Ceram. Soc. 63:574
83. Anstis GR, Chantikul P, Lawn BR, Marshall DB (1981) J. Am. Ceram. Soc. 64:533
84. Tobin AG, Pak YE (1993) In: Proc. SPIE, vol. 1916, pp. 78–86
85. Wang HY, Singh RN (1997) J. Appl. Phys. 81:7471
86. Jiang LZ, Sun CT (2001) Int. J. Solids Struct. 38:1903
87. Fu R, Zhang TY (2000) J. Am. Ceram. Soc. 83:1215
88. Shindo Y, Oka M, Horiguchi K (2001) J. Eng. Mater. Tech. 123:293
89. Popa M, Calderon-Moreno JM (2001) Mater. Sci. Eng. A 319:697
90. Huang HY, Chu WY, Su YJ, Qiao LJ, Gao KW (2005) Mater. Sci. Eng. B 122:1

Chapter 9
Mechanics of Carbon Nanotubes and Their Composites

Liang Chi Zhang

9.1 Introduction

Since the discovery of carbon nanotubes, extensive investigations have been carried out on both single-walled and multiwalled nanotubes. A major research focus in the field has been to characterize precisely the mechanical properties of nanotubes, using various methods including continuum mechanics and molecular dynamics modeling, and chemical reactivity of nanotubes with a class of matrix materials to form nanocomposites with strong interface stress transfer capability. Developing efficient techniques for fabricating nanocomposites with tailored microstructures has also been emphasized recently [3, 57].

The aim of this chapter is to discuss some fundamentals in relation to the mechanics of carbon nanotubes and their composites.

9.2 Single-Walled Carbon Nanotubes

9.2.1 Continuum Mechanics Modeling

Carbon nanotubes have been found to behave like continuum structures and possess both membrane and bending capacities. Hence, it will be beneficial if equivalent continuum theories can be established for fast analysis of their mechanical behavior. This has led to extensive studies on the equivalent or effective properties and geometrical dimensions of a carbon nanotube. Nevertheless, the mechanics modeling so far has been associated with strong assumptions, leaving many fundamentals to clarify. In this section, we will discuss some of the issues.

L.C. Zhang
School of Aerospace, Mechanical and Mechatronic Engineering, The University of Sydney, NSW 2006, Australia
zhang@aeromech.usyd.edu.au

F. Yang and J.C.M. Li (eds.), *Micro and Nano Mechanical Testing of Materials and Devices*, doi: 10.1007/978-0-387-78701-5, © Springer Science+Business Media, LLC, 2008

9.2.1.1 Effective Wall Thickness and Young's Modulus

To characterize the mechanical properties of a carbon nanotube, one often directly uses the mechanics quantities defined by continuum mechanics, such as Young's modulus. However, a carbon nanotube has a discrete molecular structure whose "wall" comprises only a number of atoms, and hence does not have continuous spatial distribution. Nevertheless, to calculate the Young's modulus of a carbon nanotube, one needs to know the wall thickness of the tube. For example, in a simple tensile loading, the Young's modulus E of a single-walled carbon nanotube (SWCNT) may be defined as $E = \sigma/\varepsilon$, where ε is the strain, and σ is the axial stress applied to the tube calculated by $F/(2\pi Rh)$ in which R is the radius of the mid-surface of the tube, h is the tube thickness, and F is the axial force applied. When R can be reasonably defined as the imagined surface radius of a nanotube through the theoretical centers of the atoms, h does not exist because a nanotube does not have a continuous wall. In the literature, mechanics modeling has been based on unreasonable assumptions of h. For example, some researchers treated a carbon nanotube as a truss member [60], a solid beam [104], or a solid cylinder [20], and some others simply let h be the inter-planar spacing of two graphite layers ($=3.4$ Å) [22, 23, 34, 47, 48, 85, 111, 112]. As a result, the Young modulus of a carbon nanotube calculated with different h values varies in a very wide range.

Vodenitcharova and Zhang introduced a concept of an effective wall thickness under the umbrella of continuum mechanics to reflect a nanotube's mechanics behavior [94]. On the basis of the consideration of force equilibrium and equivalence, they proposed a necessary condition that the effective wall thickness must be smaller than the theoretical diameter of a carbon atom, which is about 0.142 nm [93]. Their argument is that a cross-section of a nanotube contains only a number of atoms and the forces in the tube are transmitted through these atoms; but in a continuum mechanics model the same forces are transmitted through a continuous wall area. As such, the effective wall thickness cannot be greater than the theoretical diameter of a carbon atom; as otherwise, the tube equilibrium cannot be maintained. According to this necessary condition, continuum models using a wall thickness greater than or equal to the diameter of a carbon atom [4, 23, 28, 39, 47, 61, 72, 75, 89, 123] are unreasonable, while the others below that [13, 35, 63, 90, 103, 124] are possible but their validity needs to be confirmed by further study. Vodenitcharova and Zhang then proposed, using an elastic ring theory and the results from molecular dynamics analysis [77], that the effective thickness of a single-walled carbon nanotube should be $h = 0.617$ Å, which is 43.8% of the theoretical diameter of a carbon atom, and that the effective Young's modulus, E, of the tube is 4.88 TPa [94].

In an attempt to address the continuum-atomic modeling issue, Zhang et al. [115, 122] directly linked interatomic potential and atomic structure of an SWCNT with a continuum constitutive model, by equating the strain energy stored in the equivalent constitutive model to that in atomic bonds described by Tersoff-Brenner potential [8, 82]. This method seems to be reasonable because the Tersoff-Brenner potential has been shown to be appropriate for analysing SWCNTs [51]. An isotropic constitutive model was therefore derived for SWCNTs subjected to in-plane deformation with two elastic constants, the in-plane stiffness, $K_{\text{in-plane}}$, and the in-plane shear stiffness, K_{shear}. Then the following relationship can be obtained by comparing the two elastic constants of SWCNTs with their counterparts of three-dimensional (3D) thin shells of thickness h:

$$K_{\text{in-plane}} = Eh/(1 - v^2) \text{ and } K_{\text{shear}} = Gh, \tag{9.1}$$

where $G = E/2(1+v)$ is shear modulus and v is Poisson ratio of SWCNTs, which can be obtained from $K_{\text{in-plane}}/K_{\text{shear}} = (1-v)/2$. However, the effective thickness of SWCNTs cannot be determined in this way because only in-plane deformation is considered.

Huang et al. considered both the in-plane and off-plane deformation of SWCNTs [25]. By using the Tersoff-Brenner potential, V, and the modified Cauchy-Born rule [115, 122], they obtained two-dimensional isotropic constitutive relations, where the bending stiffness, D_{bending}, and the off-plane torsion stiffness, D_{torsion}, as well as $K_{\text{in-plane}}$ and K_{shear} of SWCNTs were calculated as

$$D_{\text{bending}} = \frac{\sqrt{3}}{2}\left(\frac{\partial V}{\partial \cos \theta_{ijk}}\right), \quad D_{\text{torsion}} = 0,$$

$$K_{\text{in-plane}} = \frac{1}{2\sqrt{3}}\left[\left(\frac{\partial^2 V}{\partial r_{ij}^2}\right)_0 + \frac{A}{8}\right], \quad \text{and} \quad K_{\text{shear}} = \frac{A}{16\sqrt{3}}, \tag{9.2}$$

where $\theta_{ijk}(k \neq i, j)$ is the angle between bonds $i-j$ and $i-k$, r_{ij} is the $i-j$ bond length, and A is a function of the first and second-order derivatives of V with respect to r_{ij}, θ_{ijk}, and $\theta_{ijl}(l \neq i, j, k)$. In their derivation, the bending of SWCNTs was considered as a result of the σ-bond angle change while the off-plane torsion as being independent of the deformation of the σ-bonds.

However, if we use a three-dimensional continuum thin shell theory, where bending and off-plane torsion are due to the deformation across the wall thickness, we will have a different observation about the deformation mechanics of SWCNTs, because D_{bending} and D_{torsion} of a three-dimensional thin shell are related to the shell thickness h and the in-plane material constants by

$$D_{\text{bending}} = Eh^3/\left\{12(1-v^2)\right\} \text{ and } D_{\text{torsion}} = Gh^3/12. \tag{9.3}$$

Combining (9.1) and (9.3) leads to the following condition

$$\frac{D_{\text{torsion}}}{D_{\text{bending}}} = \frac{K_{\text{shear}}}{K_{\text{in-plane}}} = \frac{1-v}{2}, \quad \text{or} \quad \frac{D_{\text{bending}}}{K_{\text{in-plane}}} = \frac{D_{\text{torsion}}}{K_{\text{shear}}} = \frac{h^2}{12}. \tag{9.4}$$

The above derivation indicates that to establish a three-dimensional elastic shell model with a well-defined effective thickness for SWCNTs, the key is to satisfy condition (9.4) when the corresponding elastic constants are obtained on the basis of an atomistic potential that accounts for the atomic structure and deformation mechanisms of SWCNTs. Unfortunately, in [25], (9.2) was obtained from potential

V with $D_{torsion} = 0$, which obviously cannot satisfy condition (9.4) and hence leads to an ill-defined effective thickness of SWCNTs.

To obtain a deeper understanding without the influence of the currently debatable wall thickness and Young's modulus, Wang and Zhang employed in-plane stiffness, $K_{in-plane}$, Poisson ratio, v, bending stiffness, $D_{bending}$, and off-plane torsion stiffness, $D_{torsion}$, as independent elastic constants in their mechanics solution to the free vibration of SWCNTs [101]. They found that the off-plane torsion stiffness cannot be zero, which is in agreement with molecular dynamics results and experimental measurements in the literature, and that the effective thickness is about 0.1 nm for (10,10) SWCNTs. Wang and Zhang then made an interesting observation, by plotting the diverse values of Young's modulus and wall thickness in the literature on a single diagram, as shown in Fig. 9.1 [102]. They found that those satisfy Vodenitcharova-Zhang's necessary condition [94] collapse very nicely along the curve of constant in-plane stiffness ($363\,J/m^2$). It means that there is something in common in predicting the mechanical properties of SWCNTs when using different methods, although the values of Young's modulus and effective wall thickness still scatter and the experimentally measured properties available correspond to different $K_{in-plane}$ values. The above observation indicates that to clarify the critical issue of effective wall thickness and Young's modulus of SWCNTs, further investigations are definitely required.

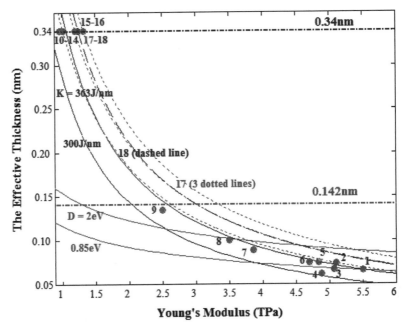

Fig. 9.1 A comparison of the scattered values of effective wall thickness and Young's modulus of SWCNTs in the literature. *Dots* numbered are from the papers listed below. 1: [109]; 2: [124]; 3: [103]; 4: [94]; 5: [63]; 6: [90]; 7: [35]; 8: [101]; 9: [72]; 10: [48]; 11: [39]; 12: [110]; 13: [62]; 14: [6]; 15: [23]; 16: [28]; 17: [34]; 18: [87]

9.2.1.2 Bending and Kinking

Carbon nanotubes are exceptionally flexible and their deformation is reversible even under a large deflection because of their high bond-breaking resistance [27, 109]. Molecular dynamics (MD) method has been used for numerical simulations [54, 109]. However, an MD simulation is time consuming and computationally expensive, and cannot bring about an analytical model to capture the inherent links of mechanics quantities. In this section, we will use continuum mechanics to establish an analytical model for characterizing the bending and kinking of an SWCNT subjected to an external bending moment M as illustrated in Fig. 9.2 (top) [95].

When M increases, an SWCNT bends and flattens, as both experiments and MD simulations in the literature have shown. Similar to the deformation of a continuous tube [113, 120], the flattening of the SWCNT's cross-section during bending before kinking can be described by a flattening ratio, $\zeta = (R_c - R)/R_c$, where R is the initial radius of the tube's mid-surface and R_c is the current radius varying with the bending angle ψ. The normal and tangential displacements are approximately determined by $w = R\zeta \cos 2\theta$ and $v = -R\zeta (\sin 2\theta)/2$, respectively. Once kinking starts, the nanotube becomes a mechanical mechanism and an analytical solution can be formulated [95]. At this deformation stage, however, the van der Waals force must be considered to account for the interaction between the opposite walls of the nanotube when they approach each other. The magnitude of the force

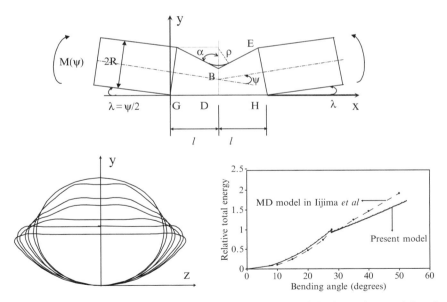

Fig. 9.2 *Top*: a continuum bending model with kinking; *Bottom left*: shape change of the tube cross-section during bending; *Bottom right*: bending angle vs. energy

depends on the distance between atoms. For large distances, the van der Waals force is attractive, but when the separation between atoms is below the equilibrium distance of 3.42 Å, it becomes strongly repulsive. With increasing the bending angle, the top and bottom parts of the kink get closer to each other, and at certain stage this distance remains unchanged upon additional bending, because there are no external normal loads applied on the walls to prevail over the repulsive van der Waals forces.

It was found that bending of a carbon nanotube without local kinking can hold up to the point of $\psi = 25.58°$. When kinking happens, the absorbed energy during bending becomes almost linear and the corresponding bending angle is almost constant. As kinking develops, the equilibrium distance of 3.42 Å is quickly reached. Iijima et al. reported that the kinking of a nanotube of radius $R = 6$ Å occurs at an angle of about 27.8° [27]. Further development of the kink is associated with an almost linear increase of the strain energy. The above mechanics model, when applied to the same nanotube, almost gives rise to the same critical bending angle of 27.9°. Comparing the strain energies, Fig. 9.2 (bottom right), the model produces only slightly higher strain energy before kinking. At the point of kinking, the model predicts the same drop in energy as the molecular dynamics simulation does.

The above mechanics analysis concludes that the prebuckling deformation of an SWCNT involves the ovalisation of the tube cross-section and the post-buckling deformation, but the process is influenced by flattening and van der Waals forces.

9.2.1.3 Deformation in Bundle Formation

MD simulations and experiments on SWCNT bundles have shown that individual SWCNTs in a bundle are already deformed in an externally unstressed state because of the intertubular van der Waals forces [11, 45, 84]. The SWCNTs are flattened, the lattice constant decreases and the distance between the SWCNTs, called equilibrium distance d, becomes smaller than the equilibrium distance between two graphite layers (3.42 Å). At a distance equal to d, the resultant forces of interaction between the neighboring SWCNTs are zero; at a distance larger than d the nanotubes attract each other; and at a distance smaller than d the intertubular interaction becomes repulsive. It has been found that the faceting of the SWCNTs in a bundle is radius-dependent [84]: the cross-sectional distortion of SWCNTs becomes noticeable when their radii are large. As the bundle polygonization is caused by van der Waals forces, it would be interesting to know their magnitude, their distribution on an SWCNT, and their dependence on the SWCNT radius.

Here, we will combine molecular dynamics and continuum mechanics to reveal the mechanical interaction between zigzag SWCNTs in a bundle without external loading [96].

Consider the deformation of the central SWCNT in a long bundle under the van der Waals forces $p(\theta)$, as illustrated in Fig. 9.3. According to the thin shell theory, a point on the mid-surface of the nanotube can undergo displacements in the longitudinal, circumferential, and radial directions, i.e., u, v, and w, measured from the undeformed geometry. Zero displacements correspond to an isolated SWCNT with no forces acting on it. As the length of an SWCNT in a bundle is much

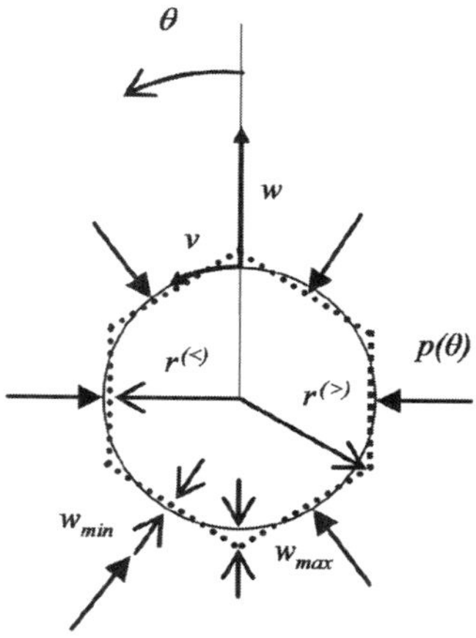

Fig. 9.3 The central SWNT in a bundle and the notations used in the continuum mechanics model

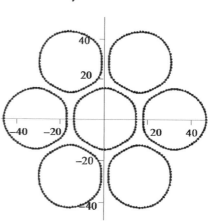

Fig. 9.4 Cross-sectional view of the deformed (36,0) SWNT bundle obtained by the MD simulation. The coordinate unit is Å

greater than its diameter and the van der Waals forces along the SWCNT axis can be considered uniform, the SWCNT can be modeled as a ring of unit length under plane-strain deformation with the longitudinal displacement u neglected. The remaining displacements are the radial displacement w (positive outward) and the tangential displacement v (positive in the direction of positive θ, Fig. 9.3). Using the ring theory of continuum mechanics, Vodenitcharova et al. obtained an analytical solution, which shows the following deformation features of an SWCNT in a bundle [96].

Figure 9.4 shows the typical deformation pattern of a (36, 0) SWCNT bundle, where each dot represents a carbon atom on the cross-section of the bundle. It is clear that the central SWCNT has been deformed significantly into a symmetrical faceting shape.

The initial van der Walls interaction energy of the circular tubes in a bundle is very small (2–4 meV per atom). This is in agreement with the results reported by [45]. The strain energy induced during flattening was calculated as the difference in the minimum total energy of the bundle and the total energy of the isolated undeformed nanotubes. It was found that the strain energy varies from 8.3 meV per

Fig. 9.5 Distribution of the
van der Waals attraction on an
SWNT as calculated from the
continuum mechanics model

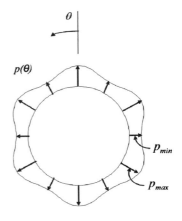

atom for a nanotube with a radius of 7.05 Å, to 7.2 meV per atom for a nanotube
with a radius of 14.1 Å. Obviously, the strain energy decreases with the increase in
radius, indicating that more energy is required to deform nanotubes of smaller radii.
The results show that the circumferential displacement v is negligible compared
with the radial displacement w, demonstrating that during the bundle formation the
nanotube atoms displace predominantly in the radial direction. Furthermore, the van
der Waals forces $p(\theta)$ are attractive (negative sign means attraction) and attain their
maximum value p_{max} in the channels (Fig. 9.5) where the attraction between the
neighboring atoms of adjacent nanotubes is more significant, and attain their minimum value p_{min} at the points of minimum intertubular distance where the attraction
is small. The radial displacement is outward even at the shortest distance between
the SWCNTs. For radii greater than 11 Å, the nonuniform inward displacement
dominates at the points of minimum intertubular distance. The bundle formation of
larger nanotubes is associated with a lower intensity of the attractive van der Waals
forces, but the values of both p_{min} and p_{max} are significant. For the nanotube radii
considered, p_{max} varies from -6 to -11 GPa and p_{min} varies from -1 to -5 GPa.

The above analysis shows that at the formation of a bundle, without external
loads, the neighboring SWCNTs attract each other, with the maximum attraction in
the channels and minimum attraction at the shortest distance between the SWCNTs.
The radius of SWCNTs plays an important role. The distance between SWCNTs
and cross-sectional distortion increase when the radius increases. An SWCNT with
radius smaller than 11 Å will experience an overall expansion in a bundle.

9.2.2 Molecular Dynamics Modeling

9.2.2.1 Fundamentals

Molecular dynamics simulation has been widely used in characterising the mechanical properties of materials and understanding their mechanisms of deformation on

the nanometer scale [116, 117]. However, the simulations must be done carefully to best represent the reality. First, it is important to select an appropriate interaction potential that can effectively describe the deformation of a nanotube correctly. Second, during a loading process, improper treatment of temperature rise can lead to fictitious results. In molecular dynamics, heat conduction is accomplished via the so-called thermostat atoms, using a thermostatting method. Adiabatic relaxation method, isokinetic thermostatting, Andersen stochastic thermostatting, and Nose-Hoover feed back thermostatting are some typical methods in the literature for temperature conversion [14]. For small systems, the adiabatic relaxation method can lead to a fluctuation of the vibrational-relaxation rate. In isokinetic thermostatting, the temperature is maintained in different ways. For example, in the Berendsen thermostat scheme with velocity scaling, the velocities of thermostat atoms are scaled to fix the total kinetic energy. In the Gaussian feedback or Evans-Hoover scheme with force scaling, however, the kinetic energy is monitored and information is fed back into the equations of motion so that the kinetic energy is kept constant to dissipate heat by controlling the thermostatting force. Velocity scaling [118, 119] has been widely used because of its simplicity in implementation. For small time steps, the Gaussian isokinetic method and velocity scaling method become identical [14]. However, on one hand, a very small time step will give an unusually high elongation speed. On the other hand, a small displacement step with a small time step will be computationally expensive. The flaw in the isokinetic-thermostatting method is that it is impossible to separate the effects of thermostatting on rate processes. The other two schemes also have this limitation to a certain extent. Third, a system has to be relaxed initially as well as during the simulation so that the velocities of the Newtonian and thermostat atoms reach equilibrium at the specified temperature of simulation; thus appropriate time step and displacement step have to be selected to get a reasonable elongation speed. A natural question is therefore: Which simulation scheme will be appropriate and effective for simulating the deformation of carbon nanotubes.

In this section of the chapter, we will discuss some necessary details that are central to a reliable simulation, such as the selection of potential, number of thermostat atoms, thermostat method, time step, displacement step, and the number of relaxation steps [51, 54].

Consider a single-walled armchair nanotube (10,10) with 100 repeat units along the axial direction and a zigzag nanotube (17, 0) with 58 repeat units along the axial direction, both having a length of about 245 Å. The interatomic forces will be described by the Tersoff (T) potential [81, 83] and the empirical bond order potential – the Tersoff-Brenner (TB) potential [8, 9]. The simulations will be carried out at 300 K with Berendsen (B) and Evans-Hoover (EH) thermostats and a time step of 0.5 fs. To examine the reliability of the simulations, we will carry out the analysis using the following schemes:

Scheme 1 (S1). In this scheme, the first two layers of atoms on both ends of a carbon nanotube will be held rigid. The next four layers will be taken as thermostat atoms and the remaining be treated as Newtonian atoms. First, the tubes will be annealed at the simulation temperature for 5,000 time steps. Then the rigid atoms on both ends will be pulled along the axial direction at an increment of 0.05 Å. After each displacement step, 1,000 relaxation steps will be done to dissipate the effect of preceding displacement step over the entire length of the tube.

Scheme 2 (S2). In this scheme, all atoms except the boundary ones rigidly held will be treated as thermostat atoms. Fifty relaxation steps will be carried out after each displacement step.

We find that simulation parameters have a remarkable influence on the results [51]. For example, different schemes lead to significantly different stress-strain curves and necking processes, as shown in Figs. 9.6 and 9.7. We conclude that a simulation using Tersoff-Brenner potential and Berendsen thermostat with all atoms as thermostat atoms (except the rigid ones) with 50 relaxation steps after

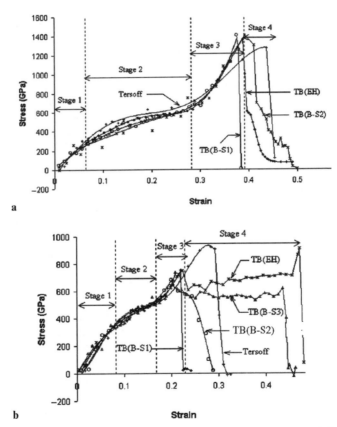

Fig. 9.6 The stress–strain curves of (**a**) a (10,10) armchair SWCNT; (**b**) a (17,0) zigzag SW-CNT using Tersoff and Tersoff-Brenner potentials. In the figure, TB(B-S1) is the calculation with Berendsen thermostat following Scheme 1, TB(B-S2) is the calculation with Berendsen thermostat following Scheme 2, and TB(B-S3) is the calculation as in TB(B-S2) but with a smaller displacement step of 0.008 Å. TB(EH) is the calculation with Evans-Hoover thermostat

Fig. 9.7 Deformation of an SWCNT when using Berendsen thermostat: (**a**) an armchair tube (**b**) a zigzag tube (**c**) necking that propagates along a zigzag carbon nanotube

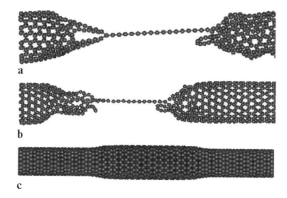

a

b

c

each displacement of 0.008 Å is a reliable and cost effective method. We also found that the Young's modulus and Poisson's ratio of (1) the armchair tube are 3.96 TPa and 0.15, respectively, and (2) the zigzag tube are 4.88 TPa and 0.19, respectively. The armchair tube can undergo a higher tensile stress compared with the zigzag tube. Under tension, both armchair and zigzag nanotubes exhibit carbon chain unravelling and one-atom chain at a strain of around 0.4.

9.2.2.2 Dynamic Properties and Ballistic Resistance

The dynamic properties of SWCNTs have not been well understood, but they are important for some potential applications such as in making light bullet proof vests and shields or explosion proof blankets [16,50], because for these applications, thinner, lighter, and flexible materials with superior dynamic mechanical properties are preferred. The best material will have a high level of elastic storage energy that will cause the projectile to bounce off or be deflected to reduce the effects of "blunt trauma" on the wearer after being struck by a bullet. Currently, the Kevlar fiber having a toughness of $\sim$33 J/gm [97] is widely used in bullet-proof vests.

In this section, we will investigate the dynamic properties and ballistic resistance of SWCNTs in terms of nanotube radius, position of bullet striking, bullet speed, and energy absorbed by a nanotube for a particular bullet size and shape [55]. Consider SWCNTs (27,0), (18,0), and (9,0) of radii 10.576, 7.051, and 3.525 Å, respectively, all having a length of $\sim$75 Å, fixed at one end to a diamond surface. A piece of diamond having 1,903 atoms with dimensions $35.67 \times 35.67 \times 7.13$ Å^3 is used as a bullet with its speed varying from 100–1,500 m/s. The bullet dimension is selected in such a way that its width is larger than that of the biggest tube considered above after flattening. The bullet is released from a target about 15 Å from the center of the tube and moves at a constant velocity in the horizontal direction (i.e., perpendicular to the nanotube axis) as shown in Fig. 9.8. The tube performance is examined for bullet released with various speeds at various heights, h, using the classical molecular

Fig. 9.8 Initial model for MD simulation

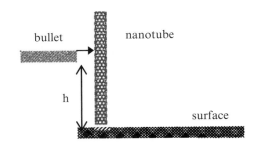

Table 9.1 The maximum bullet speed, the maximum absorption energy, and the maximum temperature of three different radii tubes when the bullet strikes the tube at various heights

h	(9,0) tube of radius 3.53 Å			(18,0) tube of radius 7.05 Å			(27,0) tube of radius 10.6 Å		
(Å)	Speed (m/s)	Absorption Energy (eV)	Temp (K)	Speed (m/s)	Absorption Energy (eV)	Temp (K)	Speed (m/s)	Absorption Energy (eV)	Temp (K)
10	190	129.1		–	–	–	–	–	–
20	360	139.9		360	225.4		340	510.1	
30	650	189.6		600	429.0		600	652.7	
42	950	298.3	523.9	870	616.0	416.7	850	781.8	381.8
50	1,080	291.2	523.0	1050	558.8	421.2	1050	752.7	387.6
60	1,400	196.3	533.2	1400	363.1	440.9	1450	670.2	416.8

dynamics (MD) method with a three-body Tersoff-Brenner potential [8, 9]. As pointed out previously [51], to minimize the heat conduction problem and improve the computational efficiency, Berendsen thermostat will be applied to all atoms except those rigidly held. At the beginning the energy of the SWCNTs is minimized by the conjugate gradient method.

We find that the maximum speeds of the bullet that the tube can bear are given in Table 9.1. When the bullet strikes the tube with a higher speed, the tube will break. As the bullet strikes, the tube is indented, and the stress developed is passed along the tube in both directions. When the bullet hits the tube below half of its length, initially the stress is released by sagging and buckling near the base as shown in Fig. 9.9a. (Hence, the bullet with high speed detaches the tube at the base because of the dynamic effect.) Further stress developed causes the tube to flatten around the bullet striking point and bend near the buckle as shown in Fig. 9.9b. The flattening and bending continues until the bullet passes away. In this period, the stress is released mainly via the top end of the tube. When the bullet hits the tube above half the length of the tube, initially the tube does not buckle near the base. This is because the stress developed is released via the top end. The upper portion of the tube bent a little and the tube flattened where the bullet struck as shown in Fig. 9.10a. Flattening and bending continued for some time; then the tube buckled near the base and started to sag (Fig. 9.10b).

During the shooting process, the tube absorbed some energy. At high bullet speeds the energy absorbed and the temperature of the tube increased with time,

Fig. 9.9 A bullet strikes the (9,0) nanotube at 20 Å with a speed of 360 m/s and traveled by (**a**) 22.5 Å in 6,250 fs and (**b**) 45 Å in 12,500 fs

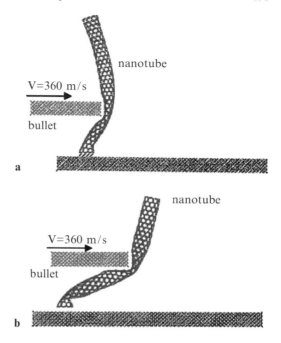

reached a maximum value, and then decreased as shown in Fig. 9.11a, b. This means that a portion of the energy is converted into thermal energy. But at lower bullet speeds and when the bullet strikes the tube somewhere below half of its length, the energy absorbed increased with time, became almost constant for some time and then decreased (Fig. 9.12a), whereas the temperature of the tube did not change except a little fluctuation within 10–15° as shown in Fig. 9.12b. The maximum energy absorbed and the maximum temperature at high bullet speeds are also tabulated in Table 9.1.

The velocity of the bullet for which the tube could be resilient, increases linearly with the height at which it strikes the tube (Fig. 9.13a). At lower heights, the bullet speed is low. As a result, the entire portion of the tube that is above the struck point bend and flattens over a period of time as the stress is developed and released at the same time. During this, the absorption energy of the tube is almost constant and this is reflected by the nearly flat portion of the energy vs. time step curve shown in Fig. 9.12a. However, all these happen in 1/100th of a nanosecond. The maximum bullet speed for which the tube could bear is almost independent of the tube radii when the bullet struck a tube at a particular height. Any one row of Table 9.1 clearly shows this. The maximum absorption energy of the tube is high when the bullet strikes the tube at half the length (Fig. 9.13b).

When the bullet strikes the tube with a high speed the maximum energy absorbed by the tube increases with the tube radius. However, the maximum temperature of the tube decreases with the tube radius (Fig. 9.14a, b). This means when the bullet

Fig. 9.10 A bullet strikes
the (9,0) nanotube at 50 Å
with a speed of 1,050 m/s
and traveled by (**a**) 52.5 Å
in 5,000 fs (**b**) 65.625 Å
in 6,250 fs

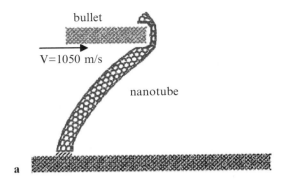

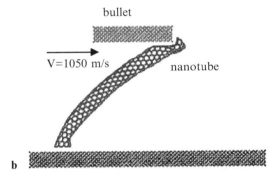

hits the small tube with high speed, more energy is converted into thermal energy. However, we find that the thermal energy kT (where k is the Boltzmann constant) is only about 0.01% of the calculated absorbed energy. Energy absorption behaviour is often characterized by the absorption efficiency ε, which can be defined as, $\varepsilon = W/(F_{max}L)$, where W is the absorbed energy at a given deflection L and F_{max} is the maximum load received by the tube up to this deflection. Although the absorption energy of the tube increases linearly with radius, the calculated absorption efficiency of all three tubes are nearly equal ($\sim$0.08) when the bullet strikes the tube at the same speed.

In summary, the above study shows that carbon nanotubes could bear projectiles travelling at comparable speeds (200–1,400 m/s) of typical rifle bullets. For a particular bullet, the bearing speed of the tube increases linearly with the height at which it strikes the nanotube, and it is independent of the tube radius when it strikes the tube at a given height. On impact, the tube absorbs more energy when the bullet is released at a height that is nearly half the length of the tube. At a given bullet speed, the absorption efficiency of a nanotube is independent of the tube radius.

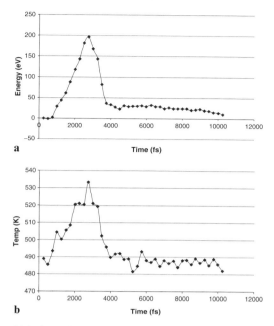

Fig. 9.11 Variation of (**a**) absorption energy and (**b**) temperature of the (9,0) CNT with time when the bullet strikes the nanotube with a speed of 1,400 m/s at a height of 60 Å

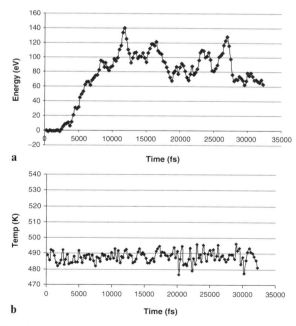

Fig. 9.12 Variation of (**a**) absorption energy and (**b**) temperature of the (9,0) CNT with time when the bullet strikes the nanotube with a speed of 360 m/s at a height of 20 Å

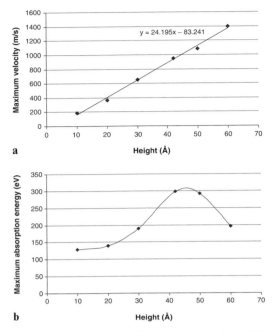

Fig. 9.13 The variation of (**a**) the maximum bullet speed for which the (9,0) tube could bear and (**b**) the maximum absorption energy of the (9,0) nanotube, with the height at which the bullet strikes

9.3 Nanotube-Reinforced Composites

9.3.1 Stress Transfer

As carbon nanotubes (CNT) have a highest axial strength and modulus among all existing whiskers [64, 74, 88, 115], they have been considered a promising reinforcing material for composites, if a large interfacial bonding strength with a matrix can be achieved for a great load transfer. Some studies on a number of CNT-reinforced polymer composites have discussed the CNT-matrix interfacial bonding strength. For example, Wager and coworkers claimed that a strong CNT-polymer adhesion and a high interfacial bonding strength in a CNT/polyurethane system were attributed to a "2 + 2" cycloaddition reaction between the tube and the polymer [46, 98–100]. Using the traditional Kelly-Tyson model [30], Wagner showed that the interfacial bonding strength in a CNT composite is higher than that in a fiber-reinforced composite, although this model is unable to demonstrate the distribution of the tensile and interfacial shear stresses along a tube under an external loading [98]. Liao and Li used molecular mechanics to simulate a pull-out process in an SWCNT-polystyrene system and reported that the interfacial bond strength could be up to 160 MPa even without considering the chemical bonding between the tube

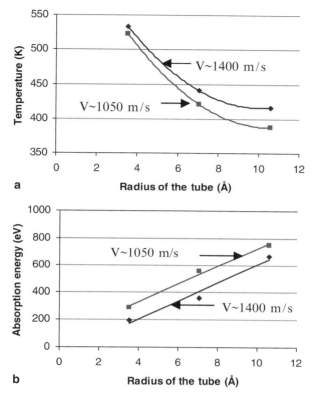

Fig. 9.14 Variation of (**a**) temperature of the tube and (**b**) absorption energy of the tube with the tube radius when the bullet speed is high

and matrix [41]. Qian and coworkers also found the significant load transfer ability of CNTs under tension [67, 68]. However, in a transmission electron microscopy study on an aligned nanotube-epoxy composite, Ajayan et al. indicated that the interfacial bonding between a CNT and epoxy matrix was weak [1]. Using Raman spectroscopy, Schadler et al. concluded that the interfacial bonding was very weak when a multiwalled CNT/epoxy composite was under tension [71].

To obtain a deeper understanding, Xiao and Zhang developed a mechanics model to investigate the effects of length and diameter of an SWCNT in an epoxy matrix on the load transfer behavior [105]. On one hand, they found that the shear stress along a tube reaches its maximum value at the two tube ends and becomes zero at the middle. Therefore theoretically, to obtain a sufficiently large load transfer, a nanotube's length should be over 200–500 nm. On the other hand, tensile stress starts to build up at the two ends of a nanotube and reaches its maximum at the middle. To have an effective reinforcement, tube length must be sufficiently long to make full use of its high tensile strength.

Because the bonding between inner and outer tubes of a multiwalled carbon nanotube (MWCNT) is weak [111, 112], inter-wall sliding may occur. Thus, an MWCNT may be approximately treated as an SWCNT in a composite system with a greater wall thickness.

Under two extreme nanotube–matrix bonding conditions, an SWCNT-reinforced composite can fail in two modes. On one hand, when the bonding is weak, the nanotube will be pulled out as the maximum shear stress at the two ends of the tube reaches the bonding strength. In this case, because a large amount of energy will be absorbed by the pull-out, the composite will exhibit good fracture toughness but a low tensile strength. On the other hand, if the interfacial bonding is very strong, the build-up of the interfacial shear stress will bring about a high tensile stress in the tube, resulting in the tube breakage as discussed before. In this case, the composite has a high tensile strength but a poor toughness. Therefore, modifying the interfacial bonding strength can lead to a property balance between tensile strength and toughness. The large stress transfer efficiency of a CNT-reinforced composite indicates that the improvement of its toughness can be achieved at a less loss of its tensile strength, and vice versa, by tailoring the interfacial bonding strength. These cannot be achieved by the usual fiber-reinforced composite [29].

The above mechanics was based on two basic assumptions: (1) the loading is in the SWCNT axis and (2) the SWCNT-matrix bonding is perfect. These are technically impossible. In the following sections, we will discuss some effects of variations to these idealised situations, such as alignment and chemical bonding.

9.3.2 Nanotube Dispersion and Alignment

It has been found that if carbon nanotubes can be well aligned and dispersed in a matrix, tailored properties of composites can be achieved. Currently, the dispersion of entangled nanotubes is ineffective, usually by sonication in a solvent such as acetone. Although a prolonged sonication can separate nanotubes by breaking them into short pieces [44], it produces defects and causes thermal degradation of the nanotubes at the same time [33]. Short nanotubes are not desirable because they cannot maximize the stress transfer ability [105], which is an important factor for making a composite of strong reinforcement. To align carbon nanotubes in polymer matrices, some methods have been proposed, including the melt-processing [21], mechanical stretching [42], slicing [1], spin coating [108], extruding [15, 86], and magnetic field-oriented processing [31].

Xiao and Zhang developed an effective mechanical method to simultaneously separate and align long, entangled carbon nanotubes in epoxy matrices by making use of the effect of matrix viscosity on the microstructures of the composites [106]. They found that to effectively disperse and align nanotubes, the matrix viscosity must be sufficiently large so that high mechanical stresses can be transferred to nanotubes. To this end, they added matrix hardener and then placed the mixture between two parallel rotating disks and applied a continuous steady shear at the rate of $0.22\,\mathrm{s}^{-1}$ at $40°C$ until the viscosity of the composite increased to above

6×10^4 Pa s. The composite was further cured in situ to ensure that the nanotubes remained in their aligned direction.

The TEM image in Fig. 9.15a shows that the nanotubes were aligned in the shear direction, as expected. The orientation distribution of the nanotubes is displayed in

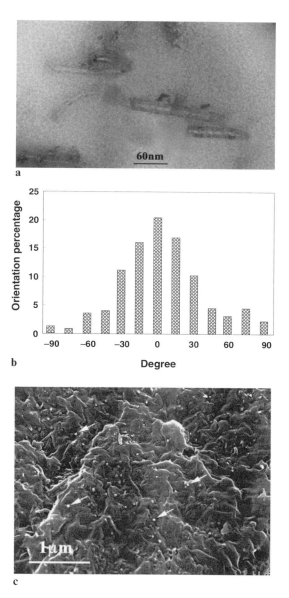

Fig. 9.15 (**a**) Nanotubes aligned in the shear direction (from *left* to *right*); (**b**) Orientation distribution histograms for the composite after alignment; (**c**) A cross-sectional view of aligned nanotubes in matrix (some are indicated by *white arrows*). Note that the nanotubes were well dispersed in the matrix

Fig. 9.15b, which shows that about 75% of the nanotubes are aligned within $\pm 30°$ along the shear direction. The cross-section of the specimen perpendicular to the nanotube alignment direction, Fig. 9.15c, clearly shows that the nanotubes, while well aligned, have been uniformed dispersed in the matrix.

9.3.3 Nanotube–Matrix Interface Bonding

Compared with carbon fibers, carbon nanotubes have many superior mechanical properties and their aspect ratio is much greater, which is preferable in making a stronger composite. As such carbon nanotubes are being considered in place of fibers for reinforcing polymers. However, reports on the relevant properties of nanotube-reinforced composites have been confusing and often conflicting. Some researchers [7, 69, 78] claimed to have obtained good improvement in fatigue resistant, fracture toughness, and tensile strength, but some others [37, 38] found that the use of nanotubes had actually reduced the composite strength. These conflicting results were mainly due to the lack of mechanism understanding of nanotube–matrix interface bonding. The theoretical rationale for the chemical bonding was not verified.

Mylvaganam and Zhang used the density functional theory to study the possible chemical bond formation between a model polyethylene chain radical and a model nanotube [52, 53]. They investigated the following four structures: Model I (C_{60}–C_5H_{11}•), a fullerene having 60 carbon atoms and a pentyl radical with unpaired electron on the third carbon atom; Model II ($C_{60}H_{10}$–C_5H_{11}•), a (5,0) nanotube segment having 60 C atoms with hydrogen atoms added to the dangling bonds of the perimeter carbons and a pentyl radical as in Model I; Model III ($C_{64}H_{20}$–C_5H_{11}•), a section of a (17,0) nanotube sidewall consisting of 23 hexagons with 20 hydrogen atoms added to the dangling bonds of the perimeter carbons (i.e., $C_{64}H_{20}$) and a pentyl radical as in Model I; and Model IV ($C_{60}H_{10}$–C_7H_{14}••), a (5,0) nanotube segment used in Model II and a C_7H_{14} biradical with unpaired electrons on the second and sixth carbon atoms.

Mylvaganam and Zhang found that the covalent bond formation between alkyl radicals and carbon nanotubes is energetically favorable, that the reaction may take place at multiple sites of nanotubes, and that the reaction is more favorable with tubes of a small diameter [52, 53]. As alkyl radicals are good representatives of polyethylene chain radicals, it may be possible to form multiple covalent bonds between polyethylene chain radicals and nanotubes. They therefore concluded that one way to improve the load transfer of carbon nanotube/HDPE composite via chemical bonds at the interface is to use free radical generators such as peroxide or incorporate nanotubes in the in situ polymerization. This offers a direct guideline on the basis of an established free radical polymerization scheme for making composites with strong carbon nanotube–polyethylene interfaces and high stress transfer ability.

Mylvaganam and Zhang [56] and Mylvaganam et al. [59] discovered that the deformation of a carbon nanotube can promote its chemical reactivity with hydrogen and alkyl radicals. A tube deformed by central loading and compression can bind atoms and radicals strongly at the ridges when the strain energy is as low as ~2 kJ/mol. Pure bending shows a much stronger binding when the strain energy is little higher than 2 kJ/mol. A nanotube under torsion also shows strong binding albeit at high strain energy and high applied force than pure bending. Thus mechanical deformations in composite blending can promote reactivity with nearby chemical species without adding specific chemical reagents to functionalize the nanotubes.

9.3.4 Wear and Friction

Relatively, very few studies have been carried out on the tribological properties of CNT-polymer composites [2, 10, 12, 19, 125]. Zoo et al. reported that an addition of 0.5 wt% CNTs to an ultra-high molecular weight polyethylene could significantly reduce the wear loss of the composite but increase its friction coefficient [125]. It was considered that this was mainly caused by the increased shear strength of the composite. An et al. acknowledged an improved wear and mechanical properties of alumina-based composites containing CNTs up to 4 wt%, though without a mechanism explanation [2]. Chen et al. believed that the favorable tribological properties of a CNT-polytetrafluoroethylene composite was due to the high strength and high aspect ratio of the CNTs [12]. They commented that the CNTs might have been released from the composite during sliding, which then prevented direct contact of worn surfaces and hence reduced both the wear rate and friction coefficient. However, no direct evidence was given. Cai et al. reported that the contribution of CNTs in a polyimide composite was to restrain the scuffing and adhesion of the polyimide matrix in sliding, providing a much better resistance than the neat polyimide [10].

In this section, we will take CNT-reinforced epoxy as an example to explore the role of CNTs in the wear of the composite. The multiwalled CNTs used were prepared by chemical vapor deposition (provided by Nanolab) with their diameters ranging from 10 to 25 nm and their lengths from 10 to 20 μm (Fig. 9.16). Details of experiments have been described by [121]. Because the dispersion of the CNTs

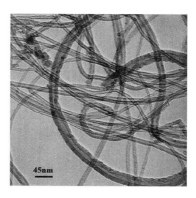

Fig. 9.16 CNTs in the initial state

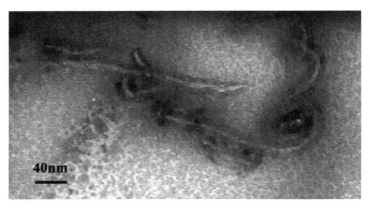

Fig. 9.17 The CNTs in a epoxy composite before a wear test

in epoxy was not absolutely uniform, particular attention was placed to monitor the coverage area ratio of CNTs to matrix on a pin surface, $R_{c/m}$. In addition to those randomly cut, a special group of pins with a specific $R_{c/m} > 0.25$ were prepared to explore the effect.

Figure 9.17 is a typical image showing the structure of the composite before wear testing. It is clear that the CNTs in the composite have the same structure as that supplied (Fig. 9.16), demonstrating that the morphology and structure of the CNTs were not altered by the composite preparation process.

Results of the wear tests are presented in Fig. 9.18a. The phenomenon of running-in wear is obvious for both the pure epoxy and the nanotube-reinforced epoxy composite. A stable wear then takes place after a sliding distance of about 2,000 m. The wear rate at this stage was $16 \times 10^{-4}\, \mathrm{mm^3/m}$ for the pure epoxy. It is clear that an addition of only 0.1 wt% nanotubes to the epoxy matrix dramatically reduces the wear rate. The samples with $R_{c/m} < 25\%$ had a wear rate of $8 \times 10^{-4}\, \mathrm{mm^3/m}$, which is over 1.7 times lower than that of the pure epoxy. These samples had rough surfaces as presented in Fig. 9.19a. Brittle fragments and multiple microcracks were evident, indicating that the material removal was via brittle fracture. More significantly, the samples with $R_{c/m} > 25\%$ had a remarkably lower wear rate (see the bottom curve with triangular dots in Fig. 9.18a), with an extremely small weight loss up to a sliding distance of 1,700 m. The wear rate upon further sliding kept low, varying in the range of 0.05×10^{-4} to $2 \times 10^{-4}\, \mathrm{mm^3/m}$, about 3.2 times smaller than that of the sample with $R_{c/m} < 25\%$ and 5.5 times lower then that of pure epoxy. The surface of these samples appeared very smooth. A more close look at the surface topography (Fig. 9.19b) did not detect a single brittle fragmentation, which seems to suggest that the mechanism of the materials removal in the wear process was ductile. The electron micrograph of the worn surface, Fig. 9.19c, clearly indicates that the CNTs covered the surface greatly slowed down the wear process of the epoxy matrix.

The coefficient of friction for the composites with $R_{c/m} > 25\%$ was much lower (Fig. 9.18b) for all sliding distances checked. Again, the surface microstructure, as

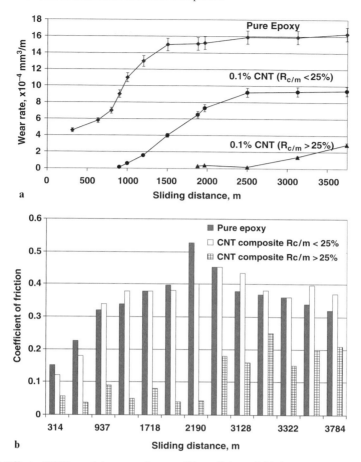

Fig. 9.18 Effect of CNTs on (**a**) wear resistance; (**b**) coefficient of friction

shown in Fig. 9.19c and to be discussed later, suggests that the CNTs exposed to
the sliding interface were acting as a solid lubricant at the sliding interface which
reduced the coefficient of friction. After a wear test, a specimen surface subjected to
sliding became very shining and smooth. For the composite with $R_{c/m} < 25\%$, the
coefficient of friction varied. A further study is necessary to clarify the mechanism.

Figure 9.19c gives the distribution of CNTs on a sample's surface with a high
$R_{c/m}$. The high concentration of CNTs, most with bamboo structures, is obvious.
The exposed length of CNTs varies from 100 to 400 nm, and it is evident that the
fragmentation of CNTs during a wear test did take place, maybe caused by the high
contact stress and abrasion at the sliding interface, because many CNTs no longer
have their caps (see CNTs 1–9 in Fig. 9.19c). The orientation of the CNTs with a
short length exposed to the sliding surface scatters, but those with longer exposed
lengths have obviously been aligned along the sliding direction of the wear test
(from right to left in Fig. 9.19c). The fragmentation of the CNTs was also coupled

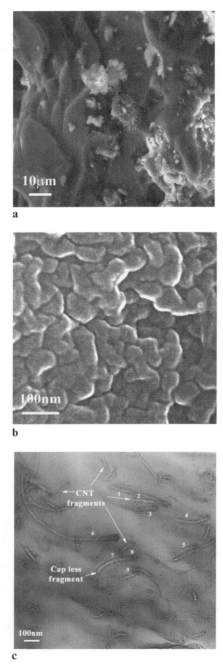

Fig. 9.19 Sample surfaces after wear test (sliding distance 2,500 m) (**a**) $R_{c/m} < 25\%$; (**b–c**) $R_{c/m} >$ 25%; (**a–b**) surface topography; (**c**) electron micrograph

with sever deformation of the walls. The removal of the CNT caps seems to occur always at the end exposed to the sliding surface. The above observation suggests the following wear mechanism. At the first stage of sliding wear, the epoxy matrix at a sample's surface was removed in a brittle mode. The CNTs were then exposed to the sliding interface. Because the CNTs have a high strength, they reduce the wear rate significantly. Nevertheless, the CNTs underwent severe deformation leading to the removal of their caps and distortion of their walls. Meanwhile, the rotation of the steel disk aligned the CNTs along the sliding direction. As a result, the structure seen in Fig. 9.19c was formed with aligned capless CNTs.

The above investigation into the wear of CNT-reinforced epoxy composites brings about the following conclusions. The surface coverage area of CNTs, $R_{c/m}$, plays a significant role in the wearability of the composites. With $R_{c/m} > 25\%$, the wear rate can be reduced by a factor of 5.5. The improvement of the wear resistance of a high $R_{c/m}$ composite was due to the CNTs exposed to the sliding interface, which protected the epoxy matrix effectively. The exposed CNTs also reduced the coefficient of friction significantly.

9.3.5 Stress-Induced Structural Change of Nanotubes

This section will discuss the deformation and structural transformation of CNTs in an epoxy composite when subjected to contact sliding [114].

The CNTs in the epoxy composite before contact sliding is presented in Fig. 9.20. The CNTs have a bamboo structure. Most of them are not deformed and have smooth walls and closed caps (Fig. 9.21a). Some CNTs are buckled (Fig. 9.21b), which could have occurred as a result of bending of CNTs above certain critical curvature [36].

The situation became different after contact sliding. Most CNTs were fragmentized to segments of 100–400 nm after a wear test. The CNT walls were no longer

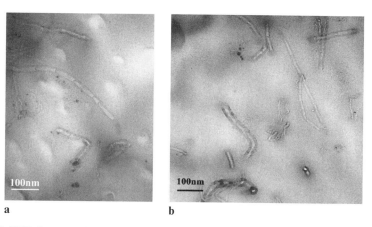

a   b

Fig. 9.20 CNTs in composite: (**a**) before wear test; (**b**) after wear test

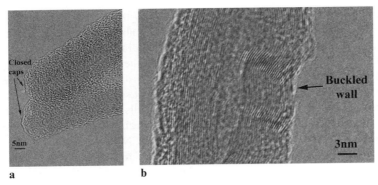

Fig. 9.21 Microstructure of CNTs before wear test: (**a**) Note closed caps of CNTs and undistorted walls; (**b**) Note buckling of some CNTs walls

smooth, and with atomic distortion and dislocations. These types of permanent deformation of the walls due to the repeated interaction of composite with the asperities of the steel disk in the contact sliding of the wear test led to the outer shell breakages of the CNTs [36]. More interestingly, nanoparticles were detected in the composite after wear tests. These nanoparticles were similar to those obtained by electron irradiation [91] or by ball milling [40]. Some nanoparticles were partially connected to CNTs proving that CNTs were their parents. For example, one side of nanoparticle marked as 1 in Fig. 9.22a is connected to the wall of a nanotube. However, the second side, marked as 2, has no connection with CNT. Rupture of this wall of CNT with following closure created nanoparticle with elliptical shape. According to Li et al. the closure of nanoparticles can be due to the deposition of small grapheme sheets, eliminating dangling bonds, on the fractured sites [40]. The nanoparticles can be seen in more details in Fig. 9.22b. It features 12 nm in outer diameter, 5 nm in inner diameter and composition of 12 shells. Both outer and inner shells are imperfect with unclosed deformed shells.

Nanoparticles having been broken away from their parent CNTs were of two main kinds. Some of them were with faceted or partly faceted walls (Fig. 9.22c). These nanopatricles had an inner diameter of 5–8 nm and a moderate number of shells 15–18. Besides that, spherical nanoparticles resembling an onion-like structure were obvious (Fig. 9.22d). The difference between nanoparticles in samples after wear tests and onion-like nanoparticles described by Ugarte was in presence of inner amorphous core of 4–5 nm and a large number of defects located in particle shells [92]. The outer and inner shells were not continuous as in the case of irradiated nanoparticles. Iijima pointed out that the onion-like nanoparticles are the most stable form of nanoparticles [26]. An irradiation treatment transforms metastable faceted nanoparticles to more stable circular onion-like particles. In this way, the system is stabilized by the energy gain from the van der Waals interaction between shells [91]. In our case the transformation from faceted particle to onion-like particle could happen not because of irradiation but due to deformation taking place during the wear test. It is possible that sharp corners between facets act as stress

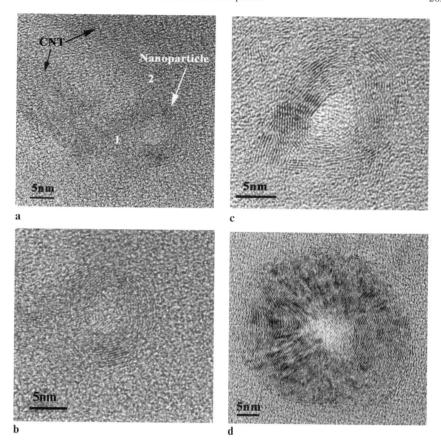

Fig. 9.22 HRTEM images of nanoparticles on the sample's surface after wear test. (**a**) A nanoparticle; note that the nanoparticle is connected with the CNT from the side marked 1 and lose its connection with the CNT at the side marked 2; (**b**) The nanoparticle in more details; (**c**) A faceted nanoparticle; (**d**) An onion-like nanoparticle

concentrators during composite-disk interaction in contact sliding and round shape particles are formed. The mechanical deformation can also destroy some shells, and as a result a lot of defects are formed in the inner structure of the nanoparticles.

9.3.6 Rheological Properties

The relationship between processability and performance is an important issue in the manufacture of CNT composites using the melt blending technique. Like the composites reinforced by whiskers or carbon fibers, the addition of CNTs to a polymer involves profound processing-related difficulties due to increased viscosity. There is usually a limited processing window, in terms of processing conditions, which gives attractive properties and hence an in-depth understanding of the rheological behavior of the materials is necessary. A few investigations on dynamic frequency sweeps [43, 49, 65, 66, 73] have been reported using melt processed thermoplastic polymer/CNT composites, whose matrices include polycarbonate [43, 65, 66], polypropylene [73], and polyamide-6 and its blends with acrylonitrile/butadiene/styrene [49]. These studies found that the complex viscosity continuously decreases with increasing frequency while storage modulus G' and loss modulus G'' increase. A characteristic change in the rheological behavior with increasing nanotube content was also found for temperatures well above the glass transition or melting temperature, referred to as the percolation threshold or gelation point in relation to fluid-to-solid, fluid-to-gel transition or a combined nanotube-polymer network.

In this section, we will discuss the rheological behavior of a low density polyethylene (LDPE) composite reinforced by multiwalled carbon nanotubes [107].

Figure 9.23 shows the variation of the complex viscosity $|\eta^*|$ of the composites as a function of sweep frequencies. Compared with the neat LDPE, the composites, in particular those with high nanotube contents, have much greater viscosities. A Newtonian plateau at low frequencies is clearly visible up to 3 wt% nanotube content. At a higher content, the plateau vanishes and the composites show a strong shear thinning effect. The viscosity almost decreases linearly with increasing the frequency, but the reduction gradient becomes greater when the CNT content increases. As a result, viscosity difference of the composites from the pure LDPE at the high frequency region becomes small. This, as to be discussed later, may imply a structural change in the composites.

Tanδ is sensitive to the structural change of the materials, as shown in Fig. 9.24. A viscoelastic peak occurs at the frequency of about 1 rad/s and disappears with

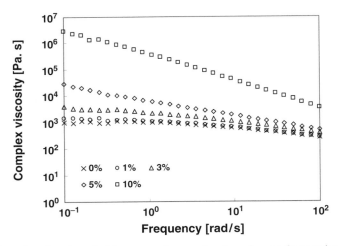

Fig. 9.23 Complex viscosity η^* of the composites as a function of sweep frequencies

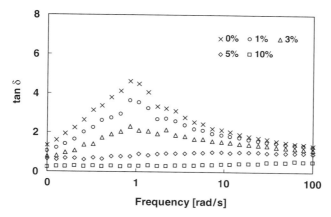

Fig. 9.24 Tanδ of the composites as a function of sweep frequencies

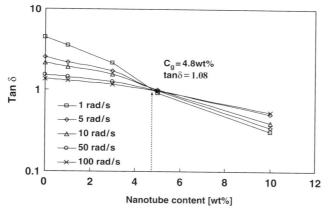

Fig. 9.25 Variation of tanδ with the nanotube content at different frequencies

increasing CNT content, showing the material becomes more elastic. This is also a characteristic when a viscoelastic fluid experiences a fluid–solid transition. Gelation as an example for such a fluid–solid transition has been extensively described by Ewen et al. [18]. At the gelation point, tanδ is expected to be independent of frequency. With the present composites, such transition occurs at CNT content between 3 and 5 wt%. In the postgel stage, tanδ increases with frequency, indicating a dominating elastic response of the material. CNT content at gelation point can be estimated more accurately from a multifrequency plot, as shown in Fig. 9.25. Tanδ decreases with frequency and all curves intersect at nearly a single point where tanδ becomes frequency-independent. Thus the gelation composition of the present composites, C_g, can be estimated to be 4.8 wt%, beyond which a continuous CNT-CNT network forms in the composites [70].

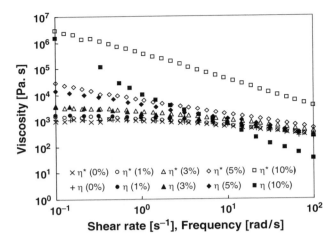

Fig. 9.26 A comparison of the complex and steady shear viscosities

The steady shear property of a material is important to the processing of a material and is often estimated by the empirical Cox-Merz rule, which states that the steady shear rate viscosity and complex viscosity are closely super-imposable for numerically equivalent values of shear rate and frequency. Kinloch et al. investigated steady shear properties of aqueous MWCNT dispersions [32], although there are no such measurements reported in polymer matrix composites. They found that Cox-Merz rule is no longer valid in these aqueous MWCNT dispersions and steady shear viscosity is a few orders of magnitude lower than complex one. The similar phenomenon was also reported in other composite systems [76]. Figure 9.26 gives a comparison of these two viscosities. It is clear that the Cox-Merz rule holds when the CNT content is low but becomes inaccurate when the content gets higher at which the composite is more solid-like. The composite with 10 wt% CNTs shows a rapid decrease in steady shear viscosity when the shear rate increases. The torque traces as a function of shear rate for the composites are displayed in Fig. 9.27. A torque peak appears at a high CNT content (e.g., 5 and 10 wt%), and reaches a steady state with increasing shear rate. Generally, the maximum torque value shifts to a low shear rate with increasing CNT content. It is, therefore, clear that the optimal processing condition changes significantly with the variation of the CNT content. On one hand, a low shear rate is preferred to achieve a low torque for composites with low CNT content. On the other hand, for a composite with a high CNT content (e.g., 10 wt%), which has a high initial viscosity (Fig. 9.26), it is easy to be processed at a high shear rate.

Figure 9.28 shows the variation of stress σ as a function of the strain ε. Below the gelation point of 4.8 wt%, σ increases with ε and levels off at high ε values. However, it should be noted that even at the upper measurement limit of $\varepsilon = 10^7$, σ does not completely reach a plateau (a typical indication of yielding). Kinloch et al. observed an obvious stress yielding in aqueous CNT dispersions [32], which occurred

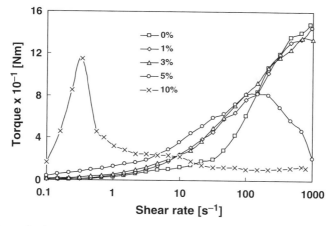

Fig. 9.27 Torque for the MWCNT-reinforced PE composites

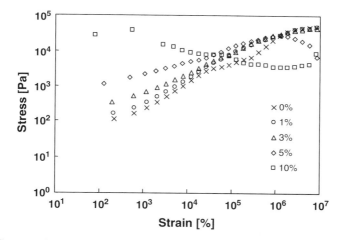

Fig. 9.28 Stress-strain relationships for the composites at various MWCNT loadings

in a shear rate range from 1 to 100/s (corresponding to a strain of 8.5×10^5 in this work) when the CNT content changed from 0.5 to 9.2 vol%. In aqueous suspensions of single wall carbon nanotubes, Hough et al. found that stress yielding is composition dependent and the higher the nanotube content, the lower the yielding strain is [24]. The yielding plateaus in these suspensions existed below $\varepsilon = 10$. They also concluded that the elasticity of nanotube networks originated from the bonding of nanotube rods rather than from stretching or bending of the tubes. A plausible explanation for this yielding behavior was proposed for flocculated networks by Barnes [5]. He argued that under shear (or strain) the network broke down into floccs. As the rate is increased, the floccs break into smaller floccs, which require less stress to flow, keeping the stress virtually constant with increasing shear rate or strain. On the basis of this explanation, it is easy to understand that in these aqueous suspensions interaction between nanotube networks dominates and it is weak. Therefore, stress yields at a quite low strain or shear rate value. In MWCNT-PE composites with tube content below C_g, where the materials are viscous fluid-like, the interaction of PE-PE molecular networks is stronger than either that of MWCNT-MWCNT networks or that of PE-MWCNT networks. Consequently, a larger strain is required to break down these PE networks. With further increasing nanotube content beyond C_g, where materials becomes solid-like and the number of weak PE-MWCNT and MWCNT-MWCNT networks increase, stress yield occurs at a lower strain value.

The above results show that the stress yielding behavior in polymeric composites is not completely the same as that in aqueous nanotube dispersions [24, 32], flocculated networks [5], the gel made of hydroxyl-12 stearic acid in nitrobenzene [80] and aqueous suspensions of steroid nanotubules [79]. The result does not show a plateau but a decrease in stress when nanotube loading is larger than C_g. This indicates that the breaking down and recovering processes of the three networks in these MWCNT loadings should be different from those with a plateau.

The above discussion shows that the dynamic mechanical moduli and viscosity increase with the incorporation of MWCNTs into PE. The materials experience a fluid–solid transition at a gelation composition of $C_g = 4.8\,wt\%$, where a continuous MWCNT network form. It was also observed that Cox-Merz rule is still valid in the MWCNT-PE composites in fluid-like composition region but becomes inaccurate beyond C_g. The processing property of the composites depends strongly on nanotube concentrations and a proper selection of shear rate can reduce torque value in the mixing of the MWCNTs with PE. It was also found that the PE-PE network is more difficult to be broken down under shear than the network of PE-MWCNT or MWCNT-MWCNT. There exists a strong composition-dependence of stress yielding behavior. However, this yielding in the composite systems studied in this section does not show a typical stress plateau as observed in many other materials.

9.4 Concluding Remarks

This chapter has discussed some fundamentals of the mechanics of carbon nanotubes and their composites, including the effective wall thickness of a single-walled carbon nanotube for continuum mechanics modeling, its mechanisms of bending and kinking, the essentials for reliable molecular dynamics simulations, the possibility of chemical bonding and deformation-promoted reactivity, dispersion and alignment, stress transfer, and the rheological behavior of nanotube-reinforced composites. It is clear that studies along the line are far from sufficient, particularly in the areas of nanomechanics modeling using continuum mechanics, improvement of interface bonding strength, tribological properties and mechanisms, and development of efficient methods for aligning nanotubes.

There are different views on the mechanical characterization of CNTs and their composites. For instance, some researchers think that there is no need to clarify the effective wall thickness and Young's modulus of CNTs, because one may use other parameters in a specific application, e.g., using bending stiffness. The author believes that the effective wall thickness and Young's modulus are the most fundamental mechanics quantities for modeling using the well-established continuum mechanics theory. Bypassing the basics in specific case studies cannot verify, at the grassroots level, the applicability of continuum mechanics theory to nanoscopic analysis [17, 58].

Acknowledgment The relevant work presented here by the author's team has been supported by the Australian Research Council.

References

1. Ajayan PM, Stephan O, Colliex C, Trauth D (1994) Science 265:1212
2. An JW, You DH, Lim DS (2003) Wear 255:677
3. Arsecularatne JA, Zhang LC (2007) Recent Patents Nanotechnol. 1:176
4. Bao WX, Zhu CC, Cui WZ (2004) Physica B 352:156
5. Barnes HA (1997) J. Non-Newton. Fluid Mech. 70:1
6. Belytschko T, Xiao SP, Schatz GC, Ruoff RS (2002) Phys. Rev. B 65:235430
7. Bower C, Rosen R, Jin L (1999) Appl. Phys. Lett. 74:3317
8. Brenner DW (1990) Phys. Rev. B 42:9458
9. Brenner DW, Shenderova OA, Harrison J, Stuart SJ, Ni B, Sinnott SB (2002) J. Phys. Condens. Matter. 14:783
10. Cai H, Yan F, Xue YQ (2004) Mater. Sci. Eng. A 364:94
11. Charlier JC, Lambin P, Ebbesen TW (1996) Phys. Rev. B 54:R8377
12. Chen WX, Li B, Han G, Wang LY, Tu JP, Xu ZD (2003) Tribol. Lett. 15:275
13. Chen X, Cao G (2006) Nanotechnology 17:1004
14. Ciccotti G, Hoover WG (1986) Molecular-Dynamics Simulation of Statistical-Mechanical Systems. Elsevier, New York
15. Cooper CA, Ravich D, Lips D, Mayer J, Wagner HD (2002) Compos. Sci. Technol. 62:1105
16. Dalton AB, Collins S, Munoz E, Razal JM, Ebron VH, Ferraris JP, Coleman JN, Kim BG, Baughman RH (2003) Nature 423:703
17. Delmotte JPS, Rubio A (2002) Carbon 40:1729
18. Ewen B, Richter D, Shiga T, Winter HH, Mours M (1997) Advances in Polymer Science. Springer, Berlin
19. Gao Y, Wang ZJ, Ma QY, Tang G, Liang J (2004) J. Mater. Sci. Technol. 20:340
20. Govindjee S, Sackman JL (1999) Solid State Commun. 110:227
21. Haggenmueller R, Gommansb HH, Rinzlerb AG, Fischera JE, Winey KI (2000) Chem. Phys. Lett. 330:219
22. Hernández E, Goze C, Bernier P, Rubio A (1998) Phys. Rev. Lett. 80:4502
23. Hernández E, Goze C, Bernier P, Rubio A (1999) Appl. Phys. A 68:287
24. Hough LA, Islam MF, Janmey PA, Yodh AG (2004) Phys. Rev. Lett. 93:168102
25. Huang Y, Wu J, Hwang KC (2006) Phys. Rev. B 74:245413
26. Iijima S (1980) J. Cryst. Growth 50:675
27. Iijima S, Brabec C, Maiti A, Bernholc J (1996) J. Chem. Phys. 104:2089
28. Jin Y, Yuan FG (2003) Compos. Sci. Technol. 63:1057
29. Kelly A, MacMillan NH (1986) Strong Solids, 3rd edn. Clarendon, Oxford

30. Kelly A, Tyson WR (1965) J. Mech. Phys. Solids 13:329
31. Kimura T, Ago H, Tobita M (2002) Adv. Mater. 14:1380
32. Kinloch IA, Roberts SA, Windle AH (2002) Polymer 43:7483
33. Koshio A, Yudasaka M, Iijima S (2001) Chem. Phys. Lett. 341:461
34. Krishnan A, Dujardin E, Ebbesen TW, Yianilos PN, Treacy MMJ (1998) Phys. Rev. B 58:14013
35. Kudin KN, Scuseria GE, Yakobson BI (2001) Phys. Rev. B 64:235406
36. Kuzumaki T, Hayashi T, Ichinose H, Miyazawa K, Ito K, Ishida Y (1998) Philos. Mag. A 77:1461
37. Lau KT, Hui D (2002) Carbon 40:1605
38. Lau KT, Shi SQ (2002) Carbon 40:2961
39. Li C, Chou TW (2003) Int. J. Solids Struct. 40:2487
40. Li YB, Wei BQ, Liang J, Yu Q, Wu DH (1999) Carbon 37:493
41. Liao K, Li S (2001) Appl. Phys. Lett. 79:4225
42. Lin J, Bower C, Zhou O (1998) Appl. Phys. Lett. 73:1197
43. Liu CY, Zhang J, He JS, Hu GH (2003) Polymer 44:7529
44. Liu J, Casavant MJ, Cox M, Walters DA, Boul P, Lu W, Rimberg AJ, Smith KA, Colbert DT, Smalley RE (1999) Chem. Phys. Lett. 303:125
45. Lopez MJ, Rubio A, Alonso JA, Qin LC, Iijima S (2001) Phys. Rev. Lett. 86:3056
46. Lourie O, Wagner HD (1999) Compos. Sci. Technol. 59:975
47. Lu JP (1997) J. Phys. Chem. Solids 58:1649
48. Lu JP (1997) Phys. Rev. Lett. 79:1297
49. Meincke O, Kaempfer D, Weickmann H, Friedrich C, Vathauer M, Warth H (2004) Polymer 45:739
50. Munoz E, Dalton AB, Collins S, Kozlov M, Razal J, Coleman JN, Kim BG, Ebron VH, Selvidge M, Ferraris JP, Baughman RH (2004) Adv. Eng. Mater. 6:801
51. Mylvaganam K, Zhang LC (2004) Carbon 42:2025
52. Mylvaganam K, Zhang LC (2004) J. Phys. Chem. B 108:5217
53. Mylvaganam K, Zhang LC (2004) J. Phys. Chem. B 108:15009
54. Mylvaganam K, Zhang LC (2005) J. Comput. Theor. Nanosci. 2:251
55. Mylvaganam K, Zhang LC (2006) Appl. Phys. Lett. 89:123127
56. Mylvaganam K, Zhang LC (2006) Nanotechnology 17:410
57. Mylvaganam K, Zhang LC (2007) Recent Patents Nanotechnol. 1:59
58. Mylvaganam K, Vodenitcharova T, Zhang LC (2006) J. Mater. Sci. 41:3341
59. Mylvaganam K, Zhang LC, Cheong WCD (2007) J. Comput. Theor. Nanosci. 4:122
60. Odegard GM, Gates TS, Nicholson LM, Wise KE (2002) NASA/TM-2002-211454
61. Odegard GM, Gates TS, Nicholson LM, Wise KE (2002) Compos. Sci. Technol. 62:1869
62. Ozaki T, Iwasa Y, Mitani T (2000) Phys. Rev. Lett. 84:1712
63. Pantano A, Parks DM, Boyce MC (2004) J. Mech. Phys. Solids 52:789
64. Peebles LH (1995) Carbon Fibres: Formation, Structure and Properties. CRC, Boca Raton
65. Pötschke P, Fornes TD, Paul DR (2002) Polymer 43:3247
66. Pötschke P, Goad MA, Alig I, Dudkin A, Lellinger D (2004) Polymer 45:8863
67. Qian D, Dickey EC, Andrews R, Rantell T (2000) Appl. Phys. Lett. 76:2868
68. Qian D, Dickey EC (2001) J. Microsc. 204:39
69. Ren Y, Li F, Cheng HM, Liao K (2003) Carbon 41:2177
70. Sandler JKW, Kirk JE, Kinloch IA, Shaffer MSP, Windle AH (2003) Polymer 44:5893
71. Schadler LS, Giannaris SC, Ajayan PM (1999) Appl. Phys. Lett. 73:3842
72. Sears A, Batra RC (2004) Phys. Rev. B 69:235406
73. Seo MK, Park SJ (2004) Chem. Phys. Lett. 395:44
74. Sheehan EWPE, Lieber CM (1997) Science 277:1971
75. Shen L, Li J (2005) Phys. Rev. B 71:165427
76. Shenoy AV (1999) Rheology of Filled Polymer System. Kluwer, Dordrecht
77. Tang J, Qin LC, Sasaki T, Yudasaka M, Matsushita A, Iijima S (2000) Phys. Rev. Lett. 85:1887

78. Tang W, Santare MH, Advani SG (2003) Carbon 41:2779
79. Terech P, Talmon Y (2002) Langmuir 18:7240
80. Terech P, Pasquier D, Bordas V, Rossat C (2000) Langmuir 16:4485
81. Tersoff J (1986) Phys. Rev. Lett. 56:632
82. Tersoff J (1988) Phys. Rev. B 37:6991
83. Tersoff J (1989) Phys. Rev. B 39:5566
84. Tersoff J, Ruoff RS (1994) Phys. Rev. Lett. 73:676
85. Thomsen C, Reich S, Jantoljak H, Loa I, Syassen K, Burghard M, Duesberg GS, Roth S (1999) Appl. Phys. A 69:309
86. Thostenson ET, Chou TW (2002) J. Phys. D Appl. Phys. 35:L77
87. Tombler TW, Zhou CW, Alexseyev L, Kong J, Dai HJ, Liu L, Jayanthi CS, Tang MJ, Wu SY (2000) Nature 405:769
88. Treacy MM, Ebbesen TW, Gibson JM (1996) Nature 381:678
89. Tserpes K, Papanikos P (2005) Compos. B 36:468
90. Tu ZC, Ou-Yang ZC (2002) Phys. Rev. B 65:233407
91. Ugarte D (1993) Europhys. Lett. 22:45
92. Ugarte D (1995) Carbon 33:989
93. Vainshtein BK, Fridkin VM, Indenbom VL (1995) VFI Atomic Radii, Structure of Crystals, 3rd edn. Springer, Berlin
94. Vodenitcharova T, Zhang LC (2003) Phys. Rev. B 68:165401
95. Vodenitcharova T, Zhang LC (2004) Phys. Rev. B 69:115410
96. Vodenitcharova T, Mylvaganam K, Zhang LC (2007) J. Mater. Sci. 42:4935
97. Vollrath F, Knight DP (2001) Nature 410:541
98. Wagner HD (2002) Chem. Phys. Lett. 361:57
99. Wagner HD, Lourie O (1998) Appl. Phys. Lett. 73:3527
100. Wagner HD, Lourie O, Feldman Y, Tenne R (1998) Appl. Phys. Lett. 72:188
101. Wang CY, Zhang LC (2008), Nanotechnology, 19: 075705
102. Wang CY, Zhang LC (2008), Nanotechnology, 19: 195704
103. Wang L, Zheng Q, Liu JZ, Jiang Q (2005) Phys. Rev. Lett. 95:105501
104. Wong EW, Sheehan PE, Lieber CM (1997) Science 277:1971
105. Xiao KQ, Zhang LC (2004) J. Mater. Sci. 39:4481
106. Xiao KQ, Zhang LC (2005) J. Mater. Sci. 40:6513
107. Xiao KQ, Zhang LC, Zarudi I (2007) Compos. Sci. Technol. 67:177
108. Xu XJ, Thwe MM, Shearwood C, Liao K (2002) Appl. Phys. Lett. 81:2833
109. Yakobson BI, Brabec CJ, Bernholc J (1996) Phys. Rev. Lett. 76:2511
110. Yao N, Lordi V (1998) J. Appl. Phys. 84:1939
111. Yu MF, Files BS, Arepalli S, Ruoff RS (2000) Phys. Rev. Lett. 84:5552
112. Yu MF, Lourie O, Dyer MJ, Moloni K, Kelly TF, Ruoff RS (2000) Science 287:637
113. Yu TX, Zhang LC (1996) Plastic Bending: Theory and Applications. World Scientific, Singapore
114. Zarudi I, Zhang LC (2006) Appl. Phys. Lett. 88:243102
115. Zhang P, Huang Y, Gebelle PH, Klein PA, Hwang KC (2002) Int. J. Solids Struct. 39:3893
116. Zhang LC (2006) In: Rieth M, Schommers W (eds) Handbook of Theoretical and Computational Nanotechnology, Volume 8: Functional Nanomaterials, Nanoparticles, and Polymer Design. American Scientific, Stevenson Ranch, CA, USA. pp. 395–456
117. Zhang LC, Mylvaganam K (2006) J. Comput. Theor. Nanosci. 3:167
118. Zhang LC, Tanaka H (1997) Wear 211:44
119. Zhang LC, Tanaka H (1999) JSME Int. J. A Solid Mech. Mater. Eng. 42:546
120. Zhang LC, Yu TX (1987) Int. J. Press. Ves. Pip. 30:77
121. Zhang LC, Zarudi I, Xiao K (2006) Wear 261:806
122. Zhang P, Jiang H, Huang Y, Geubelle PH, Hwang KC (2004) J. Mech. Phys. Solids 52:977
123. Zhou G, Duan W, Gu B (2001) Chem. Phys. Lett. 33:344
124. Zhou X, Zhou J, Ou-Yang ZC (2000) Phys. Rev. B 62:13692
125. Zoo YS, An JW, Lim DP (2004) Tribol. Lett. 16:305

Chapter 10
Microbridge Tests

Tong-Yi Zhang

10.1 Introduction

With wide applications of thin films in microelectronic devices, microelectro-mechanical systems (MEMS), and nanoelectromechanical systems (NEMS), thin films have attracted great interest from academics and industrialists. In general, thin films have different mechanical properties from their bulk counterparts [3, 16, 23]. The mechanical properties of a thin film, such as its Young's modulus and residual stress, are essential and necessary input information for detailed design and analysis of MEMS and NEMS devices. Mechanical characterization of thin films has in fact become one of the most important challenges in the development of micro/nanotechnologies and a very active area of research, as illustrated by the fact that the US Materials Research Society has organized 11 symposiums on the subject since 1988. The major difficulty encountered in the mechanical characterization of thin films is that they are not amenable to testing by conventional means because of their sizes and configurations. Therefore, various novel characterization methods have been developed, such as micro/nanoindentation tests [15, 24, 41, 44], bulge tests [35], bending tests with one fixed end [29] and with two fixed ends [22, 45], tensile tests [10], etc.

The MEMS/NEMS fabrication technique has made it possible to fabricate micro/nanobeams. If only one end of the beam is fixed, it is called a cantilever beam. A micro/nanobeam with two fixed ends is called a micro/nanobridge. A load and displacement sensing nanoindenter or atomic force microscope is a perfect loading machine to be used in the bending test on cantilever micro/nanobeams and micro/nanobridges. We call the bending test on micro/nanobridges as the microbridge test. Especially, employing a microwedge tip with its wedge size larger than the width of a tested microbeam in the bending test makes one-dimensional analysis of

T.-Y. Zhang
Department of Mechanical Engineering, Hong Kong University of Science and Technology, Clear Water Bay, Kowloon, Hong Kong, China
mezhangt@ust.hk

F. Yang and J.C.M. Li (eds.), *Micro and Nano Mechanical Testing of Materials and Devices*,
doi: 10.1007/978-0-387-78701-5, © Springer Science+Business Media, LLC, 2008

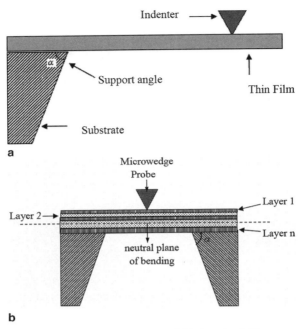

Fig. 10.1 (**a**) Depiction of the microcantilever test and (**b**) the microbridge test

the deformation behavior more appropriate [8]. Figure 10.1 schematically shows the bending test on a cantilever micro/nanobeam and the micro/nanobridge test. Homogeneous residual stresses, if any, in thin films bonded on substrates are completely released in cantilever beams. Inhomogeneous residual stresses may cause cantilever beams bent, wrinkled, or/and curled severely, therefore rendering them useless for the bending test. In contrast, homogeneous residual stresses in thin films bonded on substrates are not released in the length direction of microbridge beams because of the two fixed ends. That is why the microbridge test is able to characterize residual stress in a film.

When conducting a load–deflection test on a thin film, one has to consider the influence of substrate deformation. Fang and Wickert investigated the methods for the determination of the mean and the gradient of residual stresses in a microcantilever [12]. They considered the stress state in the still bonded film and found that the analytical ideal boundary conditions, i.e., the simply supported and clamped boundary conditions, might not completely reflect the real boundary conditions. In microcantilever bending tests, Baker and Nix found that the slope of the beam deflection vs. the applied load consisted of two terms [2]. One term was due to the beam deformation and directly proportional to the cubic of the beam length, while the other was due to the substrate deformation and directly proportional to the square of the beam length. The influence of substrate deformation makes the apparent length of the beam larger than its real length when the solution without including substrate deformation is used to fit the experimental results. Similar phenomenon was observed by Wong et al. [39] when conducting cantilever-bending tests on silicon carbide nanorods and multiwall carbon nanotubes. An extra length up to 50 nm must be added to the original distance between the loading-point and the fixed end to let the measured slope of the beam deflection vs. the applied load fit the equation without considering the substrate deformation [39]. Zhang et al. theoretically studied the effect of substrate deformation on the microcantilever beam bending test [46]. In general, there are two ways to eliminate the influence of the substrate deformation. One approach is to use a rigid substrate in experiments, which means the elastic constants of the substrate should be one or two orders higher than those of tested micro/nanobeams. An alternate and practical approach is to develop a model that considers the substrate deformation and the derived formula includes the influence of the substrate deformation.

The microbridge test technique was thus developed by employing the practical approach because experimentally measured deflection with a nanoindenter system includes the contribution from the substrate deformation [45]. The substrate deformation was modeled with three linear springs, and hence analytic solutions were derived for microbridge deformation. The analytic solutions were confirmed by finite element analysis (FEA) and experimental observations [45]. The theoretical and experimental results indicate that ignoring the substrate deformation will induce a large error in the measured elastic constant and residual stress. The microbridge test was first developed for single-layer elastic thin films [22, 45]. The microbridge test is able to simultaneously characterize the Young's modulus, film residual stress in tension or compression, the bending strength of brittle thin films [45], and the plastic behavior of ductile films [19, 47]. As microbridge samples are fabricated with the MEMS technique, they are easy to handle. Denhoff proposed a similar microbridge testing method, in which a point force provided by a surface profiler was applied at an arbitrary position along the length of the beam, and the deformation of supports was simply treated by adding an additional beam length [9]. Fitting the measured profile of the beam deflected by a given force to the theoretical formula, Denhoff determined the Young's modulus and the residual stress of PECVD silicon nitride films [9].

In comparison with single-layer thin films, mechanical characterization of multilayer films is more challenging because the mechanical property changes from layer to layer. Theoretically, Townsend et al. treated a multilayer thin film as a composite plate and provided formulas for equivalent elastic properties [34]. Klein and Miller [21] followed the theoretical framework of Townsend et al. [34] and developed applicable equations for bilayer systems, especially for systems of a thin film deposited on a thick substrate. Alshits and Kirchner studied the elasticity of multilayers, in particular, for strips, coatings, and sandwiches [1]. Finot and Suresh examined the small and large deformation of multilayer films and studied the effects of layer geometry, plasticity, and compositional gradients [13]. In a sense, we may call the theoretical study as a direct process. It is an inverse process to characterize experimentally the mechanical properties of each layer in a multilayer thin film from measured deformation behaviors.

Su et al. extended the microbridge testing method to characterize the mechanical properties of silicon nitride/silicon oxide bilayer thin films [31]. In their approach, they first tested single-layer silicon nitride films, and then assumed the Young's modulus and the residual stress of the single-layer silicon nitride films to be the same as those of the silicon nitride layers in the bilayer films in order to evaluate the mechanical properties of the silicon oxide layers in the bilayer films. This is because that a nonlinear load–deflection curve of a bilayer microbridge under large deformation reflects the response of the composite bilayer, and one can uniquely determine only two parameters from a nonlinear smooth load–deflection curve. In general, to characterize the mechanical property of one layer in an n-fold-layer thin film by a load–deflection method, it is generally required that the mechanical properties of the $n - 1$ layers in the n-fold-layer thin film are available. This approach may be called the consecutive approach, in which if the mechanical properties of the $n - 1$ layers in an n-fold-layer thin film are known, one can extract the mechanical properties of the nth layer by testing the n-fold-layer thin film. As mentioned earlier, the consecutive approach is based on the assumption that the mechanical properties of the $n - 1$ layers in an n-fold-layer thin film should be the same as that of a film made of the same $n - 1$ layers. If more information can be provided by experiments, one shall be able to determine the mechanical properties of each layer in a multilayer film.

Xu and Zhang conducted microbridge tests on two sets of symmetrical trilayer samples, consisting of $SiO_2/Si_3N_4/SiO_2$ and $Si_3N_4/SiO_2/Si_3N_4$ [40]. The experimental results were analyzed using the small deformation formula including substrate deformation. By changing the sample length, they determined the bending stiffness and the resultant residual force per unit width in each of the trilayer thin films. Then, they made a simplification that the residual stress and Young's modulus of the silicon oxide or silicon nitride layer are independent of the deposition sequence to estimate residual stress and Young's modulus in each layer. Wang et al. further extended the microbridge testing method to mechanically characterize symmetrical trilayer thin films without the previous assumption that the mechanical properties of one or two layers should be available in advance [36]. The slope of a deflection–load curve under small deformation determines the relationship between the bending stiffness and the residual force of the trilayer film. Using this additional relationship, one is able to simultaneously evaluate the Young's modulus and the average residual stress of the symmetrical trilayer film by fitting the entire nonlinear load–deflection curve under large deformation.

Wang et al. further developed the microbridge method to characterize simultaneously the mechanical properties of each layer in a trilayer thin film, which represents the microbridge testing of multilayer thin films [37]. The residual stresses in a multilayer thin film may generate a residual moment, which produces an initial deflection of a microbridge sample. Measuring the initial deflection provides more information about the mechanical property of the multilayer film. If the residual resultant force in a multilayer film is tensile, an initial deflection will be induced only by the residual moment. The initial deflection caused by a residual moment is not due to buckling. Buckling is induced by a compressive resultant force in a multilayer film when the magnitude of the compressive resultant force exceeds a critical value.

Film buckling has been studied without any applied loads [11, 14, 43]. Usually, the classic Euler beam theory is adopted in the analysis, and fitting a buckling profile with theoretical result yields residual stress in the buckled film, such as in polysilicon [14, 43] and silicon dioxide films [11]. Cao et al. conducted the microbridge test on buckled silicon oxide single-layer films [5]. The initial deflection was taken into account in the data analysis, and the unloading curve was used to determine the Young's modulus and the residual stress of the silicon dioxide film [5]. Huang and Zhang systematically studied microbridge tests on a buckled gold single-layer film [19]. In single-layer films, there is no residual moment if residual stress in the film is homogeneously distributed, which is usually assumed in the study of film stresses. In multilayer thin films, however, there may coexist a residual moment and a compressive resultant force even if residual stress in each layer of the film is homogeneously distributed. A residual moment and a compressive resultant force may jointly cause microbeams and other structures made of multilayer thin films to buckle and bend. Wang and Zhang further developed the microbridge test to characterize the mechanical properties of a multilayer thin film with a residual moment and a compressive resultant force [38]. For simplicity and clarity, a bilayer film was taken as an example to demonstrate the developed methodology.

Zhang et al. and Huang and Zhang developed a microbridge testing approach to determine the yield strength of a ductile thin film [19, 47]. The elastic deformation behavior of a ductile film under low loads should be characterized first to have the elastic properties of the film, with which one is able to build an elastically large deformation load–deflection curve as a reference. Then, using the 2% offset in the deflection, one can determine the yield strength of the film. The microbridge test was conducted on 0.48-μm-thick Au microbridge samples to illustrate the developed method.

In this chapter, the microbridge testing of thin films is summarized in detail. We start with the mechanics analysis of an n-fold-layer microbridge and its associated base. The analytic results show that only four parameters determine the elastic bending behavior of the n-fold-layer microbridge as long as the beam theory holds. The general results are then reduced to that for trilayer, bilayer, and single-layer microbridges. Experimental results are reported by following the analytic results.

10.2 Mechanics Analysis

10.2.1 Deformation of a Multilayer Laminated Beam

In the mechanics analysis, the multilayer microbridge is treated as a multilayer beam because the length of the microbridge is much larger than its width and thickness. The wedge indenter used in the experiment is wider than the sample width, which makes one-dimensional analysis appropriate. The mechanics analysis is based on the following assumptions: perfect bonding at all interfaces and validity of the

Kirchhoff hypothesis of normal-remain-normal in the multilayer beam. The residual stresses are assumed to be completely released along the beam width and thickness directions and uniformly distributed only along the beam length direction in each layer. First, we determine the neutral plane of bending of the multilayer beam, which is described in Appendix 1. The distance between the neutral plane of bending and the free surface of the first layer, d, is expressed as,

$$d = \frac{\sum\limits_{k=1}^{n} E_k h_k \sum\limits_{j=1}^{k-1} (2h_j + h_k)}{2 \sum\limits_{k=1}^{n} E_k h_k}, \tag{10.1}$$

where h_k and E_k, $(k = 1, 2, \ldots, n)$, denote the thickness and the Young's modulus of the kth layer in the multilayer film, respectively. The mechanics analysis in the following is based on the coordinate system that the x-axis lies in the neutral plane of bending, as shown in Fig. 10.2.

According to the classical laminated beam theory [33], the displacements in the kth layer of the laminated beam are expressed by

$$w_k(x,z) = w(x), \tag{10.2a}$$

$$u_k(x,z) = u(x) - z\frac{\partial w}{\partial x}, \quad (z_{k-1} \leqslant z \leqslant z_k) \tag{10.2b}$$

where z_k $(k = 0, 1, 2, \ldots, n)$ are given by,

$$z_0 = -d,$$
$$z_1 = h_1 - d,$$
$$z_2 = h_1 + h_2 - d, \tag{10.2c}$$
$$\cdots\cdots$$
$$z_n = h_1 + h_2 + \cdots\cdots + h_n - d,$$

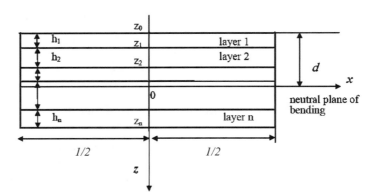

Fig. 10.2 The coordinate system of a multilayer laminated beam

and $u(x)$ and $w(x)$ denote the displacements of the neutral plane of bending along the x-axis and the z-axis, respectively. The strain is expressed by

$$\varepsilon_{xx,k} = \frac{\partial u}{\partial x} + \frac{1}{2}\left(\frac{\partial w}{\partial x}\right)^2 - z\frac{\partial^2 w}{\partial x^2}, \quad (z_{k-1} \leqslant z \leqslant z_k). \tag{10.3}$$

The stress is then deduced from Hooke's law and given by

$$\sigma_{xx,k} = E_k\left[\frac{\partial u}{\partial x} + \frac{1}{2}\left(\frac{\partial w}{\partial x}\right)^2 - z\frac{\partial^2 w}{\partial x^2}\right] + \sigma_{rk}, \quad (z_{k-1} \leqslant z \leqslant z_k) \tag{10.4}$$

where residual stress σ_{rk} $(k = 1, 2, \ldots, n)$ is added. Consequently, we calculate the resultant force per unit width, N_x, and the moment per unit width, M_x.

$$N_x = \sum_{k=1}^{n}\int_{z_{k-1}}^{z_k}\sigma_{xx,k}dz = A\left[\frac{\partial u}{\partial x} + \frac{1}{2}\left(\frac{\partial w}{\partial x}\right)^2\right] - B\frac{\partial^2 w}{\partial x^2} + N_r, \tag{10.5}$$

$$M_x = \sum_{k=1}^{n}\int_{z_{k-1}}^{z_k}\sigma_{xx,k}z\,dz = B\left[\frac{\partial u}{\partial x} + \frac{1}{2}\left(\frac{\partial w}{\partial x}\right)^2\right] - D\frac{\partial^2 w}{\partial x^2} + M_r, \tag{10.6}$$

where N_r and M_r denote residual force per unit width and residual moment per unit width, respectively, and A, B, and D are called tension stiffness, tension-bending coupling stiffness, and bending stiffness, respectively. For a multilayer microbridge, the explicit forms of A, B, and D, and N_r, and M_r are:

$$A = \sum_{k=1}^{n} E_k h_k, \tag{10.7a}$$

$$B = \sum_{k=1}^{n}\frac{1}{2}E_k(z_k^2 - z_{k-1}^2) = 0, \tag{10.7b}$$

$$D = \sum_{k=1}^{n} E_k\frac{z_k^3 - z_{k-1}^3}{3}, \tag{10.7c}$$

$$N_r = \sum_{k=1}^{n}\sigma_{rk}h_k, \tag{10.7d}$$

$$M_r = \sum_{k=1}^{n}\frac{1}{2}\sigma_{rk}(z_k^2 - z_{k-1}^2). \tag{10.7e}$$

Generally, there exists a residual moment in a multilayer microbridge, which causes an initial deflection. When the neutral plane of bending is chosen as the x-axis, the tension-bending coupling stiffness, B, becomes zero. Equation (10.7b) is consistent to (10.69) or (10.71), indicating that the neutral plane of bending can also be determined by letting $B = 0$. Equations (10.5) and (10.6) indicate that the deformation is related to the net resultant force per unit width, $\Delta N = N_x - N_r$, and the net moment per unit width, $\Delta M = M_x - M_r$. In the case of $B = 0$, (10.6) is reduced to

$$M_x = -D\frac{\partial^2 w}{\partial x^2} + M_r. \tag{10.8}$$

The equilibrium equations of the resultant force, N_x, moment, M_x, and shear force, T, are given by

$$\frac{\partial N_x}{\partial x} = 0, \tag{10.9a}$$

$$T = \frac{\partial M_x}{\partial x}, \tag{10.9b}$$

$$\frac{\partial T}{\partial x} = -q - N_x\frac{\partial^2 w}{\partial x^2}, \tag{10.9c}$$

where q denotes an applied lateral load per unit width at any location of x along the beam length. Equation (10.9a) indicates that the resultant force is constant along the length direction of the beam. Using (10.5) and (10.6), we have the governing equation for the multilayer beam

$$D_{\text{multi}}\frac{\partial^4 w}{\partial x^4} - N_x\frac{\partial^2 w}{\partial x^2} = q, \tag{10.9d}$$

where $D_{\text{multi}} = D - B^2/A$ is the equivalent bending stiffness of a multilayer beam. Again, if the neutral plane of bending is chosen as the x-axis, $B = 0$ and $D_{\text{multi}} = D$. The governing equation for bending a multilayer beam has the same form as that for bending a single-layer beam. Therefore, (10.9d) can be solved in the same way as that for a single-layer beam [45].

Figure 10.3 shows the mechanics analysis of the beam, where only a lateral load per unit width, Q, is applied at the beam center. The substrate acts a lateral force, $Q/2$, a resultant force, ΔN, and a moment, M_0, at each of the two ends. The moment boundary condition at each of the beam ends can be calculated from (10.6) and given by

$$M_0 = -D\left.\frac{\partial^2 w}{\partial x^2}\right|_{x=\pm l/2} + M_r. \tag{10.10}$$

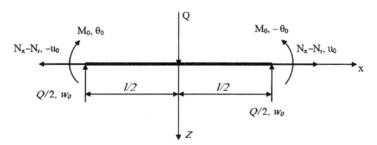

Fig. 10.3 Mechanics analysis of a multilayer laminated beam with deformable boundary conditions

In this case, integrating (10.9d) over x twice yields the governing equation for the deflection of the beam in terms of the moment balance, which is

$$D\frac{d^2w}{dx^2} - N_x(w - w_0) + \frac{Q}{2}\left(x + \frac{l}{2}\right) + \overline{M_0} = 0, \quad \text{for } -l/2 \leqslant x \leqslant 0, \quad (10.11a)$$

$$D\frac{d^2w}{dx^2} - N_x(w - w_0) + \frac{Q}{2}\left(\frac{l}{2} - x\right) + \overline{M_0} = 0, \quad \text{for } 0 \leqslant x \leqslant l/2, \quad (10.11b)$$

where $\overline{M_0} = M_0 - M_r$.

If $N_x > 0$, the general solution to (10.11) is

$$w = A_1 \cosh\left[k\left(x + \frac{l}{2}\right)\right] + B_1 \sinh\left[k\left(x + \frac{l}{2}\right)\right] + \frac{Q}{2N_x}\left(x + \frac{l}{2}\right)$$

$$+ w_0 + \frac{\overline{M_0}}{N_x}, \quad \text{for } -\frac{l}{2} \leqslant x \leqslant 0, \quad (10.12a)$$

$$w = A_r \cosh\left[k\left(\frac{l}{2} - x\right)\right] + B_r \sinh\left[k\left(\frac{l}{2} - x\right)\right] + \frac{Q}{2N_x}\left(\frac{l}{2} - x\right)$$

$$+ w_0 + \frac{\overline{M_0}}{N_x}, \quad \text{for } 0 \leqslant x \leqslant \frac{l}{2}, \quad (10.12b)$$

where $k^2 = \dfrac{N_x}{D}$.

At the two ends (i.e., $x = \pm l/2$), $w = w_0$, and at the loading point, the two portions should have the same deflection and a common tangent. These conditions yield

$$A_1 = A_r = -\frac{\overline{M_0}}{N_x}, \quad \text{for } -\frac{l}{2} \leqslant x \leqslant 0, \quad (10.13a)$$

$$B_1 = B_r = -\frac{Q}{2N_xk\cosh(kl/2)} + \frac{\overline{M_0}}{N_x}\frac{\sinh(kl/2)}{\cosh(kl/2)}, \quad (10.13b)$$

Substituting (10.15) in (10.13) yields the deflection profile. Since the deflection is symmetric about $x = 0$, we write down the result for $-l/2 \leqslant x \leqslant 0$,

$$w = \frac{\overline{M_0}}{N_x}\left\{1 + \frac{\sinh(kl/2)}{\cosh(kl/2)}\sinh\left[k\left(x + \frac{l}{2}\right)\right] - \cosh\left[k\left(x + \frac{l}{2}\right)\right]\right\}$$

$$- \frac{Q}{2N_xk\cosh(kl/2)}\sinh\left[k\left(x + \frac{l}{2}\right)\right] + \frac{Q}{2N_x}\left(x + \frac{l}{2}\right) + w_0. \quad (10.14)$$

Thus, the deflection at the loading point, $x = 0$, is given by

$$w = -\frac{Q}{2N_xk}\tanh\left(\frac{kl}{2}\right) + \frac{Ql}{4N_x} + \frac{\overline{M_0}}{N_x}\left(1 - \frac{1}{\cosh(kl/2)}\right) + w_0. \quad (10.15)$$

From (10.14), the slope of the deflection is obtained at the beam end of $x = -l/2$ to be

$$\theta_0 = \left. \frac{\partial w}{\partial x} \right|_{x=-l/2} = \frac{Q[\cosh(kl/2) - 1]}{2N_x \cosh(kl/2)} + \frac{\overline{M_0}k}{N_x} \tanh(kl/2). \qquad (10.16)$$

Because the deflection of the beam is symmetric with $x = 0$, we study the deformation of the beam in the x direction only for a half of the beam. Because of the symmetry, we have

$$u|_{x=0} = 0. \qquad (10.17)$$

From (10.5), we have

$$\frac{l}{2E_f t}(N_x - N_r) = \int_{-l/2}^{0} \left[\frac{\partial u}{\partial x} + \frac{1}{2}\left(\frac{\partial w}{\partial x}\right)^2 \right] dx. \qquad (10.18)$$

Thus, the displacement along the x-axis at the end of $x = -l/2$ is obtained

$$u_0 = \frac{1}{2} \int_{-l/2}^{0} \left(\frac{\partial w}{\partial x}\right)^2 dx - \frac{l(N_x - N_r)}{2E_f t} = I - \frac{l(N_x - N_r)}{2E_f t} \qquad (10.19a)$$

where

$$I = \frac{1}{2} \int_{-l/2}^{0} \left(\frac{\partial w}{\partial x}\right)^2 dx$$

$$= \frac{1}{32kN_x^2 \cosh^2(kl/2)} \left[-4k^3 l\overline{M_0}^2 + 4k^2\overline{M_0}^2 \sinh(kl) \right.$$

$$+ 8k\overline{M_0}Q + 2klQ^2 - 16k\overline{M_0}Q\cosh(kl/2) - 3Q^2 \sinh(kl)$$

$$\left. + kQ(8\overline{M_0} + lQ)\cosh(kl) - 4k^2 l\overline{M_0}Q\sinh(kl/2) \right]. \qquad (10.19b)$$

Equation (10.21) indicates that for a nondeformable substrate, $u_0 = 0$ and the large deformation of deflection, expressed by the first term in the right-hand side, causes a change in the resultant force, N_x, only, given by the second term in the right-hand side. For a deformable substrate, however, the large deformation of deflection causes a change in the resultant force and a displacement of the substrate at the joint point, thereby implying that substrate deformation will make the change in the resultant force small.

The above description considers positive resultant force per unit width, i.e., $N_x > 0$. However, the residual resultant force per unit width will be negative, i.e., $N_x < 0$, if residual stresses in a multilayer film are predominant by compressive stresses. A compressive residual force in a microbridge beam may cause buckling of the beam. In the case of $N_x < 0$, the general solution to (10.11) is

$$w = A_1 \cos\left[k\left(x+\frac{l}{2}\right)\right] + B_1 \sin\left[k\left(x+\frac{l}{2}\right)\right] + \frac{Q}{2N_x}\left(x+\frac{l}{2}\right)$$

$$+w_0 + \frac{\overline{M_0}}{N_x}, \quad \text{for } -\frac{l}{2} \leqslant x \leqslant 0, \tag{10.20a}$$

$$w = A_r \cos\left[k\left(\frac{l}{2}-x\right)\right] + B_r \sin\left[k\left(\frac{l}{2}-x\right)\right] + \frac{Q}{2N_x}\left(\frac{l}{2}-x\right)$$

$$+w_0 + \frac{\overline{M_0}}{N_x}, \quad \text{for } 0 \leqslant x \leqslant \frac{l}{2}, \tag{10.20b}$$

where $k^2 = -\dfrac{N_x}{D}$.

The conditions at the beam-ends and at the loading point determine the parameters:

$$A_1 = A_r = -\frac{\overline{M_0}}{N_x}, \tag{10.21a}$$

$$B_1 = B_r = -\frac{Q}{2N_x k \cos(kl/2)} - \frac{\overline{M_0}}{N_x}\frac{\sin(kl/2)}{\cos(kl/2)}. \tag{10.21b}$$

Thus, the general solution of (10.22) can be rewritten as

$$w = -\frac{Q\sin[k(l/2+x)]}{2N_x k \cos(kl/2)} + \frac{Q}{2N_x}\left(x+\frac{l}{2}\right) - \frac{\overline{M_0}}{N_x}\left[\frac{\cos(kx)}{\cos(kl/2)} - 1\right] + w_0,$$

$$\text{for } -\frac{l}{2} \leqslant x \leqslant 0, \tag{10.22a}$$

$$w = -\frac{Q\sin[k(l/2-x)]}{2N_x k \cos(kl/2)} + \frac{Q}{2N_x}\left(\frac{l}{2}-x\right) - \frac{\overline{M_0}}{N_x}\left[\frac{\cos(kx)}{\cos(kl/2)} - 1\right] + w_0,$$

$$\text{for } 0 \leqslant x \leqslant \frac{l}{2}. \tag{10.22b}$$

The slope of the deflection of the beam is obtained at the end of $x = -l/2$,

$$\theta_0 = \frac{Q}{2N_x}\left[1 - \frac{1}{\cos(kl/2)}\right] - \frac{\overline{M_0}k}{N_x}\tan(kl/2). \tag{10.23}$$

Similarly, we have the displacement along the x-axis at the end of $x = -l/2$ as

$$u_0 = I - \frac{l(N_x - N_r)}{2A} \tag{10.24a}$$

where,

$$I = \frac{1}{8k}
\begin{bmatrix}
-\dfrac{4Q\overline{M}_0 k}{N_x^2 \cos(kl/2)} + \dfrac{Q^2 kl}{4N_x^2 \cos^2(kl/2)} \\[2mm]
+\dfrac{\overline{M}_0^2 k^3 l}{N_x^2 \cos^2(kl/2)} + \dfrac{2Q^2 kl}{4N_x^2} + \dfrac{4Q\overline{M}_0 k}{N_x^2} \\[2mm]
-\dfrac{2Q^2 \sin(kl/2)}{N_x^2 \cos(kl/2)} + \dfrac{Q\overline{M}_0 k^2 l \sin(kl/2)}{N_x^2 \cos^2(kl/2)} \\[2mm]
+\dfrac{Q^2 \sin(kl)}{4N_x^2 \cos^2(kl/2)} - \dfrac{\overline{M}_0^2 k^2 \sin(kl)}{N_x^2 \cos^2(kl/2)}.
\end{bmatrix}
\qquad (10.24\text{b})$$

Actually, these equations for a compressive resultant force can be directly derived from the equations for a tensile resultant force and vice verse by using the following relations of complex functions:

$$\sinh(iz) = i\,\sin(z), \quad \cosh(iz) = \cos(z), \quad \tanh(iz) = i\,\tan(z), \quad (10.25\text{a})$$
$$\sin(iz) = i\,\sinh(z), \quad \cos(iz) = \cosh(z), \quad \tan(iz) = i\,\tanh(z), \quad (10.25\text{b})$$

where $i = \sqrt{-1}$.

10.2.2 Substrate Deformation

Figure 10.4 shows the mechanics analysis of the substrate (or base) deformation, which is modeled by three coupled springs.

Three generalized forces (force per unit width of the film), i.e., two forces, N and P, along the x and z-axis, respectively, and a moment (moment per unit width of the film), M, act on the three coupled springs and induce correspondingly three generalized displacements, u, δ, and θ. A linear constitutive equation,

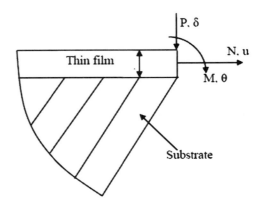

Fig. 10.4 Mechanics analysis of the substrate

$$
\begin{bmatrix} u \\ \delta \\ \theta \end{bmatrix} = \begin{bmatrix} S_{NN} & S_{NP} & S_{NM} \\ S_{PN} & S_{PP} & S_{PM} \\ S_{MN} & S_{MP} & S_{MM} \end{bmatrix} \begin{bmatrix} N \\ P \\ M \end{bmatrix} , \qquad (10.26)
$$

is applied herein, where S_{ij} are the compliances and have the symmetry of $S_{ij} = S_{ji}$. The spring compliances are determined with the consideration of the difference in the elastic constants, the dimensional analysis, and finite element calculations, which are described in detail in Appendix 2. The spring compliances depend greatly on the support angle, as shown in Fig. 10.1. Silicon wafers are widely used in MEMS and NEMS fabrication. Because of anisotropic etching, three support angles are usually formed in silicon substrates. Figure 10.5 illustrates the three support angles and the associated lattice orientations of the silicon substrates. The right support angle

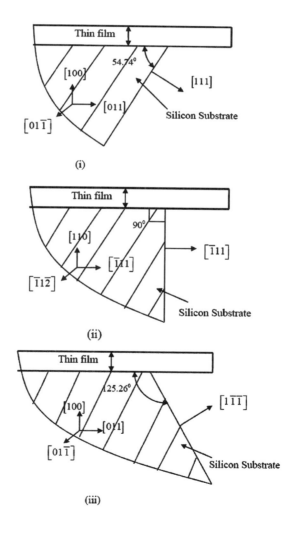

Fig. 10.5 Schematic depiction of silicon substrate support angles, (i) 54.74°, (ii) 90°, and (iii) 125.26°

of $\alpha = 90°$ is generally formed in (110) silicon wafers. For (100) silicon wafers, the backside etching generates the support angle of $\alpha = 54.74°$, whereas the front side etching forms the support angle of $\alpha = 125.26°$.

For a deformable substrate, each of the beam ends of $x = \pm l/2$ is subjected to a moment per unit width, M_0, an axial load per unit width, $N_x - N_r$, and a lateral load per unit width, $T_0 = Q/2$, and has a rotation, θ_0, an axial displacement, u_0, and a lateral displacment, w_0. At the connecting point of the film and substrate, the displacement continuity and force equilibrium require

$$
\begin{array}{lll}
u = u_0, & \delta = w_0, & \theta = \theta_0, \\
N = N_x - N_r, & P = Q/2, & M = -M_0.
\end{array}
\tag{10.27}
$$

Thus, using the boundary conditions of (10.27) and the spring compliances, we have the theoretical solution for the microbridge test.

Before the formation of the microbridge beams, the multilayer film is entirely bonded to the substrate. This state is taken here as the reference state for deformation. In the reference state, there are residual force and residual moment in the multilayer film, but deformation is treated as zero. The microbridge beams are formed by patterning, etching, and removing the underneath substrate part. After the formation of the microbridge beam, the residual stresses in the multilayer beam will be different from the original ones in the multilayer film. As described earlier, the present work assumes residual stresses being completed released in the beam width direction. In the beam length direction, the residual force will be released slightly because of the substrate deformation of the microbridge base after the removal of the underneath substrate part. Without considering buckling, the release of the residual force does not bend the beam, rather makes the beam slightly longer or shorter. The slight change in the beam length is difficult to be detected. Actually, the beam length is experimentally measured one by one after the fabrication of the microbridge samples. Therefore, the residual force is assumed unchanged before and after the removal of the underneath substrate part. That is why the net resultant force, ΔN, rather than the resultant force, N_x, is used in (10.27), to act on the substrate.

The removal of the underneath substrate part also causes bending of the microbridge beam because of the presence of the residual moment. Clearly, without the residual moment, the beam will not bend when buckling is excluded. Furthermore, if the substrate base is nondeformable, the beam will not bend even there is a residual moment in the beam. To model the bending behavior with the two features, we use M_0 rather than $\overline{M_0}$ in (10.27), indicating that the beam acts a moment, M_0, on the substrate, while the net moment acting on the beam end is $\overline{M_0}$. The value of M_0 depends on the residual moment. This means that only a residual moment can deform the substrate and then the beam from the reference state because of the removal of the underneath substrate part. If $M = -\overline{M_0}$ is used in (10.27), without any applied load, the residual moment M_r is balanced by M_0 at each of the ends such that $\overline{M_0} = 0$. In that case, no moment will exert on the substrate; and no moment will cause the bend of the beam and the beam will remain flat as that in the reference state. However, an initial bending of a multilayer microbridge induced only by a residual moment has been observed experimentally, as described in Sect. 10.3.2. Furthermore, the residual moment determined from the initial bending profile is the same as that determined from load–deflection curves under large deformation, which might confirm the rationality of the present approach.

For readers' convenience, we also provide the analysis for microcantilever bending tests in Appendix 3 [46].

10.3 Deflection of a Multilayer Microbridge for $N_x > 0$ and $M_r \neq 0$

10.3.1 Analysis

Considering the substrate deformation, we have the deflection of a multilayer microbridge at the mircobridge center, $x = 0$, where the lateral load per unit width, Q, is applied, as

$$w = -\frac{Q\tanh(kl/2)}{2N_x k} + \frac{Ql}{4N_x} + \frac{\bar{M}_0}{N_x}\left[1 - \frac{1}{\cosh(kl/2)}\right] + w_0, \qquad (10.28a)$$

where $\overline{M}_0 = M_0 - M_r$, as defined by (10.12), and w_0, M_0, and N_x are coupled in the following equations:

$$w_0 = S_{NP}(N_x - N_r) + S_{PP}\frac{Q}{2} - S_{MP}M_0, \qquad (10.28b)$$

$$\frac{Q[\cosh(kl/2)-1]}{2N_x\cosh(kl/2)} + \frac{\bar{M}_0 k}{N_x}\tanh\left(\frac{kl}{2}\right) = S_{MN}(N_x - N_r) + \frac{1}{2}S_{MP}Q - S_{MM}M_0, \quad (10.28c)$$

$$I - \frac{l(N_x - N_r)}{2A} = S_{NN}(N_x - N_r) + S_{NP}\frac{Q}{2} - S_{MN}M_0. \qquad (10.28d)$$

The expression of I is given in (10.19b). For a given load, Q, and a given residual moment, M_r, (10.28c) and (10.28d) determine the values of N_x and M_0, and consequently the deflection at the beam center is predicted by (10.28a) with (10.28b). In Appendix 4, we reduce (10.31) to solutions with the simply supported and clamped boundary conditions.

If there is a residual moment, the multilayer microbridge may initially be bent without any applied loads. The initial deflection profile should be measured to provide more information about the deformation behavior. The initial deflection caused only by the residual moment is given by

$$w_r(x) - \hat{w}_0 = \frac{\hat{M}_0 - M_r}{\hat{N}_x}\left\{\begin{array}{l} 1 + \dfrac{\sinh(\hat{k}l/2)}{\cosh(\hat{k}l/2)}\sinh\left[\hat{k}\left(\dfrac{l}{2}-x\right)\right] \\[2mm] -\cosh\left[\hat{k}\left(\dfrac{l}{2}-x\right)\right] \end{array}\right\}, \qquad (10.29a)$$

for $0 \leqslant x \leqslant l/2$. Here $\hat{w}_0$, $\hat{M}_0$, $\hat{N}_x$, and $\hat{k}$ denote the initial values of w_0, M_0, N_x and k, respectively, under purely residual moment loading, and $\hat{k} = \sqrt{\hat{N}_x/D}$. The values of $\hat{w}_0$, $\hat{M}_0$ and $\hat{N}_x$ are coupled in the following equations:

$$\hat{w}_0 = S_{NP}(\hat{N}_x - N_r) - S_{MP}\hat{M}_0, \tag{10.29b}$$

$$\frac{(\hat{M}_0 - M_r)\hat{k}}{\hat{N}_x} \tanh(\hat{k}l/2) = S_{MN}(\hat{N}_x - N_r) - S_{MN}\hat{M}_0, \tag{10.29c}$$

$$\frac{-\hat{k}l + \sinh(\hat{k}l)}{8\hat{N}_x^2\cosh^2(\hat{k}l/2)} \hat{k}\left(\hat{M}_0 - M_r\right)^2 - \frac{l(\hat{N}_x - N_r)}{2A} = S_{NN}(\hat{N}_x - N_r) - S_{MN}\hat{M}_0. \tag{10.29d}$$

For a given residual moment, M_r, (10.29c) and (10.29d) determine the values of $\hat{N}_x$ and $\hat{M}_0$. Then, (10.29a) with (10.29b) describes the profile of the initial deflection. At the center of the microbridge, $x = 0$, the deflection is calculated from (10.29a) and given by

$$w_r - \hat{w}_0 = \frac{\hat{M}_0 - M_r}{\hat{N}_x}\left(1 - \frac{1}{\cosh(\hat{k}l/2)}\right). \tag{10.29e}$$

The initial deflection induced by the residual moment may be large or small. Under small deformation, the initial values of $\hat{M}_0$ and $\hat{N}_x$ can be approximately calculated from (10.29c) and (10.29d) and are given by

$$\hat{M}_0 = \frac{k_0 \tanh(k_0 l/2)M_r + S_{MN}N_r\Delta N}{k_0 \tanh(k_0 l/2) + S_{MN}N_r}, \tag{10.30a}$$

$$\Delta N = k_0 S_{MN} \tanh(k_0 l/2)M_r/$$
$$\{S_{MN} \tanh(k_0 l/2) + (S_{MM}S_{NN} - S_{MN}S_{MN})N_r \tag{10.30b}$$
$$+ \frac{l}{4A}[k_0 \tanh(k_0 l/2) + S_{MM}N_r]\}$$

where $k_0 = \sqrt{N_r/D}$ and $\Delta N = \hat{N}_x - N_r$. Equations (10.30a) and (10.30b) indicate that the moment, $\hat{M}_0$, is proportional to the residual moment, M_r. Therefore, the initial deflection is proportional to the residual moment under small deformation. The initial deflection can be concave or convex, depending on the residual moment. Experimentally, we can measure the profile of the initial deflection by using a laser beam to scan the microbridge along its length. The profile of the initial deflection provides us with the material properties of the thin film.

When the applied load, Q, is small, the deflection is approximately linear to the load. The slope, $p = w/Q$, of the deflection–load curve under small loads can be obtained by Taylor expansion of (10.28a) at the load starting point, $Q = 0$, no matter whether the initial deflection caused by the residual moment is small or large. The slope, p, is given by

$$p = -\frac{\tanh(\hat{k}l/2)}{2\hat{N}_x\hat{k}} + \frac{l}{4\hat{N}_x} + \frac{S_{PP}}{2} - \left[S_{MP} + \frac{1}{\hat{N}_x}\left(\frac{1}{\cosh(\hat{k}l/2)} - 1\right)\right]$$

$$\times \frac{S_{MP}\hat{N}_x + [1/\cosh(\hat{k}l/2) - 1]}{2S_{MM}\hat{N}_x + 2\hat{k}\tanh(\hat{k}l/2)} + \left[S_{NP} - \frac{S_{MP} + \dfrac{1/\cosh(\hat{k}l/2) - 1}{\hat{N}_x}}{S_{MM}\hat{N}_x + \hat{k}\tanh(\hat{k}l/2)}\right] \quad (10.31)$$

$$\times \frac{S_{MN}\left[S_{MP}\hat{N}_x + (1/\cosh(\hat{k}l/2) - 1)\right] - S_{NP}\left[\hat{k}\tanh(\hat{k}l/2) + S_{MM}\hat{N}_x\right]}{2(\hat{k}\tanh(\hat{k}l/2) + S_{MM}\hat{N}_x)(S_{NN} + \frac{l}{2A}) - 2S_{MN}^2\hat{N}_x}$$

The slope, p, under small loads is experimentally measurable and depends on the bending stiffness, D, the beam length, l, the initial resultant force without any applied loads, $\hat{N}_x$, and the tension stiffness, A. For a given beam length, the slope, p, is a function of A, $\hat{N}_x$, and D or a function of A, $\hat{N}_x$, and $\hat{k}$ through $\hat{k} = \sqrt{\hat{N}_x/D}$.

As described earlier, the four parameters, A, D, N_r, and M_r, determine the bending behavior including the initial deflection. Wang et al. experimentally determined the four parameters from the slope, p, the initial displacement at the microbridge ends ($x = \pm l/2$), $\hat{w}_0$, and the profile of the initial deflection [37]. From the measured value of the slope, p, we express the initial resultant force, $\hat{N}_x$, as a function of $\hat{k}$ and A, whereas from the measured displacement of $\hat{w}_0$, we express the residual force, N_r, as a function of $\hat{M}_0$, $\hat{N}_r$, and A through the spring compliances. Since $\hat{N}_x$ is determined as a function of $\hat{k}$ and A, the residual force, N_r, can be expressed as a function of $\hat{M}_0$, $\hat{k}$, and A. Then, we can express the initial moment induced by the boundary, $\hat{M}_0$, and the residual moment, M_r, both in terms of $\hat{k}$ and A by using (10.29c) and (10.29d). As indicated in (10.29a), the profile of the initial deflection apparently depends on the parameters, $\hat{k}$, $\hat{N}_x$, $\hat{M}_0$, and M_r. With the experimental measurements of the slope, p, and the initial displacement, $\hat{w}_0$, we are able to reduce the independent parameters to A and $\hat{k}$, which can be uniquely determined from fitting the initial deflection curve with (10.29a) by using the least squares technique. When A and $\hat{k}$ are determined, we will also have the values of the initial resultant force, $\hat{N}_x$, through (10.31), and the residual force, N_r, the residual moment, M_r, the initial moment induced by the boundary, $\hat{M}_0$, by using (10.29b), (10.29c), and (10.29d). Then, the bending stiffness, D, is also determined through $D = \hat{N}_x/\hat{k}^2$. In brief, the characteristic parameters, A, D, N_r, and M_r, of an asymmetrical multilayer microbridge can be extracted from the profile of the initial deflection and the slope of deflection vs. load under small loads. Once the four parameters, A, D, N_r, and M_r, are determined, we are able to predict the nonlinear load–deflection behavior of the multilayer thin film under large loads.

Under large loads, the deflection varies nonlinearly with the load, and the deflection–load behavior depends on the parameters, the tension stiffness, A, the bending stiffness, D, the residual force, N_r, and the residual moment, M_r. All these material properties are already determined from fitting the initial deflection curve. Here, we may use the deflection–load relationship under large loads to verify the results extracted from fitting the initial deflection profile and the slope of deflection vs. load under small loads. As we can uniquely determine only two independent parameters from fitting a smooth curve, we may randomly choose two parameters as the input data from four parameters, the tension stiffness (A), the bending stiffness (D), the residual force (N_r), and the residual moment (M_r). For example, we may choose the tension stiffness (A) and the residual force (N_r), as the input data to determine the other two parameters, the bending stiffness (D) and the residual moment (M_r). Or we may choose the tension stiffness, A, and the residual moment, M_r, as the input data to determine the bending stiffness, D, and the residual force, M_r. In brief, we shall uniquely determine the values of two remaining independent parameters from fitting a nonlinear load–deflection curve with (10.31) by using the least square technique,

$$S = \sum_{i=1}^{n} [(w_i^e(Q_i) + w_r) - w_i^t(Q_i, \ X_1, \ X_2)]^2, \tag{10.32}$$

where n denotes the number of data, and $w_i^e(Q_i)$ and $w_i^t(Q_i, \ X_1, \ X_2)$ denote the experimentally observed and theoretically predicted deflections, respectively. X_1 and X_2 can be any two of the four parameters, the tension stiffness, A, the bending stiffness, D, the residual force, N_r, and the residual moment, M_r. The sign of the initial deflection at the bridge center, w_r, is negative, if the shape of the initial deflection is convex; otherwise, the sign is positive if the shape is concave. The initial deflection at the bridge center, w_r, should be added to the measured deflection, $w_i^e(Q_i)$, for the fitting purpose because the theoretical predictions are based on the bending neutral plane without any initial deflections.

In summary, we are able to measure the four properties of the tension stiffness, A, the bending stiffness, D, the residual force, N_r, and the residual moment, M_r, which determine the initial deflection and the response of a multilayer microbridge to an applied load during the microbridge test.

10.3.2 Experimental Results for Asymmetrical Trilayer Thin Films [37]

SiO_2 (300 nm)/Si_3N_4 (570 nm)/SiO_2 (140 nm) asymmetrical trilayer thin films were used as an example to demonstrate the microbridge testing method for multilayer thin films, where the thickness of each layer is indicated in the parentheses after the layer material. The SiO_2 (300 nm)/Si_3N_4 (570 nm)/SiO_2 (140 nm) asymmetrical trilayer microbridge samples were fabricated on two-side polished 4-in. p-type (100) 525-μm thick silicon wafers with the following deposition process. The first layer of SiO_2 was grown by the wet oxidation technique. After the wafer was put in a horizontally oriented diffusion furnace (ASM LB 45) at $700\,^\circ$C for 25 min, the furnace temperature was boosted up to $1,050\,^\circ$C. Then, superfluous O_2 and H_2 carried by N_2 were burned at the back end of the furnace to produce H_2O (about 95%) and O_2 (about 5%) under a total pressure of 1,000 mTorr for wet oxidation. The wet oxidation with a rate of about 333 nm/min lasted for 54 s to produce an about

300-nm thick silicon oxide layer. Sequentially, an annealing process was done at 900 °C in an O_2 atmosphere for 30 min to make the oxide dense. On the top of the oxide, a silicon nitride layer was deposited by means of the LPCVD (low pressure chemical vapor deposition) technique at 840 °C and under 200 mTorr pressure. The gas flow ratio among DCS (Dichlorosilane, SiH_2Cl_2)/NH_3/N_2 was 16:4:25. The deposition lasted for 140 min at a rate of about 4.0 nm/min to grow a 570 nm silicon nitride layer. Another silicon oxide layer was deposited on the silicon nitride layer with the LPCVD technique at 425 °C. The gas flow ratio between SiH_4 and O_2 was 4:5, and the total gas pressure was 110 mTorr, thereby resulting in a deposition rate 11.5 nm/min. The deposition lasted for 26 min, yielding a 300 nm silicon oxide layer. After that, the SiO_2/Si_3N_4/SiO_2/Si sample was also annealed at 900 °C in an O_2 atmosphere for 30 min to produce dense silicon oxide. The final annealing process might make the upper and lower oxide layers have the same Young's modulus and the same residual stress. The final thickness of each layer was determined by a film thickness measurement system (Nanometric, model 4150) at room temperature.

Figure 10.6 illustrates the fabrication process of the SiO_2/Si_3N_4/SiO_2 microbridge samples by the MEMS method. All films on the backsides of the wafers were patterned by the photolithography technique and subsequently etched away by plasma etching. The exposed silicon was etched in a tetramethyl ammonium hydroxide (TMAH) solution at a temperature of 80 °C to create rectangular windows with the designated dimensions, which yields a support angle of 54.74°. Finally, the trilayer film on the frontside of the wafer was aligned with the backside windows, patterned by photolithography, and then dry-etched by plasma to complete the microbridge structures. As the microbridges were covered by resists after dry etching, the wafers must be immersed in an organic resist stripper MS2001 at 70 °C for 5 min to strip off the resists. The bridge length and width of each microbridge sample were precisely measured one by one with the digital coordinates provided by a Hysitron Nanoindenter system. The microbridge lengths ranged from 100.67 to 143.15 μm, whereas the widths of the microbridges were all about 12 μm with a standard deviation of 0.5 μm. Figure 10.7 shows the top view of the image, taken by a WYKO NT3300 optical profiler in the vertical-scanning interferometry mode, of a microbridge sample. The gap between two adjacent bridges is 57 μm, which is large enough so that each microbridge can be treated as an isolated one. The extension of the trilayer beam on the underneath substrate, l_0, is about 25 μm, as shown in Fig. 10.7. Totally, 120 microbridge samples were tested and all of them were chosen from the center part of the wafer such that the 120 samples had the same thickness.

The initial profile of each microbridge was measured by the laser interferometry technique with the WYKO NT3300 optical profiler. The optical system has a vertical resolution of 0.1 nm. The results indicate that the microbridges were initially curved convexly, implying the existence of residual moments that bent the microbridges. As an example, Fig. 10.8 shows a laser profile of a microbridge with a length of 133.63 μm. The initial deflection at the microbridge end $(x = 0)$, w_r^0, was determined by taking a region on the substrate 30 μm far away from the microbridge end as a reference. As shown in the inset of Fig. 10.8, the experimentally measured initial deflection at the microbridge end $(x = -l/2)$, $\hat{w}_0$, has a finite nonzero

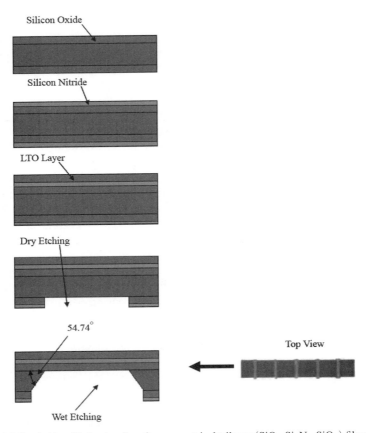

Fig. 10.6 Microbridge fabrication flow for symmetrical trilayer $(SiO_2\text{-}Si_3N_4\text{-}SiO_2)$ films

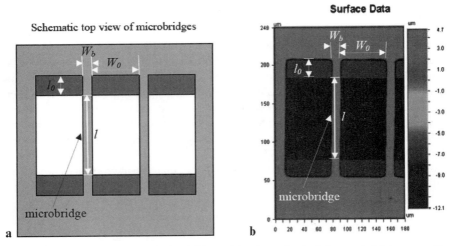

Fig. 10.7 The *top view* of the microbridges. (**a**) schematic depiction and (**b**) laser profile topography

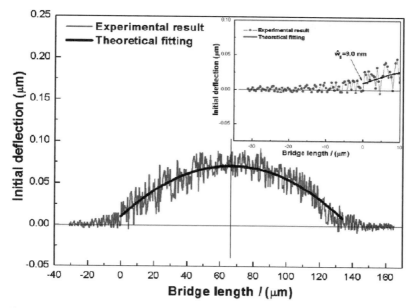

Fig. 10.8 Fitting an initial deflection profile of a microbridge with a length of $l = 133.6\,\mu m$

value, indicating that the substrate is deformed. The initial deflection is convex, as indicated in Fig. 10.8. The measured initial deflections at the microbridge center range from 60 to 85 nm for all the measured samples, and the deflection is about 70 nm in the sample shown in Fig. 10.8. The initial deformation is thus small because the magnitude of the deflection is much smaller than the thickness, 1,010 nm, of the trilayer film. For a small initial deflection, as mentioned earlier, we can use the simplified equation of (10.33) in the numerical fitting.

A Hysitron Nanoindenter system equipped with a wedge indenter was used to conduct the microbridge test, as schematically shown as Fig. 10.1. The Nanoindenter system has a load resolution of $0.1\,\mu N$ and a vertical displacement resolution of 0.1 nm. The width of the diamond wedge indenter, $20\,\mu m$, is wider than the sample width of about $12\,\mu m$, thereby making one-dimensional analysis appropriate. Calibration and alignment were performed prior to the microbridge test to ensure that the wedge indenter was set at the middle of each microbridge and perpendicular to the microbridge length direction. To prevent oscillation during the microbridge test, each sample was affixed to a steel disk and the steel disk was then adhered on the stage by a very strong magnetic force. The microbridge tests were conducted at room temperature with a preset maximum load of $1,500\,\mu N$ and at a loading rate of $250\,\mu N/s$ and the same rate during unloading.

Figure 10.9 illustrates the deflection–load curves during loading and unloading, showing that the unloading curve almost completely coincides with the loading curve. The almost completed overlap of the unloading curve over the loading curve indicates the elastic behavior of the silicon oxide/silicon nitride/silicon oxide film and the silicon substrate. Figure 10.10 shows a loading curve under small

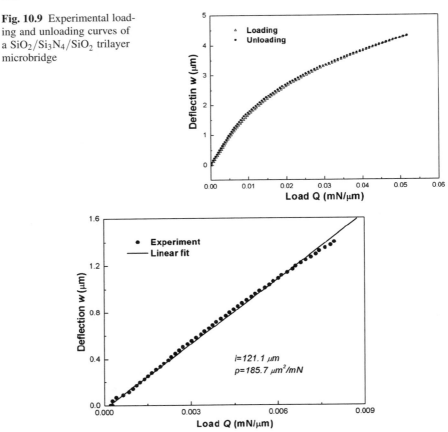

Fig. 10.9 Experimental loading and unloading curves of a $SiO_2/Si_3N_4/SiO_2$ trilayer microbridge

Fig. 10.10 A typical load–deflection curve of a microbridge under small deflection

loads where the solid circles represent the experimental data and the straight line is the linear fitting result. In the study, the deflection was regarded as small deformation if the magnitude of the deflection was less than the film thickness and the deflection within this range was fitted as a linear function of the load, which yielded the slope, p. For the example shown in Fig. 10.10, the slope was determined to be $185.7\,(\mu m)^2/mN$.

From the initial deflection profile and the slope, p, under small loads, we determined the tension stiffness, A, the bending stiffness, D, the residual moment, M_r, and the residual force, N_r, for each microbridge sample of the asymmetrical $SiO_2(300\,nm)/Si_3N_4(570\,nm)/SiO_2$ (140 nm) film. Figures 10.11–10.14 show the determined values of the tension stiffness, A, the bending stiffness, D, the residual moment, M_r, and the residual force, N_r, for all the 120 microbridge samples, respectively. As expected, the four parameters do not vary with the microbridge length, as indicated in Figs. 10.11–10.14. The average values and associated standard deviations are $A = 160.0 \pm 6.7 \times 10^3$ N/m, $D = 6.792 \pm 0.231 \times 10^{-9}$ N/m, $M_r = 3.98 \pm 0.08 \times 10^{-5}$ N/m, and $N_r = 118.41 \pm 8.94$ N/m, thereby giving relative errors of 4.19%, 3.40%, 2.01%, and 7.55% for A, D, M_r, and N_r, respectively.

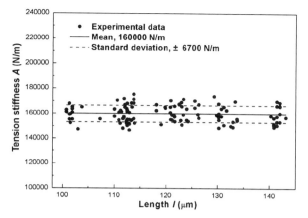

Fig. 10.11 The tension stiffness vs. the microbridge length

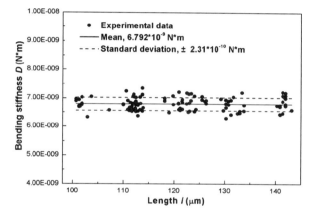

Fig. 10.12 The bending stiffness vs. the microbridge length

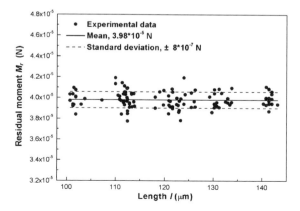

Fig. 10.13 The residual moment vs. the microbridge length

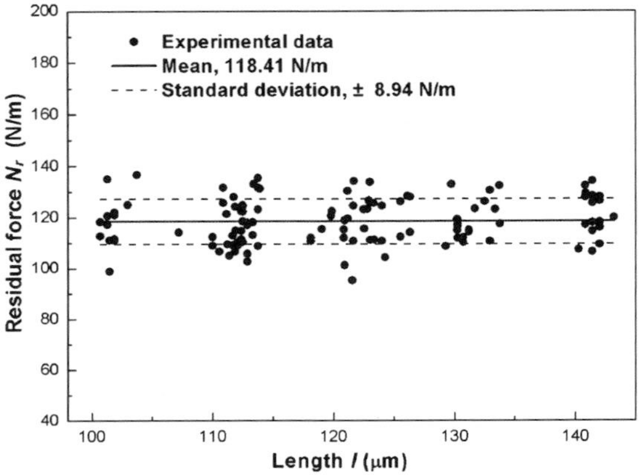

Fig. 10.14 The residual resultant force vs. the microbridge length

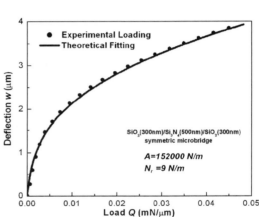

Fig. 10.15 Comparison of theoretical solution and the experimental value for the large deflection of microbridge with length $l = 121.1\,\mu m$

When the values of the tension stiffness, A, the bending stiffness, D, the residual moment, M_r, and the residual force, N_r, are available, we can verify any two of the four parameters, A, D, M_r, and N_r, from fitting the load–deflection curve under large deformation with (10.31). As an example, Fig. 10.15 illustrates a typical experimental load–deflection curve under large deformation and the theoretical fitting curve, indicating that the analytic solution fits perfectly the experimental curve. For the microbridge sample, which load–deflection curve is demonstrated in Fig. 10.15, the values of the four parameters extracted from fitting the initial deflection profile and the slope, p, are $A = 155,100\,N/m$, $D = 6.641 \times 10^{-9}\,N/m$, $M_r = 3.966 \times 10^{-5}\,N/m$, and $A = 119.51\,N/m$. If we choose the tension stiffness, A, and the residual moment, M_r, as the input data, the values of the bending stiffness, D, and the residual force, N_r, determined from fitting the load–deflection curve under large deformation are $D = 6.727 \times 10^{-9}\,N\,m$, and $N_r = 121.34\,N/m$. When comparing to the results extracted from the initial deflection and the slope, p, this gives a relative error of 1.3% and 1.5% for D and N_r, respectively. Alternately, if we choose the tension stiffness, A, and the residual force, N_r, as the input data, the values of the bending stiffness, D, and the residual moment, M_r, are $D = 6.738 \times 10^{-9}\,N\,m$ and

$M_r = 4.025 \times 10^{-5}$ N, from fitting the load–deflection curve under large deformation. This gives a relative error of 1.5% and 1.5% for D and M_r, respectively, when comparing to the results extracted from the initial deflection and the slope, p. Consistent results are expected because the four parameters completely determine the initial deflection and the response of an asymmetrical trilayer microbridge sample under an applied lateral load. In the following discussion, we shall use the results from the fitting of the initial deflection profile and the slope, p.

The final annealing of the fabricated samples at 900 °C in an O_2 atmosphere for 30 min might let the upper and lower oxide layers have the same Young's modulus and the same residual stress. On the basis of this point, we are able to determine the Young's modulus and the residual stress in each layer of the asymmetrical trilayer film from the determined values of A, D, M_r, and N_r by using (10.7). The average values and the associated standard deviations of the Young's modulus of the silicon oxide, E_1, and the Young's modulus of the silicon nitride, E_2, are 37.92 ± 1.84 GPa and 251.15 ± 11.68 GPa, respectively, for the 120 microbridge samples, which give the relative errors of 4.8% and 4.6% for E_1 and E_2, respectively. The estimated Young's modulus of the silicon oxide layers is lower than the values reported in the literature. However, the Young's modulus of a thin film is considerably dependent on its deposition process. For example, Reyntjens and Puers used the focused ion beam deposition technique, and had the Young's modulus of 45 GPa for the deposited silicon oxide film, while the Young's modulus of silicon oxide films grown by the thermal dry growth, thermal wet growth, and sputtered growth were 69 ± 14, 57 ± 11, and 92 ± 18 GPa, respectively [25, 27]. Furthermore, Zywitzki et al. found that with increasing the monomer hexamethyldisiloxane (HMDSO) in vapor stream the Young's modulus of silicon oxide coatings decreased from approximately 42 to 15 GPa [48]. The similar behavior has been found for the Young's modulus of silicon nitride films. For reference, a few reported values of Young's modulus of silicon nitride films are listed here, 222 ± 3 GPa [35], 130 GPa [26], and 370 GPa [42]. Similarly, we calculated the residual stress in the silicon oxide layers, σ_{r1}, and the residual stress in the silicon nitride layer, σ_{r2}, for each of the 120 microbridge samples. The average values and the associated standard deviations of σ_{r1} and σ_{r2} are -464.67 ± 10.52 and 566.43 ± 13.36 MPa, respectively, thereby yielding the relative error of 2.3% for σ_{r1} and the relative error of 2.4% for σ_{r2}. As the case for Young's modulus, residual stress in a thin film is highly dependent on the process conditions. Robic et al. systematically investigated the residual stresses in silicon oxide films produced by ion-assisted deposition [28]. Their results show that the residual stresses in silicon oxide films vary from -10 to -470 MPa, depending on the process conditions. For reference, a few values of residual stresses in silicon nitride films reported in the literature are listed here, 113–151 MPa [9], 110 MPa [35], and 1.0 GPa [32].

If there are only two layers in a thin film, there are only two Young's moduli, E_1 and E_2, and two residual stresses, σ_{r1} and σ_{r2}, to be determined. Therefore, no assumptions are needed in the evaluation of the four properties by the microbridge test. To demonstrate this method, we fabricated SiO_2 (400 nm)/Si_3N_4 (800 nm) bilayer microbridges and adopted the same approach as described earlier. We determined the four parameters, A, D, M_r, and N_r to be $A = 223,500 \pm 7,800$ N/m, $D = 1.625 \pm 0.0457 \times 10^{-8}$ N/m, $M_r = 1.13 \pm 0.0305 \times 10^{-4}$ N/m,

$N_r = 242.36 \pm 21.59\,\text{N/m}$. The Young's moduli and residual stresses of the two layers are $E_1 = 36.48 \pm 1.30\,\text{GPa}$, $E_2 = 261.14 \pm 9.82\,\text{GPa}$, $\sigma_{r1} = 430.60 \pm 14.72\,\text{MPa}$, and $\sigma_{r2} = 518.25 \pm 19.63\,\text{MPa}$.

10.4 Deflection of a Multilayer Microbridge for $N_x > 0$ and $M_r = 0$

If there is a tensile residual resultant force and no residual moment in a microbridge sample, the microbridge beam will not be initially bent without any applied load. In this case, the deflection of a multilayer microbridge at the mircobridge center, $x = 0$, where the lateral load per unit width, Q, is applied, is reduced from (10.31) and given by

$$w = -\frac{Q\tanh(kl/2)}{2N_x k} + \frac{Ql}{4N_x} + \frac{M_0}{N_x}\left[1 - \frac{1}{\cosh(kl/2)}\right] + w_0, \qquad (10.33a)$$

where w_0, M_0, and N_x are coupled in the following equations:

$$w_0 = S_{NP}(N_x - N_r) + S_{PP}\frac{Q}{2} - S_{MP}M_0, \qquad (10.33b)$$

$$\begin{aligned}
&\frac{Q[\cosh(kl/2) - 1]}{2N_x\cosh(kl/2)} + \frac{M_0 k}{N_x}\tanh(kl/2) \\
&= S_{MN}(N_x - N_r) + \frac{1}{2}S_{MP}Q - S_{MM}M_0,
\end{aligned} \qquad (10.33c)$$

$$I - \frac{l(N_x - N_r)}{2A} = S_{NN}(N_x - N_r) + S_{NP}\frac{Q}{2} - S_{MN}M_0, \qquad (10.33d)$$

$$\begin{aligned}
I = \frac{1}{32kN_x^2\cosh^2(kl/2)}&[-4k^3 l M_0^2 + 4k^2 M_0^2\sinh(kl) + 8kM_0 \\
&Q + 2klQ^2 - 16kM_0 Q\cosh(kl/2) - 3Q^2\sinh(kl) \\
&+ kQ(8M_0 + lQ)\cosh(kl) - 4k^2 l M_0 Q\sinh(kl/2)].
\end{aligned} \qquad (10.33e)$$

For a given load, Q, (10.33c) and (10.33d) determine the values of N_x and M_0 and consequently the deflection at the beam center is predicted by (10.33a) and (10.33b). In Appendix 5, we reduce (10.36) to the solutions with the simply supported and clamped boundary conditions.

Under small deformation, we may approximately ignore the deformation of the beam in the x direction. Thus, we render (10.33d) useless and have $N_x = N_r$ and $k = \sqrt{N_r/D}$ and

$$M_0 = \frac{\dfrac{1}{2}S_{MP}N_r + \dfrac{1}{2}\left[\dfrac{1}{\cosh(kl/2)} - 1\right]}{S_{MM}N_r + k\tanh(kl/2)}Q, \qquad (10.34)$$

from (10.33c). Equation (10.34) indicates that the moment applied by the substrate to the microbridge beam is proportional to the applied load. In this case, the deflection is proportional to the applied load, i.e., $w = pQ$, with a slope of

$$p = -\frac{\tanh(kl/2)}{2N_r k} + \frac{l}{4N_r} + \frac{S_{PP}}{2} - \left[\frac{1}{N_r}\left(\frac{1}{\cosh(kl/2)} - 1\right) + S_{MP}\right]$$
$$\times \frac{S_{MP}N_r + [1/\cosh(kl/2) - 1]}{2S_{MM}N_r + 2k\tanh(kl/2)}, \tag{10.35}$$

which depends on the bending stiffness, the beam length, and the resultant residual force. Equation (10.35) gives the relationship of the residual force, N_r, and the parameter, k, which is linked to the bending stiffness, D. If the microbridge is long enough to meet $kl > 5$, we can further approximately simplify (10.35) to

$$p = \frac{l}{4N_r} - \frac{1}{2N_r k} + \frac{S_{PP}}{2} - \frac{(1 - S_{MP}N_r)^2}{2N_r(S_{MM}N_r + k)}. \tag{10.36}$$

From (10.36), we are able to explicitly express the k parameter in terms of N_r:

$$k = \frac{-\alpha\beta + 1 + \gamma + \sqrt{(\alpha\beta - 1 - \gamma)^2 + 4\alpha\beta}}{2\alpha}, \tag{10.37}$$

where

$$\alpha = \frac{l}{2} - 2pN_r + S_{PP}N_r, \tag{10.38a}$$

$$\beta = S_{MM}N_r, \tag{10.38b}$$

$$\gamma = (1 - S_{MP}N_r)^2. \tag{10.38c}$$

Thus, the bending stiffness, D, can be explicitly expressed in terms of N_r through $k = \sqrt{N_r/D}$. Under large deformation, the deflection varies nonlinearly with the load, and the load–deflection behavior depends on the three parameters of A, D, and N_r, which we are going to extract from fitting the nonlinear load–deflection curve. With the measured slope under small deformation, we determine uniquely the parameters from a nonlinear deflection–load curve by the least square method.

10.4.1 Symmetrical Trilayer Microbridges [36]

For a symmetrical trilayer microbridge thin film with $E_1 = E_3$, $h_1 = h_3$, and $\sigma_{r1} = \sigma_{r3}$, there is no residual moment, i.e., $M_r = 0$. Hence, there will be no initial deflection. Once the values of A and D are determined, we can calculate the Young's modulus for the two materials composed in the symmetrical trilayer thin film, which are given by

$$E_1 = \frac{12D - h_2^2 A}{4h_1(h_2^2 + 2h_1^2 + 3h_1 h_2)}, \tag{10.39a}$$

$$E_2 = \frac{-12D + (4h_1^2 + 6h_1 h_2 + 3h_2^2)A}{2h_2(h_2^2 + 2h_1^2 + 3h_1 h_2)}. \tag{10.39b}$$

The average residual stress, σ_{raver}, of the symmetrical trilayer film is calculated from the residual force, N_{r}, as

$$\sigma_{\mathrm{raver}} = N_{\mathrm{r}}/(2h_1 + h_2). \qquad (10.39c)$$

In the experiment, symmetrical SiO_2 (300 nm)/Si_3N_4 (500 nm)/SiO_2 (300 nm) microbridges were fabricated. The image profile of the microbridge taken by the WYKO NT3300 optical profiler did not reveal any detectable deflection, indicating that the microbridge was initially flat and the residual moment was negligible in the samples. Again, the unloading deflection–load curves are coincided to the loading deflection–load curves, indicating the elastic deformation behavior. Under small deformation, we determined the slope, p, which provides additional information about the relationship of the residual force, N_{r}, and the bending stiffness, D. Using this relationship, we extract the tension stiffness, A, and the residual force, N_{r}, from a nonlinear load–deflection curve by using the least squares technique. The tension stiffness, A, and the residual force, N_{r}, were extracted from the fitting. Then, the corresponding value of the bending stiffness, D, for each microbridge sample, was calculated for a given value of the residual force, N_{r}. In total, 81 microbridge samples were tested and all of them were chosen from the center part of the wafer such that the 81 samples had the same thickness. The evaluated tension stiffness, A, the bending stiffness, D, and the residual force, N_{r}, of the symmetrical $SiO_2(300\,\mathrm{nm})/Si_3N_4(500\,\mathrm{nm})/SiO_2(300\,\mathrm{nm})$ films for 81 samples are $A = 162.5 \pm 17.2 \times 10^3\,\mathrm{N/m}$, $D = 6.27 \pm 0.73 \times 10^{-9}\,\mathrm{N/m}$, and $N_{\mathrm{r}} = 9.70 \pm 1.50\,\mathrm{N/m}$. Using the values of the tension stiffness, A, and the bending stiffness, D, of each microbridge sample, we calculate the Young's modulus of the silicon oxide, E_1, and the silicon nitride, E_2, for each of the microbridge samples. The average value of the Young's modulus of silicon oxide turns out to be $31 \pm 5\,\mathrm{GPa}$, and the Young's modulus of silicon nitride is $294 \pm 29\,\mathrm{GPa}$. The thickness-averaged residual stress of the symmetrical SiO_2 (300 nm)/Si_3N_4 (500 nm)/SiO_2 (300 nm) films is calculated to be 8.8 MPa.

10.4.2 Symmetrical Trilayer Microbridges under Small Deformation [40]

Equation (10.36) indicates that the slope of the deflection vs. the load is a function of the microbridge length. We can obtain more information by measuring the slopes with samples of different lengths. As many microbridge samples with different lengths can be fabricated by the MEMS technique on a substrate, the residual stress and the Young's modulus can be treated to be the same in all the samples. Thus, measuring deflection–load slopes for samples of different lengths, we can determine the bending stiffness, D, and the residual force, N_{r}. Following this methodology, Xu and Zhang conducted microbridge tests on two sets of trilayer samples, consisting of SiO_2 (300 nm)/Si_3N_4 (500 nm)/SiO_2 (300 nm) and Si_3N_4

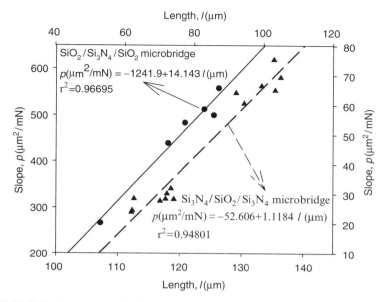

Fig. 10.16 Deflection-over-load slope, p, vs. the microbridge length, l, for the SiO$_2$ (300 nm)/Si$_3$N$_4$ (500 nm)/SiO$_2$ (300 nm) and Si$_3$N$_4$ (250 nm)/SiO$_2$ (600 nm)/ Si$_3$N$_4$ (600 nm) trilayer microbridges

(250 nm)/SiO$_2$ (600 nm)/Si$_3$N$_4$ (250 nm) [40]. The measured slope for each sample, p, vs. the microbridge length, l, is illustrated in Fig. 10.16 for the SiO$_2$/Si$_3$N$_4$/SiO$_2$ and Si$_3$N$_4$/SiO$_2$/Si$_3$N$_4$ microbridge samples. Figure 10.16 indicates that the experimental data for both sets are well fitted by the linear relationship between p and l. Using (10.7), we determine the values of N_r from the slopes in Fig. 10.16 and the values of $k = \sqrt{N_r/D_{tri}}$ from the interceptions in Fig. 10.8 and then the values of D_{tri}. Furthermore, Xu and Zhang assumed that the residual stress and Young's modulus of the silicon oxide or silicon nitride layer are independent of the deposition sequence in the SiO$_2$/Si$_3$N$_4$/SiO$_2$ and Si$_3$N$_4$/SiO$_2$/Si$_3$N$_4$ trilayer films [40]. Thus, they estimated the residual stresses of -429.49 MPa and 550.75 MPa and Young's moduli of 54.59 GPa and 270.54 GPa, respectively, for SiO$_2$ and Si$_3$N$_4$ layers.

10.4.3 Single-Layer Microbridges [45]

The microbridge test was developed first for single-layer films with residual tensile stresses. For single-layer microbridges, no residual moment exists, and hence there is no initial deflection. The Young's modulus and the residual stress are determined from fitting the nonlinearly experimental load–deflection curve with the theoretical solution by the least square technique. Zhang et al. fabricated Si$_3$N$_4$ (760 nm) singer-layer microbridge samples and conducted the test [45]. The averaged Young's modulus and residual stress of the silicon nitride film were 202.57 ± 15.80 GPa and 291.07 ± 56.17 MPa, respectively. Once the Young's modulus and residual stress are evaluated, the stress in the thin film under given loads can be determined by

$$\sigma = \frac{N_x}{t} - \frac{d^2 w}{dx^2} \frac{t}{2} E_f, \qquad (10.40)$$

where t is the thickness of the microbridge, σ denotes the maximum tensile stress, and its magnitude under the fracture load is defined as the bending strength. The bending strength of the film refers to the maximum tensile stress in the film under the fracture load, and this maximum tensile stress is almost independent of the film boundary conditions. All 33 samples were loaded to fracture during the tests, and the bending strength for each specimen was evaluated using (10.40). Figure 10.17 shows the bending strength against the bridge length, indicating that the bending strength is also independent of the bridge length. Moreover, the 33 data of the bending strength follow the Weibull distribution

$$F(\sigma) = 1 - \exp\left[-\left(\frac{\sigma}{\sigma_0}\right)^b\right], \qquad (10.41)$$

where F is the cumulative probability of failure, b is the Weibull modulus, and σ_0 is the scaling parameter. Figure 10.18 shows the experimental data on a Weibull distribution paper. From the plot, we have a Weibull modulus of 7.95 and a scaling parameter of 12.96 GPa. The average bending strength and the standard deviation are 12.26 ± 1.69 GPa, which are also plotted in Fig. 10.17 as solid and dashed lines, respectively. The bending strength depends on the process condition. For example, Cardinale and Tustison reported that the nitrogen-rich silicon nitride films have a biaxial modulus of 160 GPa and an ultimate tensile strength of 420 MPa, while the silicon-rich silicon nitride films have a biaxial modulus of 110 GPa and an ultimate tensile strength of 390 MPa [6].

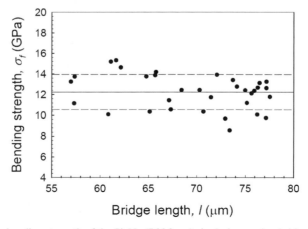

Fig. 10.17 The bending strength of the Si₃N₄ (766.2 nm) single-layer microbridges vs. the bridge length

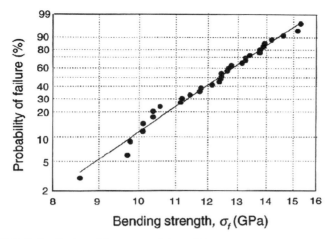

Fig. 10.18 Weibull distribution of the measured data of the bending strength shown in Fig. 10.17

10.4.4 Consecutive Approach [31]

As described earlier, the consecutive approach is based on the assumption that the mechanical properties of the $n-1$ layers in an n-fold-layer thin film should be the same as that of a film made of the same $n-1$ layers. One can extract the mechanical properties of the nth layer by testing the n-fold-layer thin film if the mechanical properties of the $n-1$ layers in an n-fold-layer thin film are known. The simple multilayer films are bilayer films. If the mechanical properties of one layer are available in a bilayer microbridge, the microbridge test is able to evaluate simultaneously the Young's modulus, and the residual stress of the other layer via the consecutive approach. In addition, the bending strength of a bilayer film can also be determined from the microbridge test. Using the consecutive approach, Su et al. conducted microbridge tests on silicon nitride(870 nm)/silicon oxide(1,080 nm) bilayer films [31]. They first tested single-layer silicon nitride films and then assumed the Young's modulus and the residual stress of the single-layer silicon nitride films to be the same as those of the silicon nitride layer in the bilayer films to evaluate the mechanical properties of the silicon oxide layer in the bilayer films. The evaluated Young's modulus and residual stress of the low-temperature silicon oxide (LTO) layer were $41.00 \pm 3.60\,\mathrm{GPa}$ and $-180.88 \pm 7.90\,\mathrm{MPa}$, respectively. Once the Young's moduli and residual stresses for both layers are known, and the stress in each layer can be determined by the following equations,

$$\sigma_1(x,z) = E_1\left[\left(\frac{B}{A}-z\right)\frac{\partial^2 w}{\partial x^2} + \frac{N_x - N_r}{A}\right] + \sigma_{r1},$$
$$\sigma_2(x,z) = E_2\left[\left(\frac{B}{A}-z\right)\frac{\partial^2 w}{\partial x^2} + \frac{N_x - N_r}{A}\right] + \sigma_{r2}.$$

(10.42)

The maximum tensile stress in the silicon nitride layer occurs at the lower surface of the bridge center, whereas the maximum tensile stress of the LTO film occurs at the upper surface of the bridge ends. The maximum tensile stress at the fracture load for each sample was calculated by using (10.42). The mean maximum tensile stresses were 0.903 ± 0.111 GPa in the LTO layer and 0.983 ± 0.087 GPa in the silicon nitride layer. Failure occurs once the maximum tensile stress in either layer reaches its bending strength. The bending strength of silicon nitride, 6.87 GPa, [45] is much larger than 0.983 GPa; therefore, the LTO layer at one of the bridge ends fractures first and then the bilayer bridge fails. This conclusion is supported by optical observations, as indicated in Fig. 10.19. Figures 10.20 and 10.21 are scanning electron microscopic (SEM) pictures of the morphology of the fracture

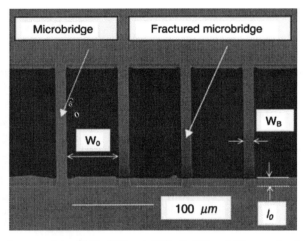

Fig. 10.19 The fracture of the Si_3N_4 (870 nm)/SiO_2 (1,080 nm) bilayer microbridges occurs at one end of the bridges

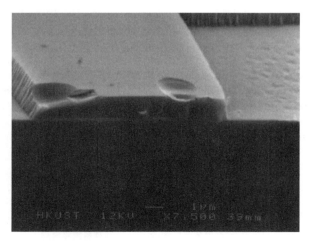

Fig. 10.20 SEM picture showing the fracture surface of a bilayer bridge, where the sample was as it was

Fig. 10.21 SEM picture showing the fracture surface of a bilayer bridge, where the sample was etched by HF solution for 5 s

surface at the bridge end, where the fractured sample in Fig. 10.20 is as it was, while the sample in Fig. 10.21 was etched by HF solution for about 5 s. Figures 10.20 and 10.21 show a flat interface between the LTO and silicon nitride films. The bonding between the LTO and silicon nitride seems good because one cannot detect the interface without etching. Figure 10.21 illustrates some surface cracks in the LTO film, indicating again that failure starts from the LTO film. Thus, the maximum tensile stress, $0.903 \pm 0.111\,\mathrm{GPa}$, represents the bending strength of the LTO layer.

10.5 Deflection of a Multilayer Microbridge for $N_x < 0$ and $M_r \neq 0$

10.5.1 Analysis

In multilayer thin films, however, there may coexist a residual moment and a compressive resultant force even if the residual stress in each layer of the film is homogeneously distributed. A residual moment and a compressive resultant force may jointly cause microbeams and other structures made of multilayer thin films to buckle and bend. From the mechanics analysis as described earlier, we have the deflection, w, at the loading point for a line load per unit width, Q, applied at the middle of a microbridge beam as

$$w = -\frac{Q}{2N_x k} \tan(kl/2) + \frac{Ql}{4N_x} - \frac{\overline{M_0}}{N_x}\left[\frac{1}{\cos(kl/2)} - 1\right] + w_0, \qquad (10.43a)$$

$$w_0 = S_{PN}(N_x - N_r) + S_{PP}\frac{Q}{2} - S_{PM}M_0, \qquad (10.43b)$$

$$\frac{Q}{2N_x}\left[1 - \frac{1}{\cos(kl/2)}\right] - \frac{\overline{M_0}k}{N_x}\tan(kl/2)$$

$$= S_{MN}(N_x - N_r) + \frac{1}{2}S_{MP}Q - S_{MM}M_0, \tag{10.43c}$$

$$I - \frac{l(N_x - N_r)}{2A} = S_{NN}(N_x - N_r) + S_{NP}\frac{Q}{2} - S_{NM}M_0, \tag{10.43d}$$

where I is given by (10.24b). For a given load, Q, a given residual force, N_r, and a given residual moment, M_r, (10.43c) and (10.43d) determine the values of N_x and M_0. Then, the deflection at the beam center is determined by (10.43a). In Appendix 6, we reduce (10.46) to the solutions with simply supported and clamped boundary conditions.

When the applied load, Q, is small, the deflection is approximately linear with the load. The slope, p, of the deflection to the load for small loads can be obtained by a Taylor expansion of (10.46) at the load starting point, $Q = 0$, no matter that the initial deflection is large or small. The slope is given by

$$p = -\frac{\tan(kl/2)}{2N_x k} + \frac{l}{4N_x} + \frac{S_{PP}}{2} - \left[S_{MP} + \frac{1}{N_x}\left(\frac{1}{\cos(kl/2)} - 1\right)\right]$$

$$\times \frac{S_{MP}N_x + [1/\cos(kl/2) - 1]}{2S_{MM}N_x + 2k\tan(kl/2)} + \left[S_{NP} - \frac{S_{MP} + \dfrac{1/\cos(kl/2) - 1}{N_x}}{S_{MM}N_x + k\tan(kl/2)}\right] \tag{10.44}$$

$$\times \frac{\left\{\begin{array}{l} S_{MN}[S_{MP}N_x + (1/\cos(kl/2) - 1)] \\ -S_{NP}[k\tan(kl/2) + S_{MM}N_x] \end{array}\right\}}{2(k\tan(kl/2) + S_{MM}N_x)(S_{NN} + \frac{l}{2A}) - 2S_{MN}^2 N_x}.$$

The slope, $p = w/Q$, under small loads is experimentally measurable and depends on the equivalent bending stiffness, D, the beam length, l, the resultant force, N_x, and the tension stiffness, A. For a given beam length, the slope, p, is a function of A, N_x, and D or a function of A, N_x, and k ($k = \sqrt{-N_x/D}$).

When there is no applied load, the buckling profile of the microbridge beam can be described as

$$w(x) = \frac{\overline{M_0}}{N_x}\left[1 - \frac{\cos(kx)}{\cos(kl/2)}\right] + w_0, \tag{10.45a}$$

$$w_0 = S_{PN}(N_x - N_r) - S_{PM}M_0, \tag{10.45b}$$

$$-\frac{k\tan(kl/2)}{N_x}\overline{M_0} = S_{MN}(N_x - N_r) - S_{MM}M_0, \tag{10.45c}$$

$$\frac{l}{8}\left[\frac{\overline{M_0}k}{N_x\cos(kl/2)}\right]^2\left[1 - \frac{\sin(kl)}{kl}\right] - \frac{l(N_x - N_r)}{2A} = S_{NN}(N_x - N_r) - S_{NM}M_0, \tag{10.45d}$$

Here again, (10.45c) and (10.45d) determine the values of N_x and M_0 for a given residual resultant force, N_r, and a given residual moment, M_r. Then, (10.45a) and (10.45b) describe the buckling profile of a microbridge.

10.5.2 Critical Resultant Force for Buckling [38]

Equations (10.48c) and (10.48d) are used to discuss the critical residual resultant force for the buckling. For convenience, we introduce the following dimensionless parameters:

$$x = kl/2, \quad x_0 = k_0l/2, \quad k_0^2 = -\frac{N_r}{D}, \quad k^2 = -\frac{N_x}{D},$$

$$\tilde{A} = \frac{A}{D/(l/2)^2}, \quad \tilde{S}_{MM} = \frac{S_{MM}D}{l/2}, \quad \tilde{S}_{MN} = \frac{S_{MN}D}{(l/2)^2},$$

$$\tilde{S}_{NN} = \frac{S_{NN}D}{(l/2)^3}, \quad \tilde{M}_0 = \frac{M_0}{D/(l/2)}, \quad \tilde{M}_r = \frac{M_r}{D/(l/2)},$$

and rewrite (10.48c) and (10.48d) in dimensionless form:

$$\tilde{M}_0\left[\tilde{S}_{MM} + \frac{1}{x}\tan(x)\right] - \frac{1}{x}\tan(x)\tilde{M}_r = \tilde{S}_{MN}(x_0^2 - x^2), \tag{10.46a}$$

$$\frac{1}{4}\left[\frac{\tilde{M}_0 - \tilde{M}_r}{-x\cos(x)}\right]^2\left[1 - \frac{\sin(2x)}{2x}\right] - \frac{1}{\tilde{A}}(x_0^2 - x^2) = \tilde{S}_{NN}(x_0^2 - x^2) - \tilde{S}_{MN}\tilde{M}_0. \tag{10.46b}$$

For simplicity, let us consider a nondeformable substrate, which means that the clamped boundary condition is applied at the two ends of the microbridge. In this case, all spring compliances are zero and (10.49) is reduced to

$$\left(\tilde{M}_0 - \tilde{M}_r\right)\frac{1}{x}\tan(x) = 0, \tag{10.47a}$$

$$\left[\frac{\tilde{M}_0 - \tilde{M}_r}{-x\cos(x)}\right]^2\left[1 - \frac{\sin(2x)}{2x}\right] = \frac{4}{\tilde{A}}(x_0^2 - x^2). \tag{10.47b}$$

Equation (10.50a) gives the critical resultant force for the buckling as

$$x_c = n\pi, \quad N_r^c = -\left(\frac{2n\pi}{l}\right)^2 D, \quad n = 1, 2, 3, \ldots, \tag{10.48}$$

which can be reduced to the well-known Euler buckling stress for clamped-clamped beams. Equation (10.50b) determines the moment if $x_0 \geqslant x = x_c$. For the first mode of buckling, the moment takes the explicit form:

$$\tilde{M}_0 - \tilde{M}_r = \pm 2\pi\sqrt{\frac{x_0^2 - \pi^2}{\tilde{A}}}. \tag{10.49}$$

It is the net moment, $\tilde{M}_{\text{net}} \equiv \tilde{M}_0 - \tilde{M}_r$, that determines the buckling behavior. Equation (10.49) indicates that with the clamped boundary condition, the upward net moment has the same magnitude as the downward net moment such that the upward and downward buckling should occur randomly if the substrate is nondeformable.

The substrate deformation affects the critical resultant force for buckling. After buckling, the magnitude of the compressive resultant force in the buckled beam should be lower than that of the original compressive resultant force. Therefore, we have $k_0 \geqslant k \geqslant 0$ and $x_0 \geqslant x \geqslant 0$. At the critical condition of buckling, the original compressive resultant force must approach a critical value and the compressive resultant force in the buckled beam must approach the same critical value. In the dimensionless variables, we have

$$x_0 \to x_c \text{ and } x \to x_c \quad \text{under the condition} \quad x_0 \geqslant x \geqslant 0, \qquad (10.50)$$

at the onset of buckling. To calculate the x value for a given x_0 value, we substitute (10.49a) into (10.49b) and then have

$$\frac{1}{4}\left[\frac{\tilde{S}_{\text{MN}}(x_0^2 - x^2) - \tilde{S}_{\text{MM}}\tilde{M}_r}{x\cos(x)\tilde{S}_{\text{MM}} + \sin(x)}\right]^2 \left[1 - \frac{\sin(2x)}{2x}\right] \qquad (10.51)$$
$$- \left[\tilde{S}_{\text{NN}} + \frac{1}{\tilde{A}} - \frac{x\tilde{S}_{\text{MN}}^2}{x\tilde{S}_{\text{MM}} + \tan(x)}\right](x_0^2 - x^2) + \frac{\tilde{S}_{\text{MN}}\tilde{M}_r\tan(x)}{x\tilde{S}_{\text{MM}} + \tan(x)} = 0.$$

For a given x_0 value, the x value is determined from solving (10.51) numerically; and there are multiple values of x corresponding to multiple modes of buckling. Taking a microbridge as an example, where $\tilde{S}_{\text{MM}} = 3.8 \times 10^{-3}$, $\tilde{S}_{\text{MN}} = 1.3427 \times 10^{-5}$, $\tilde{S}_{\text{NN}} = 3.7983 \times 10^{-7}$, $\tilde{A} = 1.329 \times 10^{22}$, and $\tilde{M}_r = 5.2071 \times 10^8$, the left-hand side of (10.51) is plotted as a function of x for a given x_0 in Fig. 10.22 to show the multiple modes of buckling. At each mode of buckling, there are two values of $x = x_1$ and $x = x_2$ for a given value of x_0, although they are very close to each other, as indicated in Fig. 10.22. Two roots of x at each mode of buckling are caused by the substrate deformation. As indicated in (10.50), we should have $x_0 \to x_c$ and $x \to x_c$ on the onset of buckling, which is illustrated in Fig. 10.23 for the example shown in Fig. 10.22, where the critical values, x_{1c} and x_{2c}, are determined to be 2.9846 and 3.2681, respectively. Clearly, the critical values of the resultant force depend strongly on the residual moment. Figure 10.24 shows that the x_{1c} value decreases with increasing $\tilde{M}_r$, whereas the x_{2c} value increases with increasing $\tilde{M}_r$. Therefore, the higher the residual moment is, the bigger the difference between the two critical values will be.

For a given value of x_0 larger than both x_{1c} and x_{2c}, there are two roots of x corresponding to a given residual moment. As an example, Fig. 10.25 shows the two roots of x as a function of the residual moment.

As the residual moment increases, one root, x_2, increases, while the other root, x_1, decreases. As mentioned earlier, the net moment, $\tilde{M}_{\text{net}}$, determines the buckling behavior. Figure 10.26 shows the variation of the net moment $\tilde{M}_{\text{net}}$ with the residual

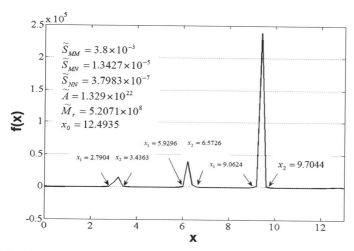

Fig. 10.22 Multivalues of x under a given x_0, showing the multi modes of buckling. At each mode of buckling, there are two values of x, very close to each other, corresponding to a given x_0

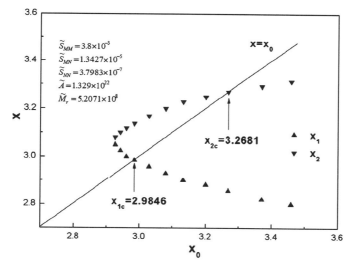

Fig. 10.23 Determine the critical buckling stress. The two intersections give the minimum values of x_0 for buckling

moment $\tilde{M}_r$ for a given value of $x_0 = 12.4935$. For a given value of the residual moment, there are two values of the net moment corresponding to the two roots of x_1 and x_2, respectively. Figure 10.26 indicates that $\tilde{M}_{\text{net}}(x_1)$ is positive and $\tilde{M}_{\text{net}}(x_2)$ is negative. For example, $\tilde{M}_{\text{net}}(x_1)$ increases from 0.36×10^7 to 0.9×10^7, 1.45×10^7, and 2.3×10^7 as the residual moment $\tilde{M}_r$ increases from 0 to 2×10^8, 4×10^8, and 8×10^8, meanwhile $\tilde{M}_{\text{net}}(x_2)$ decreases from -0.4×10^7 to -1.2×10^7, -2.1×10^7,

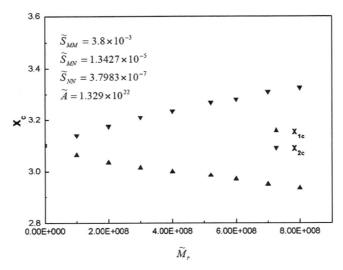

Fig. 10.24 The critical buckling stress vs. the residual moment

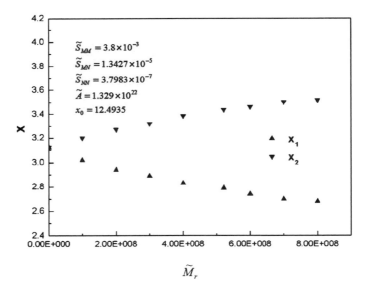

Fig. 10.25 The stress after buckling vs. the residual moment at a given residual stress

and -3×10^7. The magnitude of $\tilde{M}_{net}(x_2)$ is larger than the magnitude of $\tilde{M}_{net}(x_1)$, and the difference between the magnitudes of $\tilde{M}_{net}(x_2)$ and $\tilde{M}_{net}(x_1)$ increases from 0.04×10^7, 0.3×10^7, 0.65×10^7, and 0.7×10^7 when the residual moment $\tilde{M}_r$ increases from 0 to 2×10^8, 4×10^8, and 8×10^8. The numerical results and $\tilde{M}_{net} \equiv \tilde{M}_0 - \tilde{M}_r$ indicate that the net moment with a greater magnitude has an opposite

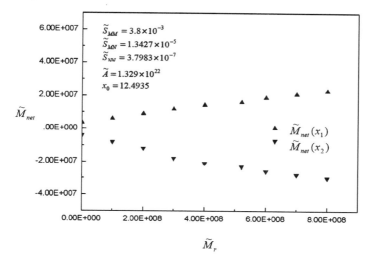

Fig. 10.26 The net moment vs. the residual moment for a given x_0. The negative net moment corresponds to upward buckling

sign in comparison to the residual moment and the microbridge will be bent by the larger absolute net moment. As shown by (10.48a), a positive net moment causes the microbridge to bend and buckle downward, while net negative moment causes the microbridge to bend and buckle upward.

As mentioned above, the four parameters, A, D, N_r and M_r, determine the bending behavior including buckling. As indicated in (10.44), we can express the resultant force, N_x, as a function of k and A, once the slope, p, is available. Additional information can be obtained by the experimentally measured displacement at the end of the microbridge, w_0, and using (10.48b), from which we can express the residual force, N_r, as a function of M_0, N_x, and A through the spring compliances. The residual force, N_r, can be expressed as a function of M_0, k, and A, since N_x is determined as a function of k and A by the slope, p. Then, we can express the moment applied by the substrate to the microbridge, M_0, and the residual moment, M_r, both in terms of k and A by using (10.48c) and (10.48d). Finally, the two independent parameters, A and k, can be uniquely determined from fitting the buckling profile with (10.48a) by using the least square technique. When A and k are determined, we will also have the values of the resultant force, N_x, through (10.44), the residual force, N_r, and the residual moment, M_r, the moment applied by the substrate to the microbridge, M_0, through (10.48b), (10.48c), and (10.48d). The bending stiffness, D, is also determined through $D = -N_x/k^2$. In brief, the characteristic parameters, A, D, N_r, and M_r, of a bilayer microbridge can be determined from the buckling profile and the slope of deflection vs. the load under small loads.

Under large loads, the deflection varies nonlinearly with the load. The load–deflection behavior is also governed by the tension stiffness, A, the bending stiffness, D, the residual force, N_r, and the residual moment, M_r, which are already determined from fitting the buckling profile. Here again, we may use the load–deflection relationship under large loads to verify the results obtained from fitting the buckling profile and the slope of deflection vs. load under small loads, as described in Sect. 10.3. Note that the buckling height at the bridge center, w_r, should be added to the measured deflection, $w_i^e(Q_i)$, for the fitting purpose because the buckling profile is taken as a reference for the measured deflection, and the theoretical predictions are based on the bending of the neutral plane without any initial deflections. The sign of the buckling height at the bridge center, w_r, is negative, if the shape of the buckle is convex, otherwise the sign is positive if the shape is concave. Furthermore, for a bilayer film, we can calculate the Young's moduli, E_1 and E_2, and the residual stresses, σ_{r1} and σ_{r2}, of each layer from the parameters of A, D, N_r, and M_r.

10.5.3 Bilayer Microbridges [38]

Experimentally, Si_3N_4 (200 nm)/SiO_2 (400 nm) bilayer microbridges were fabricated on two-side polished 4-in. p-type (100) 525-μm thick silicon wafers. The fabrication process is the same as that described previously. The final thickness of each layer was determined by a film thickness measurement system (Nanometric model 4,150) at room temperature. The bridge length and width of each microbridge sample were precisely measured one by one, using the digital coordinates provided by the Hysitron Nanoindenter system. The initial profile of each microbridge was measured by the scanning along the microbridge length direction, using the laser interferometry technique with the WYKO NT3300 optical profiler. The results indicate that the microbridges were initially curved, as shown in Fig. 10.27 and its inset, implying the existence of residual moments and/or the buckling induced by the compressive force that bent the microbridges. The microbridge tests were conducted with the Hysitron Nanoindenter system at room temperature with a preset maximum load of 1,500 μN and at a controlled loading rate of 250 μN/s and the

Fig. 10.27 Measured and fitted buckling profiles of a microbridge. The inset is an optical *top view* of a microbridge sample. The profile is obtained by the scanning along the microbridge length direction, which is indicated by a vertical line

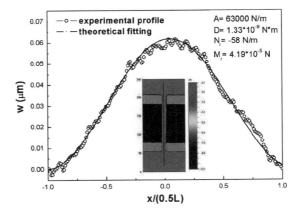

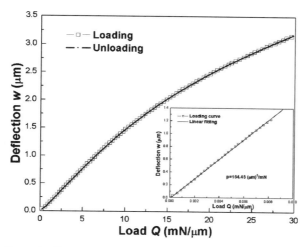

Fig. 10.28 Loading-unloading curve of a microbridge. The inset is the curve under small deformation, together with a linear fitting curve

same rate during unloading. Experimental observations found that all fabricated microbridge samples were buckling upwards. This is because the silicon nitride layer has a tensile residual stress, and the silicon oxide layer has a compressive residual stress such that the residual moment is positive. This is consistent with the theoretical prediction described earlier. As an example, Fig. 10.27 shows one measured buckling profile. The buckling profile is not ideally symmetric, which might be due to the nonuniform thickness and/or roughness $(9.5 \pm 1.5 \,\mathrm{nm})$ of the beam. The initial deflection profile was fitted with (10.48). To determine the values of the tension stiffness, A, the bending stiffness, D, the residual force, N_r, and the residual moment, M_r, from fitting the buckling profile, we need the slope of the deflection vs. the load under small loads. Figure 10.28 illustrates the load–deflection curves during loading and unloading. The unloading curve coincides with the loading curve, indicating the completely elastic behavior of the silicon nitride/silicon oxide film and the silicon substrate. The inset of Fig. 10.28 shows a loading curve under small loads, where the squares represent the experimental data and the straight line is the linear fitting result. For the example shown in Fig. 10.28, the slope was determined to be 154.45 $(\mu\mathrm{m})^2/\mathrm{mN}$.

The theoretical buckling profile was also shown in Fig. 10.27, from which the four parameters were extracted to be $A = 63,000\,\mathrm{N/m}$, $D = 1.33 \times 10^{-9}\,\mathrm{N\,m}$, $N_r = -58\,\mathrm{N/m}$, and $M_r = 4.19 \times 10^{-5}\,\mathrm{N}$, respectively. The theoretical profile did not match to the measured profile perfectly because of the asymmetry of the real buckling profile itself. Totally, 30 microbridge samples were tested. All of them were chosen from the center part of the wafer such that all the samples had the same thickness. The mean values and standard deviations of the tension stiffness, bending stiffness, residual force, and residual moment were $A = 64,000 \pm 1,700\,\mathrm{N/m}$, $D = 1.34 \pm 0.06 \times 10^{-9}\,\mathrm{N\,m}$, $N_r = -58 \pm 8\,\mathrm{N/m}$, and $M_r = 4.18 \pm 0.19 \times 10^{-5}\,\mathrm{N}$, respectively. As expected, the four parameters do not vary with the microbridge length.

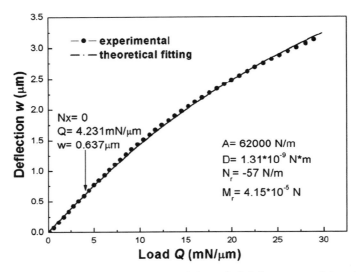

Fig. 10.29 The experimental load–deflection and theoretical fitting curves of the microbridge whose buckling profile is shown in Fig. 10.27

When the values of the tension stiffness, A, the bending stiffness, D, the residual force, N_r, and the residual moment, M_r, are available, we can verify any two of the four parameters, A, D, N_r, and M_r, from fitting the load–deflection curve under large deformation with (10.46). As an example, Fig. 10.29 illustrates a typical experimental load–deflection curve under large deformation and the theoretical fitting curve, indicating that the analytic solution fits the experimental data well. The extracted values are more or less correspondingly the same as the values obtained from the buckling profile fitting, as shown in the Figs. 10.27 and 10.29. The consistent results are expected because the four parameters completely determine the buckling profile and the response of a bilayer microbridge sample under an applied lateral load. It should be noted that the resultant force in microbridge would change gradually from compression to tension with increasing load. Thus, there must exist a particular load, at which $N_x = 0$. When the load is lower than this level, $N_x < 0$, while $N_x > 0$ if the load exceeds this level. This particular load is also determined in fitting and indicated in Fig. 10.29, where $Q = 4.231\,\text{mN/\textmu m}$ and $w = 0.637\,\text{\textmu m}$. After this load, $N_x > 0$, thus, k is a pure imaginary number and complex calculation is carried out in solving (10.46).

Using (10.7) and the determined values of A, D, N_r, and M_r, we can determine the Young's modulus and the residual stress in each layer of the bilayer film. The average value and the associated standard deviation of the Young's modulus are $E_1 = 35 \pm 2\,\text{GPa}$ for the silicon oxide layer and $E_2 = 250 \pm 7\,\text{GPa}$ for the silicon nitride layer. For silicon oxide films, the estimated values of the moduli are essentially the same as the previous value of 38 GPa [37]. For silicon nitride films, the estimated values of the moduli are smaller than the previous reported value of 290 GPa [32], but essentially the same as the values of 251 GPa [37], 222 GPa [35], 235 GPa [18].

Similarly, we calculated the residual stress in the silicon oxide layers, σ_{r1}, and the residual stress in the silicon nitride layer, σ_{r2}, for each of the 30 microbridge samples. The average value and the associated standard deviation of the residual stress are $\sigma_{r1} = -380 \pm 20$ MPa for the silicon oxide layer and $\sigma_{r2} = 470 \pm 15$ MPa for the silicon nitride layer. For silicon oxide films, the estimated magnitudes of the residual stress are larger than the reported value of -320 MPa [28]. For silicon nitride films, the estimated values of the residual stress are larger than the reported values of 120–150 MPa [35], 151 MPa [9], 420 MPa [6] in the literature.

In summary, a multilayer beam with a compressive residual resultant force will buckle if the compressive residual resultant force exceeds a critical value. If there coexists a residual moment and a compressive residual resultant force in a multilayer beam, the residual moment may change the critical value of the compressive residual resultant force for buckling due to substrate deformation. The larger absolute net moment determines the buckling direction. Measuring the buckling profile and the slope of deflection to load, we determine the four parameters of A, D, N_r and M_r, which govern the buckling and bending behavior of the multilayer beam. For bilayer beams, the Young's modulus and the residual stress of each layer can be uniquely determined from the microbridge test without any further assumptions.

10.6 Deflection of a Multilayer Microbridge for $N_x < 0$ and $M_r = 0$

10.6.1 Buckling Behavior

If the magnitude of the residual compressive force per unit width, N_r, exceeds a critical value, the microbridge beam buckles, causing partial relaxation of the residual compressive force. If N_x denotes the compressive force per unit width along the length direction in the buckled beam, both N_r and N_x are negative and $N_r < N_x$ or $|N_r| > |N_x|$. In this case, (10.48) reduces to

$$w = \frac{M_0}{N_x}\left[1 - \frac{\cos(kx)}{\cos(kl/2)}\right] + w_0. \tag{10.52a}$$

$$w_0 = S_{PN}(N_x - N_r) - S_{PM}M_0, \tag{10.52b}$$

$$-\frac{k\tan(kl/2)}{N_x}M_0 = S_{MN}(N_x - N_r) - M_0 S_{MM}, \tag{10.52c}$$

$$\frac{l}{8}\left[\frac{M_0 k}{N_x\cos(kl/2)}\right]^2\left[1 - \frac{\sin(kl)}{kl}\right] - \frac{l(N_x - N_r)}{2A} = \tag{10.52d}$$
$$S_{NN}(N_x - N_r) - S_{NM}M_0.$$

For a given value of residual compressive force, N_r, (10.55c) and (10.44d) determine the values of N_x and M_0. Then, (10.55a) with (10.55b) describes the buckling profile of a microbridge. We introduce the thickness-averaged Young's modulus, $E_f = A/t$, where t is the thickness of a multilayer film. Then, we rewrite (10.55c) and (10.55d) in dimensionless form:

$$\tilde{M}_0 \left[\tilde{S}_{MM} + \frac{1}{x} \tan(x) \right] = \tilde{S}_{MN}(x_0^2 - x^2), \tag{10.53a}$$

$$\frac{1}{4} \left[\frac{\tilde{M}_0}{x \cos(x)} \right]^2 \left[1 - \frac{\sin(2x)}{2x} \right] - \frac{\tilde{t}^2}{12}(x_0^2 - x^2) = S_{NN}(N_x - N_r) - S_{NM}M_0. \tag{10.53b}$$

where $\tilde{t} = t/(l/2)$. Substituting (10.56a) into (10.56b) yields

$$\left[1 - \frac{\sin(x)\cos(x)}{x} \right](x_0^2 - x^2) - \tilde{S}[\tilde{S}_{MM}x\cos(x) + \sin(x)]^2$$
$$+ 4x\cos(x)[\tilde{S}_{MM}x\cos(x) + \sin(x)] = 0, \tag{10.54}$$

with

$$\tilde{S} = 4 \left(\tilde{S}_{NN} + \frac{\tilde{t}^2}{12} \right) \bigg/ \tilde{S}_{MN}^2.$$

For a given x_0 value, the x value is determined by solving (10.54) numerically and there are multivalues of x corresponding to multimodes of buckling. Taking a microbridge as an example, where $\tilde{S}_{MM} = 5.84608 \times 10^{-3}$, $\tilde{S}_{MN} = 2.65162 \times 10^{-5}$, $\tilde{S}_{NN} = 8.72744 \times 10^{-7}$, and $\tilde{t} = 0.0121705$, we plot the left-hand side of (10.54) as a function of x under a given x_0 in Fig. 10.30 to show this multimodes of buckling. Apparently, at each mode of buckling, there are two values of $x = x_1$ and $x = x_2$ for a given value of x_0, although they are very close to each other, as indicated in Fig. 10.30. This phenomenon is caused by the substrate deformation. Each

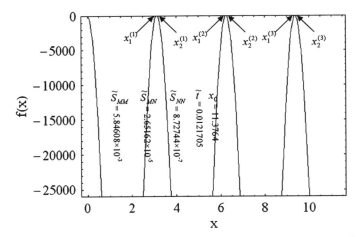

Fig. 10.30 Multivalues of x under a given x_0, showing the multimodes of buckling. At each mode of buckling, there are two values of x, very close to each other, corresponding to a given x_0

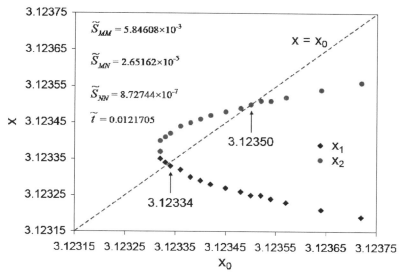

Fig. 10.31 Numerical solutions of x as function of x_0. The two intersections give the minimum values of x_0 for buckling

of the x values increases with increasing the x_0 value, indicating that a higher initial residual stress will result in a higher residual stress in the buckled beam, which is different from the prediction by the Euler buckling theory that the residual stress in a buckled beam is a constant.

As we are interested in the critical stress for buckling, we plot the x value as a function of x_0 near the critical point, at which $x_0 \to x_c$ and $x \to x_c$, in Fig. 10.31. The values located on the right side of the straight line of $x_0 = x$ satisfy $x_0 \geqslant x \geqslant 0$. Figure 10.31 shows that with decreasing x_0, x_1 and x_2 approach the critical values, x_{1c} and x_{2c}, respectively, at which $x_1 = x_0 = x_{1c}$ and $x_2 = x_0 = x_{2c}$. For the example, the critical values, x_{1c} and x_{2c}, are 3.12334 and 3.12350, respectively. In fact, the critical buckling condition of $x_0 = x$ reduces (10.54) to

$$\{-\tilde{S}[\tilde{S}_{MM}x\cos(x) + \sin(x)] + 4x\cos(x)\} [\tilde{S}_{MM}x\cos(x) + \sin(x)] = 0. \quad (10.55)$$

Equation (10.55) is satisfied if

$$\tilde{S}_{MM}x\cos(x) + \sin(x) = 0, \quad (10.56a)$$

and/or

$$\tilde{S}[\tilde{S}_{MM}x\cos(x) + \sin(x)] - 4x\cos(x) = 0. \quad (10.56b)$$

Then, taking the zero moment as the buckling criterion, we obtain (10.59b) from (10.59c).

The critical values, x_{1c} and x_{2c}, are determined by (10.59a) and (10.59b), respectively. Actually, (10.59a) can be directly deduced from (10.56a) by letting $x_0 = x$. If we substitute (10.56a) into (10.56b) to eliminate the term of $\left(x_0^2 - x^2\right)$, we have

$$\frac{1}{4}\left[\frac{1}{x\cos(x)}\right]^2 \tilde{M}_0\left[1 - \frac{\sin(2x)}{2x}\right] = \left(\frac{\tilde{t}^2}{12} + \tilde{S}_{NN}\right)\frac{\tilde{S}_{MM} + \frac{1}{x}\tan(x)}{\tilde{S}_{MN}} - \tilde{S}_{MN}. \quad (10.56c)$$

Using the values of x_1 and x_2 for a given x_0, we can calculate the moment, $\tilde{M}_0$ from (10.56) and determine the buckling direction from the moment sign. For a given set values of x_1 and x_0, or x_2 and x_0, (10.56a) gives a moment value, but (10.56b) yields two moment values:

$$\tilde{M}_0^{(1,2)} = 2\tilde{S}_{MN}[x\cos(x)]^2 \times$$

$$\frac{-\tilde{S}_{MN} \pm \sqrt{\tilde{S}_{MN}^2 + \dfrac{\left[1 - \frac{\sin(2x)}{2x}\right] \times \left(\frac{\tilde{t}^2}{12} + \tilde{S}_{NN}\right)\left(x_0^2 - x^2\right)}{[x\cos(x)]^2}}}{1 - \frac{\sin(2x)}{2x}}, \quad (10.57)$$

where $\tilde{M}_0^{(1)}$ and $\tilde{M}_0^{(2)}$ correspond to the solutions with the plus and minus signs of the square root, respectively. It should be pointed out that only one root of (10.57) is consistent with the solution of (10.56a) and this root is the true solution. For the x_1 set, $\tilde{M}_0^{(1)}$ from (10.56b) gives the same value as that determined by (10.56a), whereas for the x_2 set, $\tilde{M}_0^{(2)}$ from (10.56b) gives the same value as that determined by (10.56a). Numerical results are plotted in Fig. 10.32 to demonstrate this conclu-

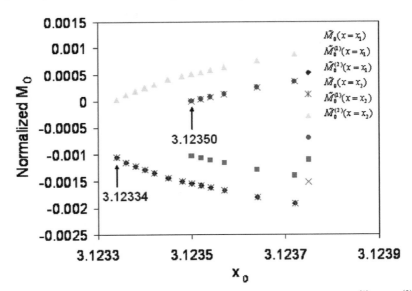

Fig. 10.32 Calculated moments for various x_0, where $\tilde{M}_0$ is from (10.12a), and $\tilde{M}_0^{(1)}$ and $\tilde{M}_0^{(2)}$ are from (10.59b). The negative moment corresponds to downward buckling

sion. Figure 10.32 shows that the moment for the x_1 set is negative and the buckling is downward, while the moment for the x_2 set is positive and the associated buckling is upward. When $x_0 \geqslant x_{2c} = 3.12350$, the microbridge could buckle upward or downward, randomly oriented, but the downward buckling will release more residuals tress, as indicated in Fig. 10.31. When $3.12350 = x_{2c} > x_0 \geqslant x_{1c} = 3.12334$, the microbridge could uniquely buckle downward. While $x_0 < x_{1c} = 3.12334$, though the moment has a nonzero value if the microbridge buckles downward, the stress in it does not release at all, as shown in Fig. 10.31. In this case, the microbridge will not buckle. If a disturbance occurs to make it buckle, the microbridge will completely recover to its flat state, at which the strain energy is lower than that at the buckled state. This situation is similar to the classic nucleation case, in which a new nucleus will grow up when it is larger than a critical size, otherwise, the new nucleus will disappear.

If the substrate is nondeformable, i.e., if the clamped boundary condition is applied at the two ends of the microbridge, the spring compliances will be zero. In this case, (10.56) reduces to

$$\tilde{M}_0 \frac{1}{x} \tan(x) = 0, \tag{10.58a}$$

$$\left[\frac{\tilde{M}_0}{x\cos(x)} \right]^2 \left[1 - \frac{\sin(2x)}{2x} \right] = \frac{\tilde{t}^2}{3}(x_0^2 - x^2). \tag{10.58b}$$

Equation (10.61a) gives the buckling stress

$$x_c = n\pi, \quad N_r^c = -\left(\frac{2n\pi}{l} \right)^2 D, \quad \text{and} \quad \sigma_r^c = -\left(\frac{2n\pi}{l} \right)^2 \frac{D}{t}, \tag{10.59}$$
$$n = 1, 2, 3, \ldots,$$

which is the well-known Euler buckling stress. Equation (10.61b) determines the moment if $x_0 = x = x_c$. For the first mode of buckling, the moment takes the explicit form:

$$\tilde{M}_0 = \pm \pi \tilde{t} \sqrt{\frac{x_0^2 - \pi^2}{3}}. \tag{10.60}$$

Equation (10.60) is a special case of (10.49) for $M_r = 0$. With the clamped boundary condition, the upward moment has the same magnitude as the downward moment, as indicated in (10.60).

10.6.2 Load–Deflection Curves under Elastic Deformation

Without any residual moment, (10.46) reduces to

$$w = -\frac{Q}{2N_x k}\tan(kl/2) + \frac{Ql}{4N_x} - \frac{M_0}{N_x}\left[\frac{1}{\cos(kl/2)} - 1\right] + w_0, \quad (10.61a)$$

$$w_0 = S_{PN}(N_x - N_r) + S_{PP}\frac{Q}{2} - S_{PM}M_0, \quad (10.61b)$$

$$M_0 = \frac{S_{MN}N_x(N_x^- N_r) + \frac{1}{2}S_{MP}QN_x + \frac{1}{2}Q\left[\frac{1}{\cos(kl/2)} - 1\right]}{S_{MM}N_x - k\tan(kl/2)}, \quad (10.61c)$$

$$S_{NN}(N_x - N_r) + S_{NP}\frac{Q}{2} - S_{NM}M_0 = I - \frac{l(N_x - N_r)}{2E_f t}, \quad (10.61d)$$

with

$$I = \frac{1}{8k}\begin{bmatrix} -\dfrac{4QM_0 k}{N_x^2\cos(kl/2)} + \dfrac{Q^2 kl}{4N_x^2\cos^2(kl/2)} \\[2mm] +\dfrac{M_0^2 k^3 l}{N_x^2\cos^2(kl/2)} + \dfrac{2Q^2 kl}{4N_x^2} + \dfrac{4QM_0 k}{N_x^2} \\[2mm] -\dfrac{2Q^2\sin(kl/2)}{N_x^2\cos(kl/2)} + \dfrac{QM_0 k^2 l\sin(kl/2)}{N_x^2\cos^2(kl/2)} \\[2mm] +\dfrac{Q^2\sin(kl)}{4N_x^2\cos^2(kl/2)} - \dfrac{M_0^2 k^2\sin(kl)}{N_x^2\cos^2(kl/2)} \end{bmatrix} \quad (10.61e)$$

Equation (10.64) reduces to (10.55) if the load, Q, is not applied. In Appendix 7, we reduce (10.64) to the solutions with the simply supported and clamped boundary conditions.

10.6.3 Experiments on Gold Thin Films [19]

Au microbridge specimens were fabricated on (100) Si wafers with double polished sides. Figure 10.33 is a schematic depiction showing fabrication steps. A layer of Si_3N_4 was first deposited on both sides of the wafer. Dry etching was then used to create windows in the bottom side of the Si_3N_4 layer to expose the underneath Si, and to remove the top side of the Si_3N_4 layer. After that, the sputtering technique was used to deposit a few nanometer thick of Cr layer on the top side of the wafer, which aids adhesion of Au to Si, and followed by depositing a few nanometer thick Au layer, which serves as a seed layer in subsequent electroplating process. Next, the photoresist was spinning-coated on the Au layer and patterned to microbridge shapes. Then, the microbridges on the wafer were further electroplated with Au to a designed thickness. The commercially available NEUTRONEX 309 gold electroplating technique, which is frequently used in "bump" plating on metallized Si wafers, was used in this step, depositing the Au layer of 99.99% purity. During plating, the pH value of electrolyte was maintained between 9.2 and 9.8, the plating bath temperature was kept at 50 °C, and the current density was set at 0.25 mA/cm^2.

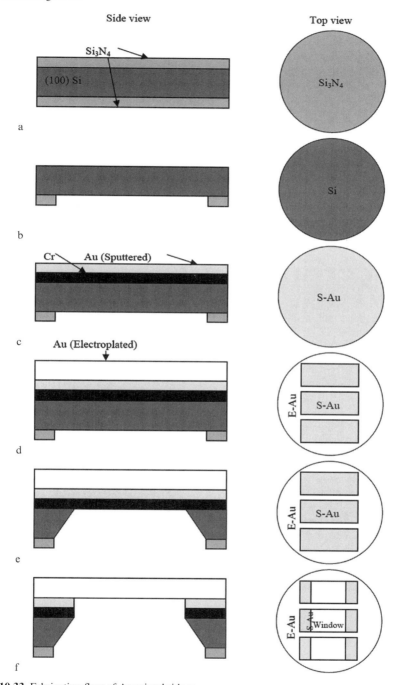

Fig. 10.33 Fabrication flow of Au microbridges

After the electroplating, the wafer was immersed in a KOH solution to etch Si from the bottom side windows with protection of the top side covered by a Teflon chuck. After the Si was etched through, windows through the wafer thickness was formed, while the sputtered Au/Cr film still sustained microbridges to suspend upon the windows. The sputtered Cr and Au layers were etched away sequentially from the bottom side by wet etching using the Cr etchant and the mixture of nitric acid and hydrochloric acid, respectively. Finally, the microbridges were made of only the electroplated Au and free-stand on the windows. Figure 10.34 shows the optical

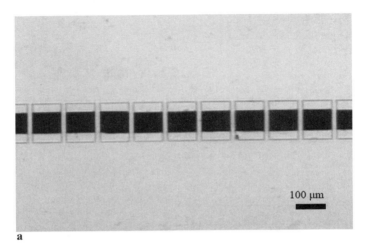

a

b

Fig. 10.34 Optical images of Au microbridge(s), (**a**) gross view and (**b**) fine view

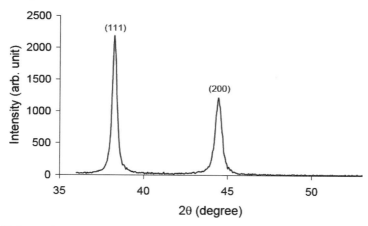

Fig. 10.35 X-ray diffraction pattern of the gold film

image of the microbridges. The dimensions, such as length, width, and thickness, of each microbridge were measured individually using the WYKO NT3300 optical profiler working in the vertical-scanning interferometry mode. From the height difference of topography of the sputtered and electroplated Au, the average thickness of microbridge was measured to be $0.481 \pm 0.005\,\mu m$. The average width of microbridge was $16.1 \pm 0.1\,\mu m$. Different lengths of microbridges were designed, which ranged from 68 to $82\,\mu m$ with an accuracy of $0.2\,\mu m$. The extension of microbridges on the supporting substrate was about $25\,\mu m$. The gap between two adjacent microbridges was $100\,\mu m$, which was large enough so that each microbridge can be treated as an isolated one.

The X-ray diffraction (PW1830, Philips) pattern, as shown in Fig. 10.35, reveals two features of the gold film. First, the peaks at $38.18°$ and $44.39°$ correspond to the Au (111) and (200) diffractions, indicating the film is polycrystalline in structure. Second, the peak intensity ratio of (200) over (111) is 0.56, which is slightly higher than the ratio of 0.52 for a random orientation distribution [20]. This exhibits that the amount of grains in the film grown along the $\langle 100 \rangle$ direction perpendicular to the film surface is larger than that with a random orientation distribution, i.e., the Au film has a gentle {100} texture. The quantitative texture analysis of the film was further carried out by using Siemens D-5000 X-ray diffraction equipment. The (111) and (200) pole figures shown in Fig. 10.36 were characterized with the reflection method under a condition of $35\,kV$ and $28\,mA$. The orientation distribution function of grains was then determined from the pole figures. With the measured grain orientation distribution function, Young's modulus of the film was calculated and compared with the experimental results. The microbridge tests were conducted with the same test procedure as that described previously. The profiles of the beams prior to and after the bending test were examined also using the WYKO NT3300 Profiler.

Fig. 10.36 Pole figures of the
thin gold films

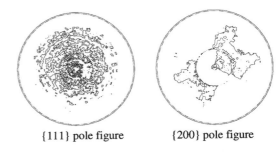

{111} pole figure   {200} pole figure

10.6.3.1 Young's Modulus of the Film Calculated from the Texture Analysis

The orientation distribution function (ODF) is expressed in terms of three Euler
angles: $g = (\varphi_1, \Phi, \varphi_2)$ (4), which were determined from the (111) and (200) pole
figures. For a single FCC crystal film, we can calculate the elastic modulus in the
direction parallel to the film surface, provided that the elastic constants of the crystal
are available

$$\frac{1}{E(g)_{//}} = S_{11} - 2(S_{11} - S_{12} - \tfrac{1}{2}S_{44})(l_1^2 l_2^2 + l_2^2 l_3^2 + l_3^2 l_1^2), \qquad (10.62)$$

where S_{11}, S_{12}, and S_{14} are the independent elastic compliances of the crystal. In the
case of Au, S_{11}, S_{12}, and S_{44} are 23.66, -10.83, and 23.81 TPa^{-1}, respectively [7].
l_1, l_2, and l_3 in (10.62) are the direction cosines to transform the crystallographic
axes to parallel direction to the film surface and read as

$$l_1 = \cos \varphi_1 \cos \varphi_2 - \sin \varphi_1 \sin \varphi_2 \cos \Phi,$$
$$l_2 = -\cos \varphi_1 \sin \varphi_2 - \sin \varphi_1 \cos \varphi_2 \cos \Phi, \qquad (10.63)$$
$$l_3 = \sin \varphi_1 \sin \Phi.$$

The average elastic modulus, $\bar{E}_{\|}$, of thin film, in parallel direction to the surface was
estimated from [17]

$$\bar{E}_{\|} = \frac{1}{2}\left[\oint E(g)_{\|} F_g dg + \frac{1}{\oint \frac{1}{E(g)_{\|}} F_g dg} \right]. \qquad (10.64)$$

where $dg = \sin \Phi d\varphi_1 d\Phi d\varphi_2 / 8\pi^2$, and F_g is the ODF of $g = (\varphi_1, \Phi, \varphi_2)$. The integral
space is the entire orientation space, i.e. $0 \leqslant \varphi_1 \leqslant 2\pi$, $0 \leqslant \Phi \leqslant \pi$, $0 \leqslant \varphi_2 \leqslant 2\pi$.

The calculated Young's moduli at two mutually orthogonal directions in the surface plane are 68.17 and 68.42 GPa, respectively. The almost identical values imply
the mechanical properties of the thin Au film are in-plane isotropic, i.e., direction-
independent in the surface plane of the film.

10.6.3.2 Young's Modulus and Residual Stress Determined from the Initial Buckling Profile Fitting

The buckling direction of the microbridges is random, could be upward or downward. Figure 10.37 shows two measured initial profiles of microbridges with opposite buckling directions. Note that in our coordinates, the downward direction is defined as positive. The buckling profile is not ideally symmetric, which might be due to the nonuniform thickness and/or roughness $(7.3 \pm 1.1 \, \text{nm})$ of the beam. Nevertheless, the buckling profile was fitted with (10.55). As a result, the theoretical profiles were also shown in Fig. 10.37, from which the Young's moduli and residual stresses were obtained to be 67.2 GPa, $-8.4 \, \text{MPa}$ and 67.4 GPa, $-7.7 \, \text{MPa}$, respectively. The theoretical profile could not match to the measured profile perfectly because of the asymmetry of the real buckling profile itself. Totally, 22 profiles of microbridges with different lengths were measured and fitted, yielding the mean values of the Young's modulus and the residual stress to be $66.2 \pm 2.9 \, \text{GPa}$ and $-8.5 \pm 1.0 \, \text{MPa}$, respectively, as shown in Fig. 10.38. It is noted that the extracted Young's modulus, $66.2 \pm 2.9 \, \text{GPa}$, is close to the value, 68 GPa, determined by the texture analysis.

10.6.3.3 Elastic Deformation Under Small Loads

Under small loads, the response of the microbridge to the load is elastic. From the elastic part of a load–deflection curve, we can extract the Young's modulus and the residual stress by fitting the experimental load–deflection curve with the theoretical prediction. Figure 10.39 is an example of such fitting. The coincidence of loading and unloading curves indicates the bending is elastic. The theoretical curve fits to

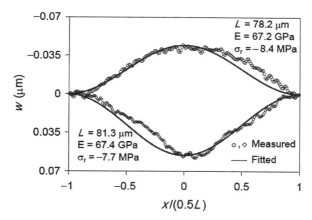

Fig. 10.37 Measured and fitted profiles of the microbridges buckling upward and downward. Young's modulus and residual compressive stress of the thin gold film are extracted from the buckling profiles and indicated inside the figure

Fig. 10.38 Young's moduli
and residual stresses determined from the profile fitting
of the buckled microbridges
of different lengths

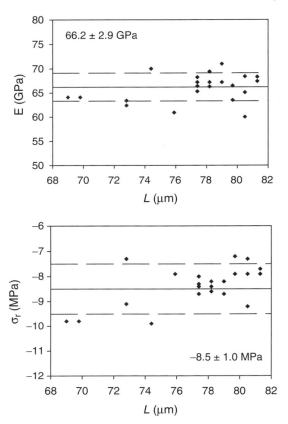

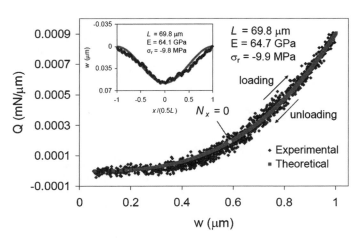

Fig. 10.39 The measured and fitted load–deflection curves of the microbridge under small loads.
The insert is the corresponding initial buckling profile and an associated fitting curve

the experimental data well, from which the Young's modulus and residual stress are evaluated to be 64.7 GPa and −9.9 MPa, respectively, which are correspondingly identical to the values of 64.1 GPa and −9.8 MPa obtained by the buckling profile fitting, as shown in the insert of Fig. 10.39.

A matter deserving to point out is that the resultant force in microbridge will change gradually from compression to tension with increasing load. Thus, there must exist a particular load, at which $N_x = 0$. When the load is lower than this level, $N_x < 0$, while $N_x > 0$ if the load exceeds this level. This particular load is also determined in fitting and indicated in Fig. 10.39, where $Q = 0.1892 \mu N/\mu m$ and $w = 0.5885 \mu m$. After this load, $N_x > 0$, thus, k is a pure imaginary number and complex calculation is carried out in conducting (10.64). The Young's modulus and residual stress were determined from fitting elastic load–deflection curves. The average values of Young's modulus and residual stress are 66.4 ± 2.8 GPa and $−8.5 \pm 1.0$ MPa, respectively, which are correspondingly the same as those from the buckling profile fitting.

10.7 Yield Strength

When the applied load is high, plastic deformation will occur in the Au beams. As an example, Fig. 10.40 shows the load–deflection curves during loading and unloading for the microbridge whose initial buckling profile is downward, as shown in Fig. 10.37. Clearly, the loading and unloading curves do not coincide with each other, as shown in Fig. 10.40, indicating plastic deformation occurred in the Au beam during the test. Figure 10.41 is a scanning interferometer image of two microbridge beams after loading–unloading. Clearly, plastic deformation occurred in

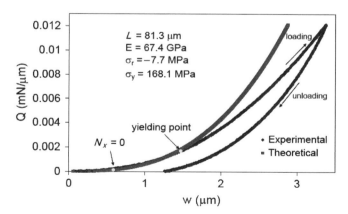

Fig. 10.40 The load–deflection curve of the microbridge whose initial buckling profile is shown in Fig. 10.37. A theoretically elastic loading curve is plotted here as a reference, which coincides with the experimental curve within the elastic range

Fig. 10.41 Scanning interferometer image of two microbridge beams after loading–unloading. Plastic deformation occurred in the lower beam

the lower beam. The smooth change of load with deflection excludes the possibility to determine the yield strength from a sudden change in the load–deflection curve. To determine the yield strength, we plot an elastic load–deflection curve with the material constants obtained from fitting elastic load–deflection curve under small loads for the same microbridge beam, as shown by the red line in Fig. 10.40. The theoretical and experimental curves coincide with each other under small loads, i.e., within the elastic range. Beyond the elastic range, the experimental curve deviates from the elastic curve due to plastic deformation. Following the 0.2% offset strain in the determination of the yield strength in a tensile test, we define a 2% offset deflection here, which means that the yield load is defined as such a load at which the normalized deflection deviation from the theoretical elastic deflection is equal to 2%, i.e., $(w^{\exp} - w^{ela})/w^{ela} = 2\%$. By this way, we determined the yield point and indicated it in Fig. 10.40. For clarity, we marked also in Fig. 10.40 the load at which $N_x = 0$. Apparently, the yield point far exceeds the point of $N_x = 0$, indicating the resultant force is tensile at the yield point. Then, from the yield point, we can determine the yield strength of the microbridge beam.

As indicated in (10.40), the stress in a microbridge changes along the thickness direction and is composed of two parts, one due to the middle plane tension (or compression) and the other due to bending. For the film with an initial residual compressive stress without residual moment, we have from (10.64) that

$$w = -\frac{Q\sin[k(l/2+x)]}{2N_x k\cos(kl/2)} + \frac{Q}{2N_x}\left(\frac{l}{2}+x\right) - \frac{M_0}{N_x}\left[\frac{\cos(kx)}{\cos(kl/2)} - 1\right] + w_0$$

$$\text{for } -\frac{l}{2} \leqslant x \leqslant 0. \tag{10.65}$$

Because of the symmetry, (10.65) gives the solution only for a half length of the beam for simplicity. Equation (10.65) is still valid after $N_x > 0$ during loading except that k is a pure imaginary number and the calculation is carried out in complex form, just the same as that in elastic fitting of load–deflection curve with (10.64) after $N_x > 0$. From (10.65), we have

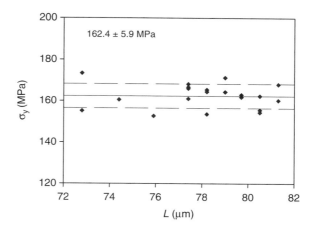

Fig. 10.42 Yield strengths determined from the 2% offset deflection approach

$$\frac{\partial^2 w}{\partial x^2} = \frac{Qk^2 \sin[k(l/2+x)]}{2N_x k \cos(kl/2)} + \frac{M_0 k^2}{N_x} \frac{\cos(kx)}{\cos(kl/2)}. \tag{10.66}$$

The maximum stress should occur at some points on the surfaces, at which the bending stress, denoted by the second term on the right-hand side of (10.40), has the same sign as the tensile stress, represented by the first term on the right-hand side of (10.40). Thus, we can substitute the absolute value of the bending stress into (10.40), which yields

$$\sigma = \frac{N_x}{t} + \frac{t}{2} E_{\mathrm{f}} \left| \frac{Qk^2 \sin[k(l/2+x)]}{2N_x k \cos(kl/2)} + \frac{M_0 k^2}{N_x} \frac{\cos(kx)}{\cos(kl/2)} \right|. \tag{10.67}$$

The maximum tensile stress occurs at the lower surface of the microbridge center. The maximum tensile stress at the offset point is defined as the yield strength of the film. By this way, we found the yield strength of the gold film to be 168.1 MPa, which is also shown in Fig. 10.40. Totally, 20 load–deflection curves of microbridges were measured and fitted. The determined values of yield strength were plotted in Fig. 10.42 with a mean value of 162.4 ± 5.9 MPa, which is close to the value of 170 MPa measured on the 0.5 μm thickness gold film with the microscale tension test method [10].

10.8 Concluding Remarks

The microbridge testing method for the mechanical characterization of thin films is based on the elastic beam theory. For brittle single-layer films, the microbridge testing method is able to evaluate simultaneously the Young's modulus, residual stress, and bending fracture strength from an experimental load–deflection curve. For multilayer films, the analysis shows that the bending of a multilayer beam is equivalent to the bending of a single-layer beam with an equivalent bending stiffness, a tension stiffness, a residual force, and a residual moment. The residual moment results in an initial deflection. Measuring the profile of an initial deflection provides more information of the multilayer film, which is therefore one aspect of the mechanical characterization of the film. With the measured value of the slope of a load–deflection curve under small loads, which gives the relationship between the equivalent bending stiffness and the initial resultant force, we are able to determine the tension stiffness, A, the bending stiffness, D, the residual moment, M_r, and the residual force, N_r, from the profile of the initial deflection. The four parameters completely determine the elastic bending behavior of multilayer microbridges. For bilayer thin films, the Young's modulus and the residual stress of each layer can be determined from the four parameters without any assumptions. If the layer number is equal to or larger than three, we may need to make more assumptions to determine the Young's modulus and the residual stress of each layer, as described above for the symmetrical and asymmetrical trilayer films. One may conduct the microbridge test on samples with different lengths to obtain more information about the mechanical behavior of a multilayer film. As microbridge samples are fabricated by the MEMS technique on a substrate, it is reasonable to assume that the microbridge samples have the same mechanical properties except for the difference in length. The consecutive approach means that one can estimate the Young's modulus and the residual stress of one layer in an n-layer film if the corresponding values of the rest of the $n-1$ layers are known. Although the consecutive approach sounds perfect in theory, it may be difficult in experiment to have the mechanical properties of the $n-1$ layers.

A multilayer beam with a compressive residual resultant force will buckle if the compressive residual resultant force exceeds a critical value. If there coexist a residual moment and a compressive residual resultant force in a multilayer beam, the residual moment may change the critical value of compressive residual resultant force for buckling because of the substrate deformation. The larger absolute net moment determines the buckling direction. Measuring the buckling profile and the slope of deflection to load determines the four parameters of A, D, N_r, and M_r, which govern the buckling and bending behavior of the multilayer beam. For bilayer beams, the Young's modulus and the residual stress of each layer can be uniquely determined from the microbridge test without any further assumptions.

The microbridge testing method is like a three-point bending test at the micrometer scale. The stress due to bending does not uniformly distribute along the thickness of a tested sample. Therefore, it is, in some way, subjective to define a yield point by the two percent deviation in the deflection. However, this definition is analogous to the definition of the offset yield strength in a tensile test, where a 0.2% plastic strain is used for a smooth stress-strain curve without any detectable yielding point. The yield strength of the Au films determined by the microbridge test is more or less the same as that reported in the literature, thereby demonstrating that the microbridge test can be utilized in the mechanical characterization of ductile thin films.

Under large deformation, the deflection is much larger than the beam thickness such that the axial stretching could dominate the deformation in the microbridge beam. On the basis of this fact, Espinosa et al. developed a tensile test on thin films by applying a lateral load [10]. The microbridge test together with the bending test of cantilever beams will allow researchers to investigate the strain gradient plasticity behavior [30].

Acknowledgment This work was supported by an RGC grant from the Hong Kong Research Grants Council, Hong Kong Special Administrative Region, China. The author thanks Professor MH Zhao, Professor CF Qian, Professor JR Li, and Dr. J Wang for their help in the theoretical analysis and in the calculation of the spring compliances. The author is also grateful to Professor YJ Su, Dr. LQ Chen, Dr. WH Xu, Dr. XS Wang, and Dr. B Huang for their experimental work. The experiments were conducted at the Microelectronics Fabrication Facility and the Design and Manufacturing Services Facility, HKUST.

Appendix 1: The Neutral Plane of Bending

The coordinate system is first set up such that the x'-axis lies in the free surface of the first layer and along the length direction, the y'-axis is along the width direction, and the z'-axis is along the thickness direction of the trilayer beam. We use h_k, z'_k, and z'_{k-1} to denote the thickness and the coordinates of two surfaces of the kth layer $(k = 1, 2, \ldots, n)$ and E_k and $\sigma_{\mathrm{r}k}$ to stand for the Young's modulus and the residual stresses of the kth layer, respectively. In this coordinate system, we have

$$
\begin{aligned}
z'_0 &= 0, \\
z'_1 &= z'_0 + h_1 = h_1, \\
z'_2 &= z'_1 + h_2 = h_1 + h_2, \\
&\cdots\cdots \\
z'_n &= z'_{n-1} + h_n = \sum_{i=1}^{n} h_i.
\end{aligned}
\tag{10.68}
$$

Figure 10.2 shows a new coordinate system in which the x-axis lies in the neutral plane of bending and along the length direction of the laminated beam, and the z-axis is along the thickness direction of the thin film. The new coordinate system is linked to the old coordinate system by transiting the x-axis from the free surface of the first layer to the neutral plane of bending, such that $z = z'-d$, where d is the distance from the neutral plane of bending to the free surface of the first layer, as shown in Fig. 10.2. The neutral plane of bending is determined by the condition

$$
\int_t \sigma_k \mathrm{d}z = 0,
\tag{10.69}
$$

where t denotes the total thickness of the n-layer film and σ_k is the stress in the kth layer induced by bending only, which is given by,

$$\sigma_k = E_k \frac{z}{R}, \tag{10.70}$$

with R being the bending curvature radius of the neutral plane of bending. Using $z = z'-d$ and (10.70), we rewrite (10.69) as

$$\int_t E_k \frac{z'-d}{R} \mathrm{d}z = 0. \tag{10.71}$$

From (10.71), we determine the value of d, as indicated in (10.1).

Appendix 2: The Spring Compliances

2.1 Dimensional Analysis of the Spring Compliances

The six compliance coefficients, S_{ij} $(i, j = M, N,$ and $P)$, are functions of the parameters, namely Young's modulus of substrate (E_s), Young's modulus of film (E_f), Poisson's ratio of substrate (v_s), Poisson's ratio of film (v_f), film thickness (t), film width (W_b), the support angle (α), and the gap between two adjacent bridges (W_0). Mathematically, we have

$$S_{ij} = f_{ij}(E_s, E_f, v_s, v_f, t, W_b, W_0, \alpha). \tag{10.72}$$

Among the eight parameters, E_s, E_f, v_s, v_f, t, W_b, α, and W_0, we may choose E_s and t to be the independent dimensions. Thus, the dimensions of E_f, v_s, v_f, t, W_b, α, W_0 and S_{ij} are given by

$$
\begin{aligned}
[E_f] &= [E_s], \\
[v_s] &= [E_s]^0 [t]^0, \\
[v_f] &= [E_s]^0 [t]^0, \\
[\alpha] &= [E_s]^0 [t]^0, \\
[W_b] &= [t]^1, \\
[W_0] &= [t]^1, \\
[S_{MM}] &= [E_s]^{-1} [t]^{-2}, \\
[S_{MN}] &= [E_s]^{-1} [t]^{-1}, \\
[S_{MP}] &= [E_s]^{-1} [t]^{-1}, \\
[S_{NN}] &= [E_s]^{-1} [t]^0, \\
[S_{NP}] &= [E_s]^{-1} [t]^0, \\
[S_{PP}] &= [E_s]^{-1} [t]^0.
\end{aligned} \tag{10.73}
$$

Using the dimensional analysis, we have

$$S_{MM} = \frac{1}{E_s t^2} f_{MM} \left(\frac{E_f}{E_s}, \nu_s, \nu_f, \frac{t}{W_b}, \frac{W_0}{W_b}, \alpha \right) ,$$

$$S_{MN} = \frac{1}{E_s t} f_{MN} \left(\frac{E_f}{E_s}, \nu_s, \nu_f, \frac{t}{W_b}, \frac{W_0}{W_b}, \alpha \right) ,$$

$$S_{MP} = \frac{1}{E_s t} f_{MP} \left(\frac{E_f}{E_s}, \nu_s, \nu_f, \frac{t}{W_b}, \frac{W_0}{W_b}, \alpha \right) ,$$

$$S_{NN} = \frac{1}{E_s} f_{NN} \left(\frac{E_f}{E_s}, \nu_s, \nu_f, \frac{t}{W_b}, \frac{W_0}{W_b}, \alpha \right) , \qquad (10.74)$$

$$S_{NP} = \frac{1}{E_s} f_{NP} \left(\frac{E_f}{E_s}, \nu_s, \nu_f, \frac{t}{W_b}, \frac{W_0}{W_b}, \alpha \right) ,$$

$$S_{PP} = \frac{1}{E_s} f_{PP} \left(\frac{E_f}{E_s}, \nu_s, \nu_f, \frac{t}{W_b}, \frac{W_0}{W_b}, \alpha \right) .$$

2.2 Simplified Expression of the Spring Compliances

In addition to Young's modulus of the substrate and the film thickness, (10.74) shows that the spring compliances of S_{ij} are functions of E_f/E_s, ν_s, ν_f, t/W_b, W_0/W_b, and α. With the commercial software ABAQUS/Standard V6.4, Wang et al. [37] conducted finite element analysis to examine the above factors.

First, Wang et al. examined the influence of W_0/W_b on the spring compliances [37]. In FEA models, they [37] changed the ratio of W_0/W_b without changing other factors and calculated the spring compliances. The used materials data in the FEA calculations were $E_f = E_s = 159\,GPa$, $\nu_f = \nu_s = 0.25$, $t = 0.8\,\mu m$, $W_b = 10\,\mu m$, and $\alpha = 54.74°$, while the ratio of W_0/W_b varied from 1 to 20. The results show that W_0/W_b has little influence on S_{ij}. Especially, when $W_0/W_b \geqslant 4$, the bridge gap is large enough to approximately neglect the interaction between bridges.

Second, Wang et al. examined the influence of Poisson's ratio of materials on the spring compliances by calculating S_{ij} with different values of Poisson's ratios [37]. In the FEA calculations, the support angle was 54.74°, $W_0/W_b = 16$, $t = 0.8\,\mu m$, and $E_f = E_s = 159\,GPa$, whereas Poisson's ratio of the film material was the same as that of the substrate material and varies from 0 to 0.5. The results show that Poisson's ratio has little influence on S_{ij}. Especially, when Poisson's ratio is in the range of 0.1–0.4, the spring compliances are almost independent of the Poisson's ratio.

Then, Wang et al. examined the influence of the film thickness (t), i.e., the influence of the ratio of the film thickness to the film width (t/W_b), on the spring compliances [37]. In the FEA calculations, the support angle was 54.74°, $W_0/W_b = 16$, $E_f = E_s = 159\,GPa$, and $\nu_f = \nu_s = 0.25$. For a fixed ratio of t/W_b, the results show that S_{MN} and S_{MP} are inversely proportional to t, S_{MM} are inversely proportional to t^2, while S_{NN}, S_{NP}, and S_{PP} remain nearly constant, thereby being consistent with the above dimensional analysis. For a given film thickness, however, the FEA results show that S_{MN}, S_{MP}, and S_{MM} are independent from the ratio of t/W_b, but

S_{NN}, S_{NP}, and S_{PP} vary with the ratio of t/W_b. In practice, Poisson's ratio of most materials is among the range of 0.1–0.4. The normalized gap between two adjacent bridges, W_0/W_b, is designed to be larger than 4. From the FEA results, Wang et al. [37] further simplified the expressions of the compliance coefficients as,

$$S_{MM} = \frac{1}{E_s t^2} f_{MM} \left(\frac{E_f}{E_s}, \alpha \right),$$

$$S_{MN} = \frac{1}{E_s t} f_{MN} \left(\frac{E_f}{E_s}, \alpha \right),$$

$$S_{MP} = \frac{1}{E_s t} f_{MP} \left(\frac{E_f}{E_s}, \alpha \right),$$

$$S_{NN} = \frac{1}{E_s} f_{NN} \left(\frac{E_f}{E_s}, \frac{t}{W_b}, \alpha \right),$$

$$S_{NP} = \frac{1}{E_s} f_{NP} \left(\frac{E_f}{E_s}, \frac{t}{W_b}, \alpha \right),$$

$$S_{PP} = \frac{1}{E_s} f_{PP} \left(\frac{E_f}{E_s}, \frac{t}{W_b}, \alpha \right).$$

(10.75)

2.3 The Final Expression of the Spring Compliances for the Base Support Angle of 54.74°

As mentioned in the text, the support angle of 54.74° is due to anisotropic etching of (100) Si wafers. Wang et al. numerically calculated the spring compliances for the support angle, $\alpha = 54.74°$ [37]. The Reuss average value of 159 GPa of the silicon Young's modulus was used in the FEA calculations for the substrate. Using $W_0/W_b = 16$, $W_b = 10\,\mu m$ and $v_f = v_s = 0.25$, Wang et al. calculated the spring compliances for different Young's modulus of the film from 19.875 GPa to 1,276 GPa and various film thickness from 0.2 to 12.8 μm. Fitting the FEA results as functions of E_f/E_s and t/W_b, Wang et al. had the spring compliances [37]:

$$S_{MM} = \frac{7.48}{E_s t^2} \left(\frac{E_f}{E_s} \right)^{-0.692},$$

$$S_{MN} = \frac{2.47}{E_s t} \left(\frac{E_f}{E_s} \right)^{-0.835},$$

$$S_{MP} = \frac{4.85}{E_s t} \left(\frac{E_f}{E_s} \right)^{-0.298},$$

$$S_{NN} = \frac{3.19}{E_s} \left(\frac{t}{W_b} \right)^{-0.318} \left(\frac{E_f}{E_s} \right)^{-0.645(t/W_b)^{0.149}},$$

$$S_{NP} = \frac{3.04}{E_s} \left(\frac{t}{W_b} \right)^{-0.371} \left(\frac{E_f}{E_s} \right)^{-0.502-0.0557\ln(t/W_b)},$$

$$S_{PP} = \frac{7.41}{E_s} \left(\frac{t}{W_b} \right)^{-0.367} \left(\frac{E_f}{E_s} \right)^{-0.367(t/W_b)^{0.182}}.$$

(10.76)

In the work, thin films were deposited on (100) silicon wafers. The Reuss average value of 159 GPa of the silicon Young's modulus was used in the fitting as the substrate Young's modulus.

Appendix 3: Mechanics Analysis of the Microcantilever Beam-Bending Test

Figure 10.43a illustrates the mechanical analysis of a cantilever beam subjected to a load P (per unit width) and the associated coordinate system. The governing equation for the beam bending is

$$D\frac{d^2w}{dx^2} = P(L-x),$$ (10.77)

and its general solution has the form of

$$Dw = \frac{P}{6}(L-x)^3 + Ax + B,$$ (10.78)

where A and B are constants to be determined by the boundary conditions. The moment, M_0, the displacement, w_0, and the rotation, θ_0, at the supported end, i.e., at $x = 0$, are, respectively, given by

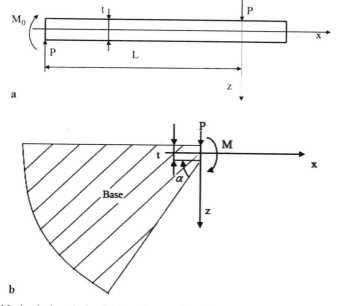

Fig. 10.43 Mechanical analysis of (**a**) a microcantilever beam under a lateral load, P, and (**b**) its substrate

$$M_0 = -PL, \tag{10.79}$$

$$Dw_0 = \frac{P}{6}L^3 + B, \tag{10.80}$$

$$D\theta_0 = -\frac{P}{2}L^2 + A. \tag{10.81}$$

Figure 10.43b shows the mechanical analysis of the substrate or the base. For simplicity, the substrate includes the top part of the deposited thin film, and this part of the thin film is assumed to have the same mechanical properties as the substrate. The substrate deformation is modeled by two coupled springs subjected to a moment, M, and the same load, P. The substrate deformation is described by a local rotation, θ, and a local displacement, δ. The constitutive equation of the two coupled springs is proposed to be

$$\begin{bmatrix} \delta \\ \theta \end{bmatrix} = \begin{bmatrix} S_{PP} & S_{PM} \\ S_{MP} & S_{MM} \end{bmatrix} \begin{bmatrix} P \\ M \end{bmatrix}. \tag{10.82}$$

Clearly, (10.82) can be reduced from (10.26). Zhang et al. conducted finite element analysis (FEA) with the commercial software ABAQUS to calculate the compliance tensor for the silicon substrates [46]. For simplicity, silicon substrates were assumed to be mechanically isotropic with a Reuss Young's modulus of 159 GPa and a Poisson's ratio of 0.25 in the calculations. The compliance values for three supported angles were listed in the previous publication on the basis of the FEA results [46].

At the connecting point of the film and substrate, the displacement continuity and Newton's third law require

$$\begin{aligned} \delta &= w_0, \\ \theta &= \theta_0, \\ M &= -M_0. \end{aligned} \tag{10.83}$$

Using (10.83), we determine the constants A and B as

$$A = \left[\frac{L^2}{2} + D(S_{PM} + S_{MM}L) \right] P, \tag{10.84}$$

$$B = \left[-\frac{L^3}{6} + D(S_{PP} + S_{PM}L) \right] P. \tag{10.85}$$

Then, (10.78) is rewritten as

$$w = \left(\frac{1}{3D}L^3 + S_{MM}L^2 + 2S_{PM}L + S_{PP} \right) \cdot P. \tag{10.86}$$

Equation (10.86) shows that the deflection at the loading point is proportional to the load, and the bending compliance is

$$C = \frac{1}{3D}L^3 + S_{MM}L^2 + 2S_{PM}L + S_{PP} = \frac{1}{3D}L^3 + S_{MM}L^2 \left(1 + \frac{2S_{PM}}{S_{MM}L} + \frac{S_{PP}}{S_{MM}L^2} \right). \tag{10.87}$$

The bending compliance is a third-order polynomial of the length from the loading point to the supported beam end. The first term is the same as that for the clamped boundary conditions, representing the beam compliance. The second, third, and fourth terms are attributed to the substrate deformation. The second is due to the local rotation; the third is due to the coupling between local rotation and displacement; and the fourth is due to the local displacement. The analytic results were verified by the FEA [46]. If the value of the last two terms in the parentheses of (10.87) is much smaller than unity, (10.87) reduces to

$$C = \frac{1}{3D}L^3 + S_{MM}L^2,$$
(10.88)

with sufficient accuracy. Equation (10.88) is identical to the equation proposed by Baker and Nix [2]. If adding an extra length, ΔL, to the original length, we have

$$C = \frac{1}{3D}(L + \Delta L)^3 = \frac{L^3}{3D}\left[1 + 3\frac{\Delta L}{L} + 3\left(\frac{\Delta L}{L}\right)^2 + \left(\frac{\Delta L}{L}\right)^3\right].$$
(10.89)

When $(\Delta L/L) \ll 1$, we may ignore the high order terms of $\Delta L/L$. Thus, the extra length is estimated as

$$\Delta L = D S_{MM}.$$
(10.90)

Note that the extra length depends not only on the substrate properties but also on the beam thickness and Young's modulus of the beam.

Appendix 4: Solutions with the Simply Supported and Clamped Boundary Conditions for $N_x > 0$ and $M_{(r)} \neq 0$

The clamped boundary conditions give

$$w|_{x=\pm l/2} = 0,$$
(10.91)

$$\left.\frac{\partial w}{\partial x}\right|_{x=\pm l/2} = 0,$$
(10.92)

$$u|_{x=\pm l/2} = 0.$$
(10.93)

Comparing (10.91)–(10.93) with (10.31b)–(10.31d) indicates that the clamped boundary conditions treat the substrate to be nondeformable with zero values of all spring compliances S_{ij}. Then, it is a straightforward matter to obtain expressions of the deflection at the beam center for the clamped boundary conditions from (10.31). Letting all spring compliances be zero leads to

$$w = -\frac{Q\tanh(kl/2)}{2N_x k} + \frac{Ql}{4N_x} + \frac{\bar{M}_0}{N_x}\left[1 - \frac{1}{\cosh(kl/2)}\right],$$
(10.94)

where $\overline{M}_0 = M_0 - M^{(r)}$ and N_x are coupled in the following equations:

$$\overline{M}_0 = -\frac{Q[\cosh(kl/2) - 1]}{2k\sinh(kl/2)},\tag{10.95}$$

$$\frac{l(N_x - N^{(r)})}{2A} = \frac{1}{32kN_x^2\cosh^2(kl/2)}\left[4k^2\overline{M}_0^2\sinh(kl) - 16k\overline{M}_0Q\cosh(kl/2)\right.$$

$$+ kQ(8\overline{M}_0 + lQ)\cosh(kl) - 4k^3l\overline{M}_0^2$$

$$- 3Q^2\sinh(kl) - 4k^2l\overline{M}_0Q\sinh(kl/2) + 2klQ^2$$

$$\left. + 8k\overline{M}_0Q\right].\tag{10.96}$$

One assumption in the simply supported boundary conditions is that there is no constraint to the slope of the deflection, namely to the rotation angle, at the boundary. Therefore, the rotation angle equation, i.e., (10.28c) is surrendered if the simply supported boundary conditions are applied. When there is a residual moment, the simply supported boundary conditions require

$$w\big|_{x=\pm l/2} = 0,\tag{10.97}$$

$$M^{(r)} - D\frac{\partial^2 w}{\partial x^2}\bigg|_{x=\pm l/2} = 0.\tag{10.98}$$

$$u\big|_{x=\pm l/2} = 0.\tag{10.99}$$

Equation (10.98) means $M_0 = 0$ and $\overline{M}_0 = -M^{(r)}$. Then, considering the symmetry of the spring compliances indicates that letting all spring compliances and M_0 be zero reduces (10.31) to that for simply supported ends, which gives

$$w = -\frac{Q\tanh(kl/2)}{2N_x k} + \frac{Ql}{4N_x} - \frac{M_r}{N_x}\left[1 - \frac{1}{\cosh(kl/2)}\right],\tag{10.100}$$

$$\frac{l(N_x - N_r)}{2A} = \frac{1}{32kN_x^2\cosh^2(kl/2)}\left[-4k^3lM_r^2 + 2klQ^2\right.$$

$$4k^2\,M_r^2\sinh(kl) + 16k\,M_rQ\cosh(kl/2)$$

$$- 3Q^2\sinh(kl) + kQ(-8M_r + lQ)\cosh(kl)$$

$$\left. + 4k^2lM_rQ\sinh(kl/2) - 8k\,M_rQ\right].\tag{10.101}$$

As described in the text, only a residual moment can bend the microbridge beam if the deformable boundary conditions are used. If the base is nondeformable, i.e., letting all spring compliances be zero in (10.32), we shall have $N_x = N_r$, $M_0 = M_r$ and $w = 0$. Thereby this indicates that only a residual moment cannot bend the microbridge beam if the clamped boundary conditions are used. For clamped and simply supported boundary conditions, we may assume that small deformation is satisfied if there is no stretch in the neutral plane of bending, and thus $N_x = N_r$. With this assumption, the deflection is independent of the residual moment under the clamped boundary conditions and is given by

$$w = -\frac{Q\tanh(kl/2)}{2N_r k} + \frac{Ql}{4N_r} - \frac{Q[\cosh(kl/2)-1]}{2kN_r\sinh(kl/2)}\left[1 - \frac{1}{\cosh(kl/2)}\right], \quad (10.102)$$

For the simply supported boundary ends, there is no constraint to the rotation angle at the boundary and no moment is applied by the base to the microbridge beam. In this case, the deflection under small deformation is

$$w = -\frac{Q\tanh(kl/2)}{2N_r k} + \frac{Ql}{4N_r} - \frac{M_r}{N_r}\left[1 - \frac{1}{\cosh(kl/2)}\right]. \quad (10.103)$$

If the residual resultant force, N_r, is zero, the k parameter will be zero. Then, using

$$\sinh(x) \approx x + \frac{x^3}{6}, \quad \cosh(x) \approx 1 + \frac{x^2}{2}, \quad \tanh(x) \approx x - \frac{x^3}{3} \quad \text{for} \quad x \to 0,$$

we reduce (10.102) and (10.103), respectively, to

$$w = \frac{Ql^3}{192D} \quad \text{for clamped ends}, \quad (10.104)$$

$$w = \frac{Ql^3}{48D} - \frac{M_r l^2}{8D} \quad \text{for simply supported ends}. \quad (10.105)$$

Appendix 5: Solutions with the Simply Supported and Clamped Boundary Conditions for $N_x > 0$ and $M_r = 0$

Without any residual moment, we have $\overline{M_0} = M_0$. The solutions for the clamped and simply supported boundary conditions can be easily obtained from (10.94–10.96) and (10.101). For the clamped ends, the solutions are

$$w = -\frac{Q\tanh(kl/2)}{2N_x k} + \frac{Ql}{4N_x} + \frac{M_0}{N_x}\left[1 - \frac{1}{\cosh(kl/2)}\right], \quad (10.106)$$

where M_0 and N_x are coupled in the following equations:

$$M_0 = -\frac{Q[\cosh(kl/2)-1]}{2k\sinh(kl/2)}, \quad (10.107)$$

$$\frac{l(N_x - N^{(r)})}{2A} = \frac{1}{32kN_x^2\cosh^2(kl/2)}\left[-4k^3 l M_0^2 + 8kM_0Q^2\right.$$

$$+4k^2 M_0^2\sinh(kl) - 4k^2 l M_0 Q\sinh(kl/2)$$

$$-3Q^2\sinh(kl) - 16kM_0Q\cosh(kl/2) + 2klQ$$

$$\left. +kQ(8M_0 + lQ)\cosh(kl)\right]. \quad (10.108)$$

For the simply supported ends, the solution takes the form of

$$w = -\frac{Q\tanh(kl/2)}{2N_x k} + \frac{Ql}{4N_x}, \tag{10.109}$$

and N_x is calculated from

$$\frac{l(N_x - N_r)}{2A} = \frac{2kl + kl\cosh(kl) - 3\sinh(kl)}{32kN_x^2\cosh^2(kl/2)}Q^2. \tag{10.110}$$

For the clamped ends, the solutions for small deformation are the same as those of (10.102) with a finite nonzero value of N_r and (10.104) with a zero N_r. For the simply supported ends, however, the solutions are further simplified by letting $M_r = 0$, which are

$$w = -\frac{Q\tanh(kl/2)}{2N_r k} + \frac{Ql}{4N_r} \quad \text{for } N_r \neq 0 \tag{10.111}$$

$$w = \frac{Ql^3}{48D} \quad \text{for } N_r = 0. \tag{10.112}$$

Appendix 6: Solutions with the Simply Supported and Clamped Boundary Conditions for $N_x < 0$ and $M_{(r)} \neq 0$

From the same arguments described in Appendix 4, letting all spring compliances in (10.46) be zero leads to expressions of the deflection at the beam center for the clamped boundary conditions. They are

$$w = -\frac{Q}{2N_x k}\tan(kl/2) + \frac{Ql}{4N_x} - \frac{\overline{M}_0}{N_x}\left[\frac{1}{\cos(kl/2)} - 1\right], \tag{10.113}$$

$$\frac{Q}{2N_x}\left[1 - \frac{1}{\cos(kl/2)}\right] = \frac{\overline{M}_0 k}{N_x}\tan(kl/2), \tag{10.114}$$

$$\frac{l(N_x - N_r)}{2A} = \frac{1}{8k}\left[\begin{array}{c} -\dfrac{4Q\overline{M}_0 k}{N_x^2\cos(kl/2)} + \dfrac{Q^2 kl}{4N_x^2\cos^2(kl/2)} \\[2ex] +\dfrac{\overline{M}_0^2 k^3 l}{N_x^2\cos^2(kl/2)} + \dfrac{2Q^2 kl}{4N_x^2} + \dfrac{4Q\overline{M}_0 k}{N_x^2} \\[2ex] -\dfrac{2Q^2\sin(kl/2)}{N_x^2\cos(kl/2)} + \dfrac{Q\overline{M}_0 k^2 l\sin(kl/2)}{N_x^2\cos^2(kl/2)} \\[2ex] +\dfrac{Q^2\sin(kl)}{4N_x^2\cos^2(kl/2)} - \dfrac{\overline{M}_0^2 k^2\sin(kl)}{N_x^2\cos^2(kl/2)} \end{array}\right] \tag{10.115}$$

The deflection at the beam center for the simply supported boundary conditions is obtained by letting all spring compliances and M_0 in (10.46) be zero. The solution is

$$w = -\frac{Q}{2N_xk}\tan(kl/2) + \frac{Ql}{4N_x} + \frac{M_\mathrm{r}}{N_x}\left[\frac{1}{\cos(kl/2)} - 1\right], \tag{10.116}$$

$$\frac{l(N_x - N_\mathrm{r})}{2A} = \frac{1}{8k}\left[\begin{array}{c} \dfrac{4QM_\mathrm{r}k}{N_x^2\cos(kl/2)} + \dfrac{Q^2kl}{4N_x^2\cos^2(kl/2)} \\[3mm] + \dfrac{M_\mathrm{r}^2k^3l}{N_x^2\cos^2(kl/2)} + \dfrac{2Q^2kl}{4N_x^2} - \dfrac{4QM_\mathrm{r}k}{N_x^2} \\[3mm] - \dfrac{2Q^2\sin(kl/2)}{N_x^2\cos(kl/2)} - \dfrac{QM_\mathrm{r}k^2l\sin(kl/2)}{N_x^2\cos^2(kl/2)} \\[3mm] + \dfrac{Q^2\sin(kl)}{4N_x^2\cos^2(kl/2)} - \dfrac{M_\mathrm{r}^2k^2\sin(kl)}{N_x^2\cos^2(kl/2)} \end{array}\right] \tag{10.117}$$

Without any applied loads, as described in Sect. 10.5.1, a residual moment and a compressive resultant force may jointly bend the microbridge beam and reduce the magnitude of the resultant force. Buckling is of large deformation. In general, small deformation is based on the assumption that there is no stretch in the neutral plane of bending, which gives $N_x = N_\mathrm{r}$. In the discussion of small deformation, we will not consider buckling. Ignoring (10.115) and replacing N_x by N_r in (10.113) and (10.114) give the deflection under small deformation for the clamped boundary conditions,

$$\begin{aligned} w = {} & \frac{Ql}{4N_\mathrm{r}} - \frac{Q\tan(kl/2)}{2N_\mathrm{r}k} - \frac{Q}{2N_\mathrm{r}k\tan(kl/2)} \\ & \times \left[1 - \frac{1}{\cos(kl/2)}\right]\left[\frac{1}{\cos(kl/2)} - 1\right]. \end{aligned} \tag{10.118}$$

Similarly, the small deformation solution for the clamped boundary conditions is

$$w = \frac{Ql}{4N_\mathrm{r}} - \frac{Q}{2N_\mathrm{r}k}\tan(kl/2) + \frac{M_\mathrm{r}}{N_\mathrm{r}}\left[\frac{1}{\cos(kl/2)} - 1\right]. \tag{10.119}$$

When there is no residual resultant force, N_r and the k parameter are zero. Then, using

$$\sin(x) \approx x - \frac{x^3}{6}, \quad \cos(x) \approx 1 - \frac{x^2}{2}, \quad \tan(x) \approx x + \frac{x^3}{3} \quad \text{for } x \to 0,$$

we reduce (10.118) and (10.119), respectively, to

$$w = \frac{Ql^3}{192D} \quad \text{for clamped ends,} \tag{10.120}$$

$$w = \frac{Ql^3}{48D} - \frac{M_r l^2}{8D} \quad \text{for simply supported ends.} \tag{10.121}$$

As expected, (10.120) and (10.121) are identical to (10.104) and (10.105), respectively.

Appendix 7: Solutions with the Simply Supported and Clamped Boundary Conditions for $N_x < 0$ and $M_{(r)} = 0$

Without any residual moment, we have $\overline{M_0} = M_0$. The solutions for the clamped and simply supported boundary conditions can be easily obtained from (10.113)–(10.115) and (10.116)–(10.117). For the clamped ends, the solutions are

$$w = -\frac{Q}{2N_x k} \tan(kl/2) + \frac{Ql}{4N_x} - \frac{M_0}{N_x} \left[\frac{1}{\cos(kl/2)} - 1 \right], \tag{10.122}$$

$$\frac{Q}{2N_x} \left[1 - \frac{1}{\cos(kl/2)} \right] = \frac{M_0 k}{N_x} \tan(kl/2), \tag{10.123}$$

$$\frac{l(N_x - N_r)}{2A} = \frac{1}{8k} \left[\begin{array}{l} -\dfrac{4QM_0 k}{N_x^2 \cos(kl/2)} + \dfrac{Q^2 kl}{4N_x^2 \cos^2(kl/2)} \\[2mm] +\dfrac{M_0^2 k^3 l}{N_x^2 \cos^2(kl/2)} + \dfrac{2Q^2 kl}{4N_x^2} + \dfrac{4QM_0 k}{N_x^2} \\[2mm] -\dfrac{2Q^2 \sin(kl/2)}{N_x^2 \cos(kl/2)} + \dfrac{QM_0 k^2 l \sin(kl/2)}{N_x^2 \cos^2(kl/2)} \\[2mm] +\dfrac{Q^2 \sin(kl)}{4N_x^2 \cos^2(kl/2)} - \dfrac{M_0^2 k^2 \sin(kl)}{N_x^2 \cos^2(kl/2)} \end{array} \right] \tag{10.124}$$

For the clamped ends, the solution is

$$w = -\frac{Q}{2N_x k} \tan(kl/2) + \frac{Ql}{4N_x}, \tag{10.125}$$

$$\frac{l(N_x - N_r)}{2A} = \frac{Q^2}{8kN_x^2} \left[\frac{kl + \sin(kl)}{4\cos^2(kl/2)} + \frac{2kl}{4} - \frac{2\sin(kl/2)}{\cos(kl/2)} \right]. \tag{10.126}$$

For the clamped ends, the solutions for small deformation are the same as those of (10.118) with a finite nonzero value of N_r and (10.120) with a zero N_r. For the simply supported ends, however, the solutions are further simplified by letting $M_r = 0$, which are

$$w = \frac{Ql}{4N_r} - \frac{Q}{2N_r k} \tan(kl/2) \quad \text{for } N_r \neq 0 \tag{10.127}$$

$$w = \frac{Ql^3}{48D} \quad \text{for } N_r = 0. \tag{10.128}$$

References

1. Alshits VI, Kirchner HOK (1995) Philos. Mag. A 72:1431
2. Baker SP, Nix WD (1994) J. Mater. Res. 9:3131
3. Brotzen FR (1994) Int. Mater. Rev. 39:24
4. Bunge HJ (1982) Texture Analysis in Materials Science. Butterworths, London
5. Cao ZQ, Zhang TY, Zhang X (2005) J. Appl. Phys. 97:1049091
6. Cardinale GF, Tustison RW (1992) Thin Solid Films 207:126
7. Courtney TH (2000) Mechanical Behavior of Materials. McGraw-Hill, Singapore
8. De Boer MP, Gerberich WW (1996) Acta Mater. 44:3169
9. Denhoff MW (2003) J. Micromech. Microeng. 13:686
10. Espinosa HD, Prorok BC, Fischer M (2003) J. Mech. Phys. Solids 51:47
11. Fang W, Wickert JA (1994) J. Micromech. Microeng. 4:116
12. Fang W, Wickert JA (1996) J. Micromech. Microeng. 6:301
13. Finot M, Suresh S (1996) J. Mech. Phys. Solids 44:683
14. Guckel H, Randazzo T, Burns DW (1985) J. Appl. Phys. 57:1671
15. Han SM, Saha R, Nix WD (2006) Acta Mater. 54:1571
16. Hardwick DA (1987) Thin Solid Films 154:109
17. Hill R (1952) Proc. Phys. Soc. A 65:349
18. Hong S, Weihs TP, Bravman JC, Nix WD (1990) J. Electron. Mater. 19:903
19. Huang B, Zhang TY (2006) J. Micromech. Microeng. 16:134
20. Diffraction pattern of Au (JCPDS 4-784), International Center for Diffraction Data (1998)
21. Klein CA, Miller RP (2000) J. Appl. Phys. 87:2265
22. Kobrinsky MJ, Deutsch ER, Senturia SD (2000) IEEE J. Microelectromech. Syst. 9:361
23. Nix WD (1989) Metall. Trans. A 20:2217
24. Pharr GM, Oliver WC (1992) MRS Bull. 17:28
25. Petersen KE (1978) IEEE Trans. Electron Devices 25:1241
26. Petersen KE, Guarnieri CR (1979) J. Appl. Phys. 50:6761
27. Reyntjens S, Puers R (2000) J. Micromech. Microeng. 10:181
28. Robic JY, Leplan H, Payleau Y, Rafin B (1996) Thin Solid Films 291:34
29. Schweitz J (1992) MRS Bull. 17:34
30. Shi ZF, Huang B, Tan H, Huang Y, Zhang TY, Wu PD, Hwang KC, Gao HJ (2008) Int. J. Plasticity 24:1606
31. Su YJ, Qian CF, Zhao MH, Zhang TY (2000) Acta Mater. 48:4901
32. Tabata GO, Kawahata K, Sugiyama S, Igarashi I (1989) Sens. Actuators 20:135
33. Timoshenko SP, Woinowsky-krieger S (1959) Theory of Plates and Shells, 2nd edn. McGraw-Hill, New York
34. Townsend PH, Barnett DM, Brunner TA (1987) J. Appl. Phys. 62:4438
35. Vlassak JJ, Nix WD (1992) J. Mater. Res. 7:3242
36. Wang XS, Wang J, Zhao MH, Zhang TY (2005) IEEE J. Microelectromech. Syst. 14(3):634
37. Wang XS, Li JR, Zhang TY (2006) J. Micromech. Microeng. 16:122
38. Wang XS, Zhang TY (2007) Metall. Mater. Trans. A 38:2273
39. Wong EW, Sheehan PE, Lieber CM (1997) Science 277(5334):1971
40. Xu WH, Zhang TY (2003) Appl. Phys. Lett. 83:1731
41. Yang FQ, Li JCM (2001) Langmuir 17:6524
42. Yoshioka T, Ando T, Shikida M, Sato K (2000) Sens. Actuators A 82:291
43. Zhang TY, Zhang X, Zohar Y (1998) J. Micromech. Microeng. 8:243
44. Zhang TY, Chen LQ, Fu R (1999) Acta Mater. 47:3869
45. Zhang TY, Su YJ, Qian CF, Zhao MH, Chen LQ (2000) Acta Mater. 48:2843
46. Zhang TY, Zhao MH, Qian CF (2000) J. Mater. Res. 15:1868
47. Zhang TY, Wang XS, Huang B (2005) Mater. Sci. Eng. A 409:329
48. Zywitzki O, Sahm H, Krug M, Morgner H, Neumann M (2000) Surf. Coat. Technol. 133–134:555

Chapter 11
Nanoscale Testing of One-Dimensional Nanostructures

Bei Peng, Yugang Sun, Yong Zhu, Hsien-Hau Wang, and Horacio Dante Espinosa

11.1 Introduction

The emergence of numerous nanoscale materials and structures such as nanowires (NWs), nanorods, nanotubes, and nanobelts of various materials in the past decade has prompted a need for methods to characterize their unique mechanical properties. These one-dimensional (1D) nanostructures possess superior mechanical properties [1, 2]; hence, applications of these structures ranging from nanoelectromechanical systems (NEMS) [3] to nanocomposites [4] are envisioned.

Two overarching questions have spurred the development of nanomechanical testing techniques and the modeling of materials behavior at the nanoscale: how superior is material behavior at the nanoscale as compared to its bulk counterpart, and what are the underlying mechanisms that dictate this? Due to the limited number of atoms present in these nanostructures, they provide an excellent opportunity to couple experimentation and atomistic modeling on a one-to-one basis. This approach has the potential to greatly advance our understanding of material deformation and failure, as well as to validate the various assumptions employed in multiscale models proposed in the literature. A large number of atomistic simulations have been performed to predict nanostructure properties and reveal their deformation mechanisms [2]. However, due to the minute scale of these nanostructures, it has proven quite challenging to conduct well-instrumented mechanical testing and validate computational predictions.

B. Peng, and H.D. Espinosa
Mechanical Engineering, Northwestern University, 2145 Sheridan Rd., Evanston, IL 60208-3111
bpeng@northwestern.edu, espinosa@northwestern.edu

Y.G. Sun and H.-H. Wang
Center for Nanoscale Materials, Argonne National Laboratory, 9700 South Cass Ave.,
Argonne, IL 60439
ygsun@anl.gov, hau.wang@anl.gov

Y. Zhu
Mechanical and Aerospace Engineering, North Carolina State University, Raleigh, NC 27695
yong_zhu@ncsu.edu

F. Yang and J.C.M. Li (eds.), *Micro and Nano Mechanical Testing of Materials and Devices*,
doi: 10.1007/978-0-387-78701-5, © Springer Science+Business Media, LLC, 2008

Earlier nanomechanical testing techniques include thermally- or electrically induced vibration of cantilevered nanostructures inside a transmission electron microscope (TEM) [5, 6]; lateral bending of suspended nanostructures using an atomic force microscope (AFM) [7, 8]; radial compression, by means of atomic force microscopy (AFM) probes, or nanoindentation of nanostructures on substrates [9, 10]; and tensile testing of freestanding nanostructures between two AFM cantilevers within a scanning electron microscope (SEM) [11]. Despite the exciting progress achieved by these methods, they are in general not well controlled in terms of loading, boundary conditions, and force–displacement measurements. In some instances, the nanostructure properties must be inferred from assumed models of the experimental setup.

Recent advances in mechanical characterization of thin films have been remarkable and provide good insight for the testing of nanostructures. Among numerous techniques, two categories are particularly fascinating: in situ testing and on-chip testing. In situ testing provides a powerful means to obtain the deformation field and to observe the deformation mechanisms though real-time imaging, for example, by SEM. The SEM chamber is large enough to accommodate a microscale testing setup, and has been used for in situ tensile testing [12,13]. Another example of in situ testing involves an AFM to record the surface profile during a tensile test [14]. TEM is ideal for in situ testing since it provides direct evidence of the defects nucleation and reaction. Although most in situ TEM setups do not measure or control stresses in the specimen [15, 16], Haque and Saif recently incorporated a load sensor in the TEM [17, 18]. An on-chip testing system consists of micromachined elements, such as comb-drive actuators and force (load) sensors that can be integrated on a chip. One of the early attempts used electrostatic actuation and sensing for fatigue testing of silicon cantilever beams [19]. Osterberg and Senturia used electrostatic actuation for chip-level testing of cantilever beams, fixed–fixed beams, and clamped circular diaphragms to extract material properties [20]. Kahn et al. [21] used electrostatic actuators integrated with microfracture specimens to study fracture properties of polysilicon films. Owing to the capability of generating and measuring small-scale forces and displacements with high resolution, on-chip testing has the potential to impact the small-scale testing field profoundly. Note that these two concepts, in situ and on-chip, are related but different. On-chip testing can be performed in situ or ex situ. In situ testing does not necessarily utilize an on-chip device. But due to the small size of on-chip devices and integrated loading and force sensing capabilities, they conveniently facilitate in situ testing.

In this chapter, the nanomechanical characterization of 1D nanostructures is reviewed. In Sect. 11.2, we summarize the challenges for mechanical characterization of 1D nanostructures. In Sect. 11.3, an overview of the existing experimental techniques is presented. In Sect. 11.4, a newly developed nanoscale material testing system for characterizing nanostructures is described in detail. In Sect. 11.5, some experimental results [22, 23] are summarized with emphasis on the in situ electromechanical testing of nanostructures, which complements our nanomechanical testing.

11.2 Challenges for Mechanical Characterization of 1D Nanostructures

The proper measurement of loads and displacements applied to 1D nanostructures, such as NWs and CNTs, is extremely challenging because of the miniscule size. Earlier studies on mechanical properties of nanostructures focused on theoretical analyses and numerical simulations [2, 24]. Owing to advancement in various microscopic techniques, nanoscale experimental tools have been developed to exploit the capabilities of high-resolution microscopes. Major challenges in the experimental study of 1D nanostructures include (1) constructing appropriate tools to manipulate and position specimens; (2) applying and measuring forces with nano-Newton resolution; and (3) measuring local mechanical deformation with subnanometer resolution.

11.2.1 Manipulation and Positioning

The first important step in testing is to manipulate and position the nanostructures at the desired location with nanometer resolution. For tensile testing, this becomes even more challenging, as the specimens must be freestanding and clamped at both ends. Furthermore, they should be well-aligned with the tensile direction. In the following section, we review the available methods to mount the specimens meeting the requirements for the tensile tests.

11.2.1.1 Random Dispersion

After purification, a small aliquot of the nanostructure suspension is dropped onto the gap between two surfaces. By random distribution, some of the nanostructures are suspended across the gap with random orientation. After an appropriate nanostructure is identified, a technique called electron-beam-induced deposition (EBID) is then used to fix the two ends. This is done by depositing foreign materials [11], such as residual carbon in an SEM chamber or external precursors, such as trimethylcyclopentadienyl-platinum $(CH_3)_3CH_3C_5H_4Pt)$. Random dispersion is the simplest method for most of the mechanical testing experiments to date, but it is only modestly effective.

11.2.1.2 Nanomanipulation

AFM can be used to both image and manipulate carbon nanotubes. A "NanoManipulator" AFM system, comprising an advanced visual interface for manual control of the AFM tip and tactile presentation of the AFM data, has been developed [25].

In addition, Veeco Instruments (Woodbury, New York) developed the "NanoMan" system for high-resolution imaging, high-definition nanolithography, and direct nanoscale in-plane manipulation.

Electron microscopy provides the imaging capability for manipulation of nanostructures with nanometer resolution. Various nanomanipulators working under either SEM [11, 26] or TEM [6, 27] have been developed. These manipulators are generally composed of both a coarse micrometer-resolution translation stage and a fine nanometer-resolution translation stage, the latter being based on piezo-driven mechanisms. These manipulators typically have three linear degrees of freedom, and some even have rotational capabilities.

11.2.1.3 External Field Alignment

DC and AC/DC electric fields have been used for the alignment of nanowires [28], nanotubes [29], and bioparticles [30]. Microfabricated electrodes in close proximity are typically used to create an electric field in the gap between them. A droplet containing nanostructures in suspension is dispensed into the gap with a micropipette. The applied electric field aligns the nanostructures, due to the dielectrophoretic effect, which results in the bridging of the electrodes by a single nanostructure. Electric circuits may be used to ensure the manipulation of one single nanostructure by immediately switching off the field upon bridging (i.e., shorting of the electrodes) [30].

Huang et al. [31] demonstrated another method for aligning nanowires. A laminar flow was employed to achieve preferential orientation of nanowires on chemically patterned surfaces. Magnetic fields have also been used to align carbon nanotubes [32].

11.2.1.4 Directed Self-Assembly

Self-assembly is a method of constructing nanostructures by forming stable bonds between the organic or nonorganic molecules and substrates. Rao and colleagues [33] reported an approach in large scale assembly of carbon nanotubes with high throughput. Dip Pen Nanolithography (DPN) [34] was employed to functionalize the specific surface regions either with polar chemical groups such as amino $-NH_2/-NH_3^+$ or carboxyl $(-COOH/-COO^-)$, or with nonpolar groups such as methyl $(-CH_3)$. When the substrate with functionalized surfaces was dipped into a liquid suspension of carbon nanotubes, the nanotubes were attracted toward the polar regions and the tubes self-assembled to form predesigned structures, usually within 10 s, with a yield higher than 90%. This method is scalable to large arrays of nanotube devices by using high-throughput patterning methods such as photolithography, stamping, or massively parallel DPN.

11.2.1.5 Direct Growth

Rather than manipulating and aligning nanostructures postsynthesis, researchers also examined methods for controlled direct growth. Dai and co-workers [35, 36] reported several patterned growth approaches for CNTs. The idea is to pattern the catalyst in an arrayed fashion and control the growth of CNTs between specific catalytic sites. He et al. [37] recently succeeded in direct growth of Si nanowires between two preexisting single-crystal Si microelectrodes with <111> sidewalls. Si catalysts are deposited on the sidewalls of the electrodes and epitaxially grown perpendicularly to the <111> surfaces.

Direct growth is a very promising method to prepare specimens for nanomechanical characterization. It does not involve the nano-welding steps using EBID technique, which is advantageous because EBID brings foreign materials, e.g., carbon and platinum, onto the surface of nanostructures, which might cause some secondary effects on the property measurement. In addition, it is a rapid process, which does not require tedious manual nanomanipulation.

11.2.2 High Resolution Displacement and Force Measurements

SEM, TEM, and AFM have been widely used in characterizing nanostructures. These instruments provide effective ways of measuring dimensions and deformations with nanometer resolution. Electron microscopy uses high-energy electron beams for scattering (SEM) and diffraction (TEM). A field emission gun SEM has a resolution of about 1 nm and the TEM is capable of achieving a point-to-point resolution of 0.1–0.2 nm. The resolution of SEM is limited by the interaction volume between the electron beam and the sample surface. The resolution of TEM is limited by the spread in energy of the electron beam and the quality of the microscope optics.

Commercial force sensors usually cannot reach nano-Newton resolution. Therefore, AFM cantilevers have been effectively employed as force sensors [11, 38], provided that their spring constant has been accurately calibrated. Alternatively, microelectromechanical systems (MEMS) offer the capability to measure force with nano-Newton resolution. This MEMS-based methodology will be further discussed in Sect. 11.4.

11.3 Nanomechanical Testing of 1D Nanostructures

To date, the experimental techniques employed in the mechanical testing of 1D nanostructures can be roughly grouped into three major categories: dynamic vibration, bending, and tensile testing.

11.3.1 Dynamic Vibration

Treacy et al. [5] estimated the Young's modulus of MWNTs by measuring the amplitude of their thermal vibrations during in situ TEM imaging. The nanotubes were attached to the edge of a hole in 3-mm-diameter nickel rings for TEM observation, with one end clamped and the other free. The TEM images were blurred at the free ends, and the blurring was significantly increased with the temperature increase of the CNTs. This indicated that the vibration was of thermal origin. Blurring occurs because the vibration cycle is much shorter than the integration time needed for capturing the TEM image. The Young's modulus was estimated from the envelope of the thermal vibration.

Poncharal et al. [6] measured the Young's modulus of MWNTs by inducing mechanical resonance. The actuation was achieved utilizing an AC electrostatic field within a TEM (Fig. 11.1a). In the experiment, the nanotubes were attached to a fine

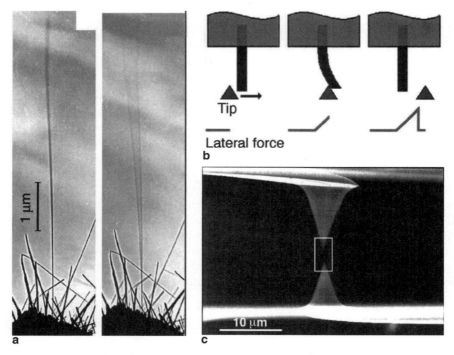

Fig. 11.1 Techniques for mechanical characterization of nanostructures. (**a**) Dynamic responses of an individual CNT to alternate applied potentials, (*left*) absence of a potential, and (*right*) at fundamental mode. Reprinted with permission from Poncharal et al., *Science* 283, 1513 (1999). © 1999, American Association for the Advancement of Science. (**b**) A CNT with one end clamped is deflected by an AFM in lateral force mode. Reprinted with permission from Wong et al., *Science* 277, 1971 (1997). © 1997, American Association for the Advancement of Science. (**c**) An individual MWNT mounted between two opposing AFM tips and stretched uniaxially by moving one tip. Reprinted with permission from Yu et al., *Science* 287, 637 (2000). © 2000, American Association for the Advancement of Science

gold wire, on which a potential was applied. In order to precisely position the wire near the grounded electrode, a special TEM holder with a piezo-driven translation stage and a micrometer-resolution translation stage were used. Application of an AC voltage to the nanotubes caused a time-dependent deflection. The elastic modulus was then estimated from the observed resonance frequencies.

11.3.2 Bending Test

Wong et al. [8] measured the Young's modulus, strength and toughness of MWNTs and SiC nanorods using an AFM in lateral force mode (Fig. 11.1b). The nanostructures were dispersed randomly on a flat surface and pinned to the substrate by means of microfabricated patches. AFM was then used to bend the cantilevered nanostructures transversely. At a certain location along the length of each structure, the force vs. deflection $(F-d)$ curve was recorded to obtain the spring constant of the system. Multiple $F-d$ curves were recorded at various locations along the structure. By considering the nanostructure as a beam, the $F-d$ data were used to estimate the Young's modulus.

Bending of nanostructures resting on a substrate is straightforward to implement. Nevertheless, it cannot eliminate the effect of adhesion and friction from the substrate. To solve the friction issue, Walters et al. [7] suspended the nanotube over a microfabricated trench and bent the nanotube using AFM lateral force mode. Salvetat et al. [38] introduced a similar method by dispersing MWNTs on an alumina ultrafiltration membrane with 200 nm pores. The adhesion between the nanotubes and the membrane was found sufficiently strong to fix the two ends. Using AFM in contact mode, the authors deflected the suspended NTs vertically and measured $F-d$ curves.

11.3.3 Tensile Test

Pan et al. [39] used a stress–strain rig to pull a very long ($\sim$2 mm) MWNT rope containing tens of thousands of parallel tubes. Yu et al. [11] conducted in situ SEM tensile testing of MWNTs with the aid of a nanomanipulator (Fig. 11.1c). A single nanotube was clamped to the AFM tips by localized EBID of carbonaceous material inside the SEM chamber. The experimental setup consisted of three components: a soft AFM probe (force constant less than 0.1 N m^{-1}) as a load sensor, a rigid AFM probe as an actuator, which was driven by a linear motor, and the nanotubes mounted between the two AFM tips. Following the motion of the rigid cantilever, the soft cantilever was bent due to the tensile load, equal to the force applied on the nanotube. The nanotube deformation was recorded by SEM imaging, and the force was measured by recording the deflection of the soft cantilever. The Young's modulus and failure strength of the MWNTs was then obtained using the acquired data.

Although significant progress has been made, the mechanical testing of nanostructures is still quite rudimental. This is mainly due to the lack of control in experimental conditions and the lack of accuracy in force and displacement measurements. Recent advances in thin film characterization have been remarkable, which have provided some guidance for nanostructure characterization.

11.4 A MEMS-Based Material Testing Stage

MEMS lend themselves naturally to material testing at the nanoscale. These systems consist of a combination of micromachined elements, including strain sensors and actuators, integrated on a single chip. Due to their intermediate size, MEMS serve as an excellent interface between the macro- and nanoworlds. Their extremely fine force and displacement resolution allows accurate measurement and transduction of forces and displacements relevant at the nanometer scale. At the same time, the larger feature sizes and signal levels of MEMS allow handling and addressing by macroscale tools. Furthermore, many of the sensing and actuation schemes are employed in MEMS scale favorably. For example, the time response, sensitivity, and power consumption of electrostatic displacement sensors improve as their dimensions shrink.

Electrostatic comb-drive actuators are often used in MEMS-based testing systems to apply time-dependent forces. van Arsdell and Brown [19] repeatedly stressed a micrometer-scale specimen in bending using a comb-drive actuator fixed to one end of the specimen. This comb-drive actuator swept in an arc-like motion while the opposite end of the specimen was fixed, causing bending stresses in the specimen. The comb-drive was also used to measure displacement, and collect fracture and fatigue data while testing to the point of failure. Kahn et al. [20, 21] determined fracture toughness by controlling crack propagation in a notched specimen using a comb-drive. One end of the specimen was fixed while the other was attached to a perpendicularly oriented comb-drive.

Electrothermal actuation schemes have also been used to apply loading [22,23]. In these actuators, Joule heating induces localized thermal expansion of regions in the actuators and an overall displacement. The resulting strains are often measured using an integrated capacitive sensor and may be verified through digital image correlation.

This section presents a detailed description of the design and modeling of a MEMS-based material testing system [24] for in situ electron microscopy mechanical testing of nanostructures. The device allows for continuous observation of specimen deformation and failure with subnanometer resolution by SEM or TEM while simultaneously measuring the applied load electronically with nano-Newton resolution. First, an analytical model of the thermal actuator used to apply tensile loading is analyzed. It includes an electrothermal analysis to determine the temperature distribution in the actuator, followed by a thermomechanical analysis to determine the resulting displacement. A coupled-field finite element analysis compliments the analytical model. Next, the differential capacitive load sensor is analyzed to determine the output voltage for a given displacement. Finally, a set of design criteria are established based on the analysis as guidelines for design of similar devices. In Sect. 11.5, experimental results using the device are described.

11.4.1 Device Description

The MEMS-based material testing system described here has previously been reported in detail [22, 23, 40]. For the purpose of this chapter, a brief review of the device is presented so that the following experimental results and the way in which they were obtained can be more clearly understood. The device consists of three parts: an actuator, a load sensor, and a gap between them for placement of nanostructures, as shown in Fig. 11.2. The devices were fabricated at MEMSCAP (Durham, NC) using the Multi-User MEMS Processes (MUMPs). Two types of actuators, an electrostatic (comb-drive) actuator [19–21] and an electrothermal actuator [41, 42] were used to apply time-dependent forces or displacements, respectively. The load sensor operates on the basis of differential capacitive sensing [22,23,40]. The sensor displacement is determined by the measured change in capacitance. This displacement is used to compute the applied force based on the known stiffness of the sensor. The design of the actuator and load sensor is described later in detail.

Figure 11.2a shows the entire device with a thermal actuator, load sensor, and space to position the specimen. The electrothermal actuator acts as a "displacement-controlled actuator" in the sense that it applies a prescribed displacement to the specimen regardless of the force required to achieve this displacement (within the functional range of the device). The load sensor is suspended on a set of folded beams of known stiffness and measures the corresponding tensile force applied to the specimen. Figure 11.2b shows an alternative loading stage using an electrostatic rather than a thermal actuator. The electrostatic actuator works as a "force controlled actuator," applying a prescribed force regardless of the resulting displacement (again within a functional range).

The electrothermal actuator has the capability of testing stiff structures, e.g., nanoscale thin films and large diameter NWs, while the comb-drive actuator is better-suited for relatively compliant structures. Both the electrothermal and electrostatic actuators can be readily made using standard microfabrication techniques. The remainder of this section focuses on using the electrothermal actuator as a case study from the perspective of its design, modeling, and integration with the rest of

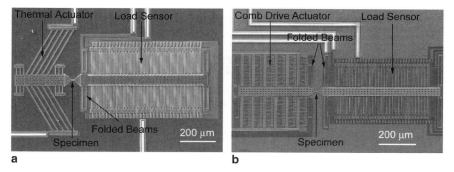

Fig. 11.2 Two variations of the MEMS-based material testing stage. (**a**) "Displacement-controlled" device using a thermal actuator and differential capacitive load sensor. (**b**) "Force-controlled" device using an electrostatic comb-drive actuator and differential capacitive load sensor

the device. Electrostatic actuators have been thoroughly described elsewhere in the literature [43–45].

In this design for in situ SEM testing, 20 devices with different types of actuators and load sensors are arranged on a $10 \times 10\,mm^2$ chip. To make electrical connection, there are 100 gold pads fabricated around the periphery of the chip. The chip is glued to the cavity of a ceramic pin grid array package, and the gold pads are wire-bonded to the 100 leads around the cavity as shown in Fig. 11.3a. The corresponding pins in the back of the package make the electric connection to a printed circuit board, which in turn is connected to electronic actuation and measuring instrumentation.

At this size scale, the changes in capacitance of the load sensor are on the order of femtofarads, which is quite challenging to measure. Fortunately, a method to measure charge that mitigates the effect of parasitic capacitance has been developed by the MEMS community [46, 47]. A commercially available integrated circuit based on this method, Universal Capacitive Readout MS3110 (MicroSensors, Costa Mesa, CA), is used here. The MEMS device chip is positioned close to the

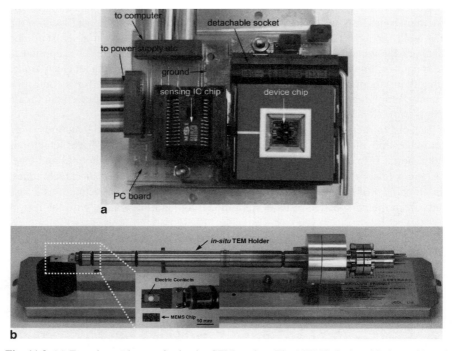

Fig. 11.3 (**a**) Experimental setup for in situ SEM testing. The MEMS device chip is positioned near the MS3110 chip on a printed circuit board. The setup is connected to a power supply, a digital multimeter, and a computer outside the SEM by means of a chamber feedthrough. (**b**) The upper image shows the In situ TEM holder containing a feedthrough, the lower one shows the magnified view of the eight electric contact pads [36] along with a $5 \times 10\,mm$ MEMS chip. In an actual experiment, the MEMS chip is flipped, placed in the TEM holder, and fixed by the left and right clamps

integrated sensing chip on the circuit board to minimize amplified electromagnetic interference (Fig. 11.3a). The output voltage of the integrated circuit is proportional to the capacitance change.

In addition to the in situ SEM measurements reported here, this device has the potential to impact other nanoscale characterization techniques. For instance, in situ TEM testing of nanostructures is possible with the addition of a microfabricated window beneath specimen gap to allow the imaging beam to pass through the device substrate. The major challenge here is to etch the window, from the back of the silicon wafer, without damaging the previously fabricated structures. This was accomplished by deep reactive ion etching of the window before releasing the devices [22, 23]. Figure 11.3b shows a MEMS chip (5×10 mm) containing four MEMS devices. The two devices in the center are used for in situ TEM testing, while the other two devices are used for calibration purposes. The chip has eight contact pads for electrical actuation/sensing. The chip is designed to be directly mounted on a specially designed TEM holder containing a feed-through and interconnects to electrically operate the devices (Fig. 11.3b) [22, 23]. In this case, the sensing integrated circuit chip (MS3110) used in the capacitance measurement is located outside of the TEM.

11.4.2 Electrothermal Actuator Design

Electrothermal actuation complements electrostatic schemes as a compact, stable, high-force actuation technique [48]. Various forms of thermal actuators have been employed in systems ranging from linear and rotary microengines [49] to two-dimensional nanoscale positioners [50], optical benches [51], and instrumentation for material characterization [52]. By incorporating compliant mechanisms, larger displacements can be achieved [50].

Modeling of the thermal actuators generally takes one of two approaches:

1. A sequential electrothermal and thermostructural analysis [53–55]
2. A completely coupled three-dimensional FEA [56]

Additional analyses include characterization of the temperature-dependent electrothermal properties [48–56] of these devices. A schematic of the thermal actuator analyzed in this section is shown in Fig. 11.4. The thermal actuator consists of a series of inclined polysilicon beams supporting a free-standing shuttle. One end of each of the inclined beams is anchored to the substrate while the other end connects to the shuttle. Thermal expansion of the inclined beams, induced by Joule heating, causes the shuttle to move forward. The heating is the result of current flowing through the beams because of a voltage bias applied across the two anchor points [57]. Modeling of these actuators requires a two-step analysis; first an electrothermal analysis to determine the temperature distribution in the device, followed by a thermostructural analysis to determine the resulting displacement field.

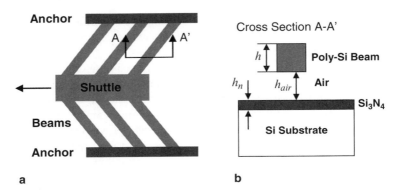

Fig. 11.4 (**a**) Schematic of the thermal actuator. (**b**) Cross section of a single beam suspended over the substrate

An electrothermal model of the device is developed to determine the temperature distribution as a function of the applied voltage. This is highly dependent upon the operating environment. When operating in air, the dominant heat transfer mechanism is heat conduction between the actuator and substrate through the air-filled gap between them [54,55,57]. In this scenario, the governing equation is

$$k_p \frac{\mathrm{d}^2 T}{\mathrm{d}x^2} + J^2 \rho = \frac{S}{h} \frac{T - T_s}{R_T}, \tag{11.1}$$

where k_p and ρ are the thermal conductivity and resistivity of the polysilicon beams, respectively; T is the temperature; J is the current density; $S = (h/w)((2h_{air}/h)+1)$ $+1$ is a shape factor accounting for the effect of element shape on heat conduction to the substrate; $R_T = (h_{air}/k_{air}) + (h_n/k_n) + (h_s/k_s)$ is the thermal resistance between the polysilicon beam and substrate; h and w are the thickness and width of a single beam, respectively; h_{air} is the gap between the beam and silicon nitride layer on the substrate; h_n is the thickness of the silicon nitride; h_s is the representative thickness of the substrate; k_{air}, k_n, k_s are the thermal conductivities of air, silicon nitride, and the substrate, respectively; and T_s is the temperature of the substrate.

The thermal conductivities k_p and k_{air} are both temperature dependent. However, the assumption of a constant k_p yields results similar to those using a temperature-dependent value of k_p [53]. Assuming a constant k_p and temperature dependent k_{air}, the finite difference method is implemented to solve (11.1) by writing the second-order differential equation in the form $\mathrm{d}^2 T / \mathrm{d}x^2 = b(x, T)$, and approximating it as

$$\frac{\mathrm{d}^2 T}{\mathrm{d}x^2} \approx \frac{1}{(\Delta x)^2} (T_{i+1} - 2T_i + T_{i-1}). \tag{11.2}$$

Figure 11.5a shows the steady-state temperature profile obtained for a two-leg (one pair of inclined beams) thermal actuator operating in air. The temperature of the shuttle is significantly lower than that of each of the beams. This is due to the relatively low current density in the shuttle, resulting in a lower rate of heat generation

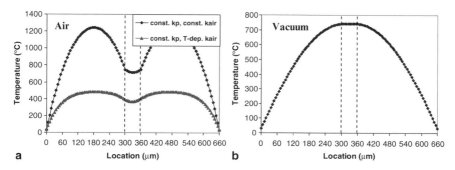

Fig. 11.5 Steady-state temperature profile (with respect to the substrate) along a pair of inclined beams and the shuttle while operated (**a**) in air for both constant and temperature-dependent values of k_{air} and (**b**) in vacuum with an input current of 10 mA. In both cases, locations 0–300 μm and 360–660 μm correspond to the thermal beams while the center region is the shuttle between the beams. The beams are anchored to the substrate at 0 and 660 μm. In air, the highest temperature occurs in the beams

as compared to that of the beams. Furthermore the relatively large area of the shuttle results in greater heat dissipation through the air as compared to the beams.

The thermal conductivity of air has a significant effect on the actuator behavior [54]. This strong dependence is clearly seen in the two curves plotted in Fig. 11.5a – in one case (lower curve) temperature dependence of thermal conductivity of air is taken into account. k_{air} increases with temperature, increasing the heat flow between the beams and shuttle and the substrate. Consequently the temperature of the beams and shuttle is lower for a given current flow. It is clear that decreasing heat conduction through the air increases the temperature of the beams. Ultimately, operation in vacuum maximizes the beam temperature for a given current flow, making the device more efficient.

In contrast, heat dissipation by conduction through the anchors to the substrate dominates in vacuum [54, 57]. Assuming each beam is thermally independent, an electrothermal model based on a single beam is presented [54]. Heat transfer within the beam is treated as a one-dimensional problem since the length dimension is significantly larger than either of the cross-sectional dimensions. To analyze the case where the thermal actuator operates in vacuum, the term for heat conduction through the air is removed from (11.1),

$$k_p \frac{d^2 T}{dx^2} + J^2 \rho = 0. \tag{11.3}$$

Figure 11.5b shows that the highest temperature now occurs in the shuttle rather than in the beams. Here the temperature depends mostly upon the distance from the anchor points which are now assumed to be the only source of heat dissipation. Since the shuttle is furthest from the anchors, it reaches the highest temperature. With the temperature distribution known from this electrothermal analysis, the thermomechanical behavior of the actuator is modeled to determine the resulting displacement [22, 23].

While the displacement of the actuator in vacuum is easily characterized experimentally [40], the temperature distribution is more difficult to obtain. Therefore a coupled-field simulation is particularly necessary. Coupled analysis also helps to assess the temperature at the actuator–specimen interface and to examine the effectiveness of the thin heat sink beams in controlling the temperature increase of the specimen during actuation.

The MEMS-based tensile stage is intended to operate within the SEM or TEM. Thus the following finite element electrothermal analysis is carried out for the case where the device operates in vacuum. The actuation voltage applied across the anchor points serves as the input while the output includes both the actuator temperature and displacement fields. Displacements at the anchor points are held fixed as applied mechanical boundary conditions. The thermal boundary conditions dictate a constant temperature at the anchors.

Figure 11.6a, b depict the temperature and displacement fields observed in the thermal actuator for an actuation voltage of 1 V. As previously mentioned, heat dissipation through the anchors is the dominant dissipation mechanism. Since the shuttle is furthest from the anchors, the highest temperature occurs in the shuttle. Due to the nonuniformity of the temperature distribution, the displacement is also nonuniform.

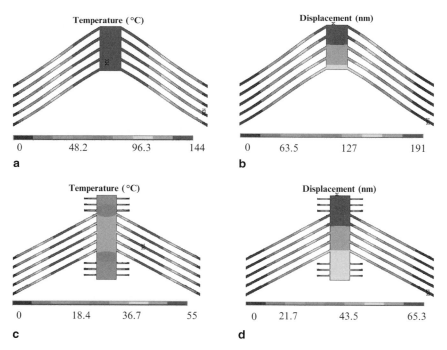

Fig. 11.6 (**a**) Temperature increase (°C) and (**b**) displacement field (nm) in the thermal actuator. The displacement component plotted is in the shuttle axial direction. (**c**) Temperature (°C) and (**d**) displacement field (nm) in the thermal actuator with three pairs of heat sink beams at the specimen end. In this analysis, the heat sink beams are 40 μm in length and 4-μm wide with 16 μm spacing between them. ANSYS Multiphysics, version 6.1 was used for this analysis

Heating of the specimen during actuation is unavoidable as a result of the increased temperature of the shuttle to which the sample is attached. However, this effect is minimized with the addition of a series of heat sink beams running between the shuttle and substrate near the shuttle-specimen interface as shown in Fig. 11.6c, d. To avoid out-of-plane bending, another three pairs of heat sink beams are placed at the opposite end of the shuttle. Comparing the two cases – with and without heat sink beams – it is found that similar displacement can be obtained at shuttle end for significantly smaller temperatures at the shuttle–specimen interface with heat sink beams. For highly temperature sensitive materials, the problem of specimen heating can be further mitigated with the addition of a thermal isolation layer between the actuator and specimen following the custom microfabrication process [57].

The thermal actuator is calibrated experimentally to verify the analytical and FEA models described above. Figure 11.7 [40] shows a comparison of experimentally measured results with the analytical and FEA predictions of the actuator displacement for a given current input. The displacement of the actuators was measured with a SEM [22], having spatial resolution of better than 5 nm. Using the analytical model, the displacement is computed based on experimentally measured temperatures in the actuator [40]. In order to obtain the current, resistance of the actuator is computed using the output temperature and a value of resistivity corresponding to the average temperature of the device.

The models agree well with the experimentally measured actuator displacements as shown in Fig. 11.7. This suggests that the models are useful in predicting the behavior of thermal actuators of other geometry. At large currents (more than ∼12 mA), both analytical and FEA models deviate slightly from the experimental results. This can be explained largely by inaccuracies in material parameters such as resistivity and thermal conductivity at high temperatures [48, 54]. Furthermore, the microstructure of polysilicon starts getting transformed at these high current levels and elevated temperatures [40].

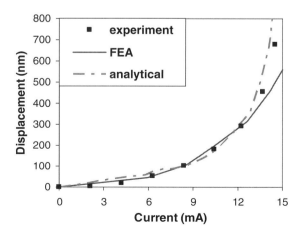

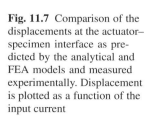

Fig. 11.7 Comparison of the displacements at the actuator–specimen interface as predicted by the analytical and FEA models and measured experimentally. Displacement is plotted as a function of the input current

11.4.3 Load Sensor Design

The load sensing mechanism consists of a differential capacitive displacement sensor suspended on a set of elastic members of known stiffness. By calibrating the stiffness of the elastic members [22, 23], the load is computed based on the measured displacement. The differential capacitive displacement sensor [45, 46, 58] is chosen for its sensitivity and linear behavior over a range of displacements appropriate for tensile testing of nanostructures.

The differential capacitive sensor is comprised of a movable rigid shuttle with electrodes (or "fingers") pointing outward as shown schematically in Fig.11.8b [22]. These fingers are interdigitated between pairs of stationary fingers (Fig. 11.8b) fixed to the substrate. Under load-free condition each movable finger sits in the middle of two stationary fingers. Each set of fingers (one movable and two stationary ones on the sides) forms two capacitors. The entire capacitance sensor is equivalent to two combined capacitances, C_1 and C_2, as shown in Fig. 11.8a.

$$C_1 = C_2 = C_0 = \varepsilon N \frac{A}{d_0} (1 + f), \tag{11.4}$$

where ε is the electric permittivity; N is the number of unit movable fingers; A and d_0 are the area of overlap and initial gap, respectively between the movable finger and each stationary finger; and $f = 0.65 d_0 / h$ is the fringing field correction factor with h being the beam height [59].

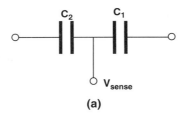

(a)

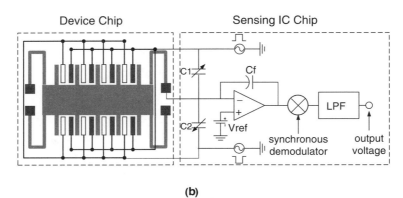

(b)

Fig. 11.8 (a) A simple model of the differential capacitor. (b) Double chip architecture used for measuring capacitance change. The capacitance change is proportional to the output-voltage change

The movable fingers are attached to the folded beams via the rigid movable shuttle, such that the displacement of shuttle and the beams is the same. This displacement yields a change in capacitance given by,

$$\Delta C = C_1 - C_2 = N\varepsilon A \left(\frac{1}{d_0 - \Delta d} - \frac{1}{d_0 + \Delta d} \right) \approx \frac{2N\varepsilon A}{d_0^2} \Delta d, \qquad (11.5)$$

where Δd is the displacement of the load sensor. Note that the fringing effect factor cancels. For displacements Δd with 50% of the initial gap d_0, the capacitance changes approximately linearly with the sensor displacement. This relatively large range of linear sensing is a major advantage of differential capacitance sensing over direct capacitance sensing which uses a single fixed beam for each movable beam.

A variety of circuit configurations may be used in measuring capacitance [45,46]. Figure 11.8b shows schematically the charge sensing method used in the device. This method mitigates the effects of parasitic capacitances that generally occur in electrostatic MEMS devices. Here the change in output voltage ΔV_{sense} is proportional to the capacitance change [22],

$$\Delta V_{\text{sense}} = \frac{V_0}{C_f} \Delta C, \qquad (11.6)$$

where V_0 is the amplitude of an AC voltage signal applied to the stationary fingers and C_f the feedback capacitor as shown in Fig. 11.8b.

Minimizing stray capacitance and electromagnetic interference is critical in high resolution capacitance measurements. In this case, integrating the MEMS differential capacitor and sensing electronics on a single chip minimizes these effects, allowing detection of changes in capacitance at the atto-Farad level [46]. However, this greatly increases the fabrication complexity. The double chip architecture depicted in Fig. 11.8 is an alternative to the single chip scheme. Here the MEMS-based system is fabricated on one chip while a commercial integrated circuit chip (for e.g., Universal Capacitive Readout MS3110, Microsensors, Costa Mesa, CA) is used to measure changes in capacitance. Both chips are housed on a single printed circuit board.

11.4.4 System Analysis

With the mechanical response of the thermal actuator known for a given current input, it is now possible to formulate a set of equations governing the behavior of the entire device [40]. A lumped model of the entire device is shown in Fig. 11.9. Here K_S is the stiffness of the tensile specimen, K_{LS} is the stiffness of the load sensor corresponding to the folding beams by which it is suspended, K_{TA} is the stiffness of the thermal actuator computed before, and U_{LS} is the displacement of the load sensor. The central shuttle is assumed to be rigid.

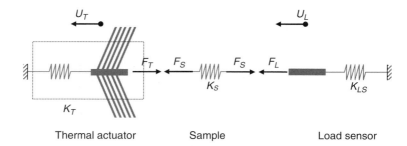

Fig. 11.9 Lumped model of the entire tensile loading device with internal forces and displacements shown in free body form

The following are the governing equations for the lumped system [40]:

$$\Delta U_S = U_{TA} - U_{LS}$$
$$U_{TA} = \frac{2m\alpha\Delta T E A s - F_{TA}}{K_{TA}}$$
$$F_{TA} = F_S = F_{LS} \qquad , \qquad (11.7)$$
$$F_S = K_S \Delta U_S$$
$$F_{LS} = K_{LS} U_{LS}$$

where $s = \sin\theta$ and ΔU_S is the elongation of the specimen, E is the Young's modulus, α is the coefficient of thermal expansion of the beam material, and m is the number of pairs of beams. Solving the system of equations (11.7), one obtains the displacement of the thermal actuator U_{TA}, the tensile force on the specimen F_S, the elongation of the specimen ΔU_S, and the corresponding displacement of the load sensor U_{LS} as,

$$U_{TA} = \frac{2m\alpha\Delta T E A s}{(K_{TA} + K_{TA}K_{LS}/K_S + K_{LS})} + \frac{2m\alpha\Delta T E A s}{(K_{TA} + K_S + K_{TA}K_S/K_{LS})}$$
$$F_S = \frac{2m\alpha\Delta T E A s}{(K_{TA}/K_S + 1 + K_{TA}/K_{LS})}$$
$$\Delta U_S = \frac{2m\alpha\Delta T E A s}{(K_{TA} + K_S + K_{TA}K_S/K_{LS})} \qquad (11.8)$$
$$U_{LS} = \frac{2m\alpha\Delta T E A s}{(K_{TA} + K_{TA}K_{LS}/K_S + K_{LS})}$$

where A is the cross-sectional area of the beam. These represent the critical parameters in obtaining force–displacement data using the MEMS-based tensile testing device.

Considering the above analyses, the following design criteria are set to achieve an effective and reliable material testing system:

1. Large load sensor displacements to maximize load resolution
2. Low temperature at the actuator–specimen interface to avoid artificial heating of the specimen
3. Displacement-controlled testing i.e., the stiffness of the thermal actuator is significantly higher than that of the specimen and load sensor

The specimen stiffness, failure load, and elongation at failure (ΔU_S) dictate the choice of actuator geometry and the number and dimensions of the beams. Consequently optimization of the device design requires some preliminary knowledge of the specimen behavior as is customary in experimental mechanics.

11.5 Experimental Results

11.5.1 Device Calibration

To obtain the relationship between the measured capacitance and corresponding displacement, an identical device with a solid connection between the actuator and load sensor (i.e., no gap for specimen mounting) was fabricated on the same chip. Real time high-resolution images were employed to calibrate the capacitance measurements [23] within a field emission SEM (Leo Gemini 1525). The device was actuated with a series of stepwise increasing voltages, applied sequentially in six ON–OFF cycles. A high contrast feature on the movable shuttle was selected for capturing images at high magnification ($\times$183k). The device state during successive ON–OFF actuation cycles was captured in a single SEM image, as shown in Fig. 11.10a. Simultaneously, the output voltage V_{sense} was recorded by a digital

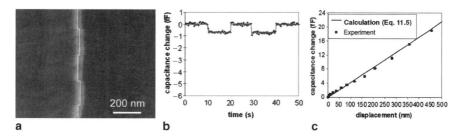

a   b   c

Fig. 11.10 Calibration of the load sensor showing the relationship between the capacitance change and measured displacement from SEM images at a series of actuation voltage. (**a**) and (**b**) Signatures when actuator is at 2 V; reference feature in the SEM image showing a motion of 15 nm due to four ON–OFF actuations (**a**), and plot corresponding to a 0.7 fF capacitance change resulting from the same actuation (**b**). Both raw data and fitted data are shown in the plot of capacitance measurements. (**c**) Plot of displacement vs. capacitance change resulting from the calibration

multimeter and converted to the capacitance change using (11.6). Figure 11.10b shows the raw data of V_{sense} at an actuation voltage of 5.5 V.

Figure 11.10c shows a plot of experimental data correlating the displacements/loads and the capacitance changes (in dots). The relationship between displacement and capacitance change as predicted in (11.5) is plotted in solid line [40] is evident that the experimental data agrees with the predictions very well. The achievable resolution of the measured capacitance change is 0.05 fF and the corresponding displacement resolution is 1 nm.

Another important step in the calibration procedure is the accurate measurement of the load sensor stiffness. This can be accomplished in one of two ways (1) by resonance methods, a common procedure in MEMS research [22] or (2) by identifying the Young's modulus of the material, E, and then using FEA with accurate metrology to determine the structural stiffness. For the parallel beams in the load sensor, the resonating voltage cannot be larger than the *pull-in* voltage so the second methodology is employed. Since the load sensor and actuator are comprised of the same material, the accuracy of the load sensor stiffness prediction is assessed by determining the comb-drive actuator stiffness using the resonance method and then comparing this result with the one calculated by FEA. In the resonance method, the stiffness is calculated by $K = (2\pi f_r)^2 (M_s + 0.3714 M_b)$ [22], where f_r is the resonant frequency, M_s and M_b the masses of the shuttle and the folded beams, respectively. For the comb-drive actuator, we measured a resonant frequency of 17.2 ± 0.1 kHz. The corresponding stiffness is 20.3 N m^{-1}, while the computed stiffness based on the *measured* folded beam geometry, using $E = 170$ GPa [60], was 20.7 N m^{-1}. This clearly shows that the stiffness computed based on the fabricated geometry and the known value of Young's modulus is in good agreement with that identified from the resonance experiment. Following this procedure, the stiffness of the load sensor designed for the testing CNTs was computed to be 11.8 N m^{-1}, which corresponds to a load resolution of 11.8 nN [22]. Likewise, the stiffness of the load sensor designed for testing NWs was 48.5 N m^{-1} with a load resolution of 48.5 nN. It should be noted that by reducing the stiffness of the load sensor, the load resolution can be achieved below 1 nN.

11.5.2 Tensile Tests of Co-Fabricated Polysilicon Thin Films

The size and fragile nature of nanostructures demands specialized techniques for preparation and mounting on the aforementioned MEMS device. Thin films may be co-fabricated with the MEMS device. This eliminates any handling or nanomanipulation of the specimen. For example, freestanding polysilicon films were co-fabricated with the MEMS device between the actuator and the load sensor (Fig. 11.11a) [23]. Due to limitations in the resolution of the photolithography used to make the devices, the initial specimen width could not be made thinner than approximately 2 μm. To reduce this dimension, the polysilicon specimen was further machined by focused ion beam (FIB) down to 350–450 nm.

Thin film specimens co-fabricated with the MEMS device were then tested in in situ SEM. The results of a tensile test of a polysilicon specimen prepared as

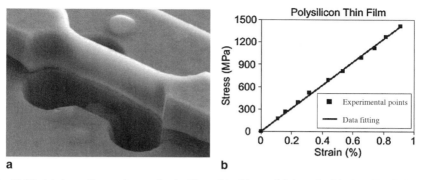

Fig. 11.11 (**a**) A tensile specimen of polysilicon thin film co-fabricated with the MEMS device and further thinned by FIB machining. (**b**) Measured stress–strain data [40]

described above are shown in Fig. 11.11b. Here the stress–strain curve shows strong linearity with a Young's modulus of 156 ± 17 GPa [40]. This result is consistent with other reported values for polysilicon films [60–62], which provides confidence in the overall experimental protocol and data reduction.

11.5.3 Tensile Tests of Nanowires

It is known that nanowires posses a relatively large surface area-to-volume ratio. Consequently interfaces, interfacial energy, and surface topography play an increasingly important role in their deformation and failure processes. In larger structures, generation and propagation of defects dictate material behavior. As grain sizes or structural dimensions fall below 50–100 nm, interatomic reorganization near surfaces gain influence over bulk material behavior. Therefore understanding the deformation and fracture mechanics of these new one-dimensional nanostructures is essential.

Testing of individual nanowires and nanotubes with the MEMS device as described above may be achieved either by growing the specimen across the gap between actuator and load sensor or by placement of the specimen in the device by nanomanipulation [23]. The latter procedure involves use of a nanomanipulator operated inside an SEM to pick up and place the individual nanostructure across the gap, followed by EBID of platinum to weld the ends.

In situ SEM tensile tests of silver nanowires using the MEMS device were performed to identify the material stress–strain behavior and failure. Figure 11.12a shows sequential SEM images obtained during the testing of a silver nanowire. The nanowire specimen began to deform at an actuation voltage of 0.8 V and failed at 6.6 V. The strain–stress data, shown in Fig. 11.12b, reveal key features. The nanowires were stressed to about 2–2.5 GPa, which is significantly higher than the tensile strength (1.7 GPa) of bulk silver [63], before necking. This phenomenon,

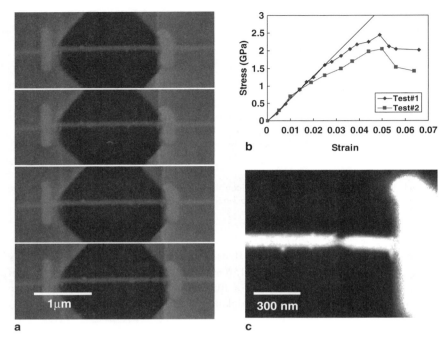

Fig. 11.12 In situ SEM tensile testing of a silver nanowire. (**a**) Sequential SEM images of a tensile loading process. The nanowire is 1.9 μm in length and 100 nm in width. (**b**) The strain–stress data, and (**c**) the fracture surface of the nanowire showing the necking

which is attributed to the absence of defects and high stress threshold for the nucleation of defects [64, 65], agrees with the hypothesis that the strength of the material increases as the specimen's characteristic dimension decreases. The nanowires underwent large localized plastic deformation leading to necking and fracture (Fig. 11.12c). A failure strain of about 6% was identified in the experiment. It is interesting to note that the softening regime captured in the experiments was possible only because of the displacement control characteristic of the thermal actuator.

11.6 Summary

The first 1D nanostructure was synthesized more than a decade ago. Since then, a number of nanodevices using these nanostructures have been developed. Mechanical characterizations at the nanoscale become an important activity to predict and assess device integrity and durability as well as to gain fundamental understanding of deformation processes at the nanoscale. Most of the developed nanomechaninical testing techniques did not possess a well-controlled loading condition and electronic load–displacement measurement. As such, in situ mechanical testing relied on indirect measurement of load through imaging of deforming beams. This chapter described the modeling and analysis involved in the design of a MEMS-based material testing system, which allows *simultaneous* load–displacement measurements combined with *real-time high resolution* SEM or TEM imaging of the specimen without the need for shifting the imaging beam. The system uses a thermal actuator to apply load and a differential capacitive displacement sensor, of known stiffness, to electronically measure the applied load. An analytical model of the thermal actuator involving electro-thermal-mechanical analysis was developed to determine the temperature distribution in the actuator, displacement and force fields. A coupled-field FEA was used to verify the analytical model and obtained further insight on the field variables. A set of design criteria were then established. Finally, examples of application of the MEMS-based material testing system to characterize co-fabricated polysilicon thin films and Ag nanowires were presented.

Acknowledgments The authors thank Alberto Corigliano for valuable discussions in the modeling of the thermal actuator. A special thank is also due to Ivan Petrov, E. Olson, and J.-G. Wen for their guidance in the development of the in situ TEM holder. This work was supported by National Science Foundation Grants DMR-0315561, CMS-00304472 and CMMI-0555734. Nanomanipulation was carried out in the Center for Micro-analysis of Materials at the University of Illinois, which is partly supported by the US Department of Energy under Grant DEFG0296-ER45439. Use of the Center for Nanoscale Materials was supported by the U.S. Department of Energy, Office of Science, Office of Basic Energy Sciences, under Contract No. DE-AC02-06CH11357.

References

1. Iijima S (1991) Nature 354(6348):56
2. Yakobson BI, Avouris P (2001) Carbon Nanotubes. Topics in Applied Physics, vol. 80. Springer, Berlin, p. 287
3. Fennimore AM, Yuzvinsky TD, Han WQ, Fuhrer MS, Cumings J, Zettl A (2003) Nature 424(6947):408
4. Dalton AB, Collins S, Munoz E, Razal JM, Ebron VH, Ferraris JP, Coleman JN, Kim BG, Baughman RH (2003) Nature 423(6941):703
5. Treacy MMJ, Ebbesen TW, Gibson JM (1996) Nature 381(6584):678
6. Poncharal P, Wang ZL, Ugarte D, de Heer WA (1999) Science 283(5407):1513
7. Walters DA, Ericson LM, Casavant MJ, Liu J, Colbert DT, Smith KA, Smalley RE (1999) Appl. Phys. Lett. 74(25):3803
8. Wong EW, Sheehan PE, Lieber CM (1997) Science 277(5334):1971
9. Shen WD, Jiang B, Han BS, Xie SS (2000) Phys. Rev. Lett. 84(16):3634
10. Li XD, Hao HS, Murphy CJ, Caswell KK (2003) Nano Lett. 3(11):1495
11. Yu MF, Lourie O, Dyer MJ, Moloni K, Kelly TF, Ruoff RS (2000) Science 287(5453):637
12. Tsuchiya T, Tabata O, Sakata J, Taga Y (1998) J. Microelectromech. Syst. 7(1):106
13. Greek S, Ericson F, Johansson S, Furtsch M, Rump A (1999) J. Micromech. Microeng. 9(3):245
14. Chasiotis I, Knauss WG (2002) Exp. Mech. 42(1):51
15. Hugo RC, Kung H, Weertman JR, Mitra R, Knapp JA, Follstaedt DM (2003) Acta Mater. 51(7):1937
16. Robertson IM, Lee TC, Birnbaum HK (1992) Ultramicroscopy 40(3):330
17. Haque MA, Saif MTA (2002) Exp. Mech. 42(1):123

18. Haque MA, Saif MTA (2004) Proc. Natl. Acad. Sci. USA 101(17):6335
19. van Arsdell WW, Brown SB (1999) J. Microelectromech. Syst. 8(3):319
20. Osterberg PM, Senturia SD (1997) J. Microelectromech. Syst. 6(2):107
21. Kahn H, Ballarini R, Mullen RL, Heuer AH (1999) Proc. R. Soc. Lond. A Math. Phys. Eng. Sci. 455(1990):3807
22. Zhu Y, Moldovan N, Espinosa HD (2005) Appl. Phys. Lett. 86(1):013506
23. Zhu Y, Espinosa HD (2005) Proc. Natl. Acad. Sci. USA 102(41):14503
24. Lu JP (1997) Phys. Rev. Lett. 79(7):1297
25. Falvo MR, Clary GJ, Taylor RM, Chi V, Brooks FP, Washburn S, Superfine R (1997) Nature 389(6651):582
26. Williams PA, Papadakis SJ, Falvo MR, Patel AM, Sinclair M, Seeger A, Helser A, Taylor RM, Washburn S, Superfine R (2002) Appl. Phys. Lett. 80(14):2574
27. Cumings J, Zettl A (2000) Science 289(5479):602
28. Smith PA, Nordquist CD, Jackson TN, Mayer TS, Martin BR, Mbindyo J, Mallouk TE (2000) Appl. Phys. Lett. 77(9):1399
29. Chen XQ, Saito T, Yamada H, Matsushige K (2001) Appl. Phys. Lett. 78(23):3714
30. Hughes MP, Morgan H (1998) J. Phys. D Appl. Phys. 31(17):2205
31. Huang Y, Duan XF, Wei QQ, Lieber CM (2001) Science 291(5504):630
32. Fujiwara M, Oki E, Hamada M, Tanimoto Y, Mukouda I, Shimomura Y (2001) J. Phys. Chem. A 105(18):4383
33. Rao SG, Huang L, Setyawan W, Hong SH (2003) Nature 425(6953):36
34. Piner RD, Zhu J, Xu F, Hong SH, Mirkin CA (1999) Science 283(5402):661
35. Dai HJ (2000) Phys. World 13(6):43
36. Kong J, Soh HT, Cassell AM, Quate CF, Dai HJ (1998) Nature 395(6705):878
37. He RR, Gao D, Fan R, Hochbaum AI, Carraro C, Maboudian R, Yang PD (2005) Adv. Mater. 17(17):2098
38. Salvetat JP, Briggs GAD, Bonard JM, Bacsa RR, Kulik AJ, Stockli T, Burnham NA, Forro L (1999) Phys. Rev. Lett. 82(5):944
39. Pan ZW, Xie SS, Lu L, Chang BH, Sun LF, Zhou WY, Wang G, Zhang DL (1999) Appl. Phys. Lett. 74(21):3152
40. Espinosa HD, Zhu Y, Moldovan N (2007) J. Microelectromech. Syst. 16:1219
41. Chu L, Que L, Gianchandani Y (2002) J. Microelectromech. Syst. 11:489
42. Fischer E, Labossiere P (2002) In: Proceedings of the SEM Annual Conference on Experimental and Applied Mechanics, Milwaukee, WI
43. Saif MTA, MacDonald NC (1996) Sens. Actuators A 52:65
44. Tang WC, Nguyen TCH, Howe RT (1989) Sens. Actuators A 20:53
45. Legtenberg R, Groeneveld AW, Elwenspoek M (1996) J. Micromech. Microeng. 6:320
46. Senturia SD (2002) Microsystem Design. Kluwer, Boston
47. Boser PE (1997)In: Proc. Transducers, Chicago, IL
48. Geisberger AA, Sarkar N, Ellis M, Skidmore GD (2003) J. Microelectromech. Syst. 12:513
49. Park JS, Chu LL, Oliver AD, Gianchandani YB (2001) J. Microelectromech. Syst. 10:255
50. Chu LL, Gianchandani YB (2003) J. Micromech. Microeng. 13:279
51. Pai MF, Tien NC (2000) Sens. Actuators A 83:237
52. Kapels H, Aigner R, Binder J (2000) IEEE Trans. Electron. Devices 47:1522
53. Chiao M, Lin LW (2000) J. Microelectromech. Syst. 9:146
54. Lott CD, Mclain TW, Harb JN, Howell LL (2002) Sens. Actuators A 101:239
55. Huang QA, Lee NKS (1999) Microsyst. Technol. 5:133
56. Mankame ND, Ananthasuresh GK (2001) J. Micromech. Microeng. 11:452
57. Que L, Park JS, Gianchandani YB (2001) J. Microelectromech. Syst. 10:247
58. Boser BE, Howe RT (1996) IEEE J. Solid-State Circuits 31:366
59. Huang JM, Liew KM, Wong CH, Rajendran S, Tan MJ, Liu AQ (2001) Sens. Actuators A 93:273
60. Espinosa HD, Peng B, Prorok BC, Moldovan N, Auciello O, Carlisle JA, Gruen DM, Mancini DC (2003) J. Appl. Phys. 94:6076

61. Sharpe WN, Jackson KM, Hemker KJ, Xie Z (2001) J. Microelectromech. Syst. 10:317
62. Corigliano A, De Masi B, Frangi A, Comi C, Villa A, Marchi M (2004) J. Microelectromech. Syst. 13:200
63. Giancoli D (2000) Physics for Scientists and Engineers, 3rd edn. Prentice-Hall, Upper Saddle River
64. Espinosa HD, Prorok B, Peng B (2004) J. Mech. Phys. Solids 52:667
65. Barber A, Kaplan-Ashiri I, Cohen S, Tenne R, Wagner H (2005) Compos. Sci. Technol. 65:2380

Chapter 12
Metrologies for Mechanical Response of Micro- and Nanoscale Systems

Robert R. Keller, Donna C. Hurley, David T. Read, and Paul Rice

12.1 Introduction

Thin films and nanomaterials lie at the heart of the burgeoning fields of nano-electronics and nanotechnology, and the accurate measurement of their properties provides a basis for consistent manufacturing, fair trade, and reliable performance. Introduction of such materials into current and future technologies has opened an entirely new suite of both materials science and measurement science challenges – effects of dimensional scaling play a stronger role in the reliability of thin films and nanomaterials than in any other materials previously known. Surfaces and interfaces can dominate and change behaviors and properties known to develop in bulk materials of the same chemical composition. As a result, extrapolation of bulk responses to the nanoscale is often inaccurate.

This chapter describes metrologies developed by NIST scientists and collaborators for mechanical properties of dimensionally constrained materials; these approaches make use of methods inherently sensitive to small volumes. Attention is focused on emerging test methods for several key forms of mechanical response – elasticity, strength, and fatigue. We attempt to avoid excessive duplication of discussion contained in other chapters. Emphasis is placed on structures that are volume constrained, with one or more dimensions less than 1 µm and often less than 100 nm.

The first metrology is contact-resonance atomic force microscopy (AFM), wherein the mechanical resonance properties of an AFM cantilever in contact with a thin-film or nanoparticle-based specimen are shown to depend sensitively on the elastic properties of a small volume of material beneath the tip. The second metrology is microtensile testing, wherein thin-film specimens are patterned into

R.R. Keller, D.C. Hurley, D.T. Read, and P. Rice
National Institute of Standards and Technology, Materials Reliability Division, Boulder, CO 80305
keller@boulder.nist.gov, hurley@boulder.nist.gov, read@boulder.nist.gov,
paul.rice@colorado.edu

F. Yang and J.C.M. Li (eds.), *Micro and Nano Mechanical Testing of Materials and Devices*,   313
doi: 10.1007/978-0-387-78701-5, © Springer Science+Business Media, LLC, 2008

free-standing tensile specimens by use of integrated circuit fabrication methods, and force–displacement response measured. The third metrology is based on application of controlled Joule heating to patterned thin films on substrates by alternating current, for the purpose of measuring fatigue and strength of thin patterned films on substrates.

12.2 Contact-Resonance AFM Methods for Nanoscale Mechanical Properties

Many emerging nanotechnology applications require nanoscale spatial resolution in more than one dimension. Furthermore, in many new applications it is desirable to visualize the spatial distribution in properties. This is because new systems increasingly involve several disparate materials integrated on the micro- or nanoscale (e.g., electronic interconnect, nanocomposites). Failure in such heterogeneous systems frequently occurs due to a localized variation or divergence in properties (void formation, fracture, etc.). Engineering these complex systems therefore requires quantitative nanomechanical imaging to better predict reliability and performance.

Methods utilizing the atomic force microscope (AFM) [1] present an attractive solution for characterizing mechanical properties with true nanoscale resolution. The AFM's scanning capabilities and the small radius of the cantilever tip (typically 5–50 nm) enable rapid, in situ imaging with nanoscale spatial resolution. The AFM was originally created to measure surface topography with atomic spatial resolution [1]. Since then, several AFM techniques to sense mechanical properties have been demonstrated [2–6]. For relatively stiff materials such as ceramics or metals, the most promising AFM methods for quantitative measurements are dynamic approaches, in which the cantilever is vibrated at or near its resonant frequencies [7]. These methods are often called "acoustic" or "ultrasonic" due to the frequency of vibration involved ($\sim$100 kHz $-$ 3 MHz). Among them are ultrasonic force microscopy (UFM) [8], heterodyne force microscopy [9], ultrasonic atomic force microscopy (UAFM) [10], and atomic force acoustic microscopy (AFAM) [11]. A general name for all of these approaches is "contact-resonance-spectroscopy AFM" or simply "contact-resonance AFM."

Of these methods, AFAM has arguably achieved the most progress in quantitative measurements. In this section, we describe how contact-resonance AFM methods can be used for quantitative measurements and imaging of nanoscale mechanical properties. We present the basic physical concepts, and then explain how they can be used to measure elastic properties. Techniques are described to obtain measurements at a single sample position as well as by both qualitative and quantitative imaging. Results are shown for specific material systems in order to demonstrate the potential of the techniques for nanoscale materials characterization.

12.2.1 Principles of Contact-Resonance AFM

Contact-resonance AFM methods make use of resonant modes of the AFM cantilever in order to evaluate near-surface mechanical properties. Several groups have developed approaches based on the basic principles described here [11–16]. Resonant vibrational modes of the cantilever are excited by either a piezoelectric element attached to the AFM cantilever holder or an external actuator such as an ultrasonic transducer. Typically, the two lowest-order flexural (bending) modes of the cantilever are used. When the tip of the cantilever is in free space, the resonant modes occur at the free or natural frequencies of the cantilever. The exact values of the resonant frequencies are determined by the geometry and material properties of the cantilever. When the tip is brought into contact with a specimen by applying a static load, the frequencies of the resonant modes are higher than the corresponding free-space values, due to tip–sample forces that stiffen the system. Contact-resonance AFM methods involve exciting these "free-space" and "contact-resonance" modes, and measuring the frequencies at which they occur. The frequency data are then interpreted with appropriate models in order to deduce the mechanical properties of the sample.

After acquiring contact-resonance spectra in the manner described below, the frequency information is first analyzed with a model for the dynamics of the vibrating cantilever. Both analytical [15, 17] and finite-element [14, 18] analysis approaches have been used. The simplest model to describe the interaction is shown in Fig. 12.1a. The cantilever is modeled as a rectangular beam of length L and stiffness k_c (spring constant). The cantilever is clamped at one end. It is coupled to the sample by a tip located a distance $L_1 < L$ from the clamped end. The coupling between the tip and sample is assumed to be purely elastic and is therefore represented by a spring of stiffness k. This assumption is valid if the applied load F_N

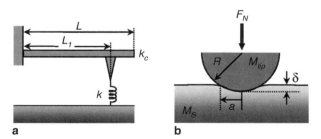

Fig. 12.1 Concepts of the contact-resonance AFM. (**a**) Model for cantilever dynamics. A rectangular cantilever beam with stiffness k_c (spring constant) is clamped at one end and has a total length L. It is coupled to the surface through a spring of stiffness k (contact stiffness) located at a position L_1 with respect to the clamped end. (**b**) Model for contact mechanics. A hemispherical tip of radius R is brought into contact with a flat surface under a normal applied force F_N. The resulting deformation of the surface is δ and the radius of contact is a. The elastic properties of the tip and sample are indicated by the indentation moduli M_{tip} and M_s, respectively

is much greater than the adhesive force, but low enough to avoid plastic deformation of the sample. These conditions are valid under typical experimental conditions involving relatively stiff materials (e.g., metals, ceramics) and stiff cantilevers ($k_c \approx 40$–50 N m^{-1}), for which $F_N \approx 0.4$–2 µN.

The analytical model for cantilever dynamics provides a characteristic equation that links the measured resonant frequencies to the tip–sample contact stiffness k. If the model assumes that the AFM tip is located at the very end of the cantilever, the values of k calculated with this equation for different resonant modes are usually not equal. The adjustable tip position parameter L_1 is used to ensure that the value of k is the same regardless of mode. Contact-resonance spectra are acquired for the two lowest-order flexural modes. Values of k are then calculated as a function of the relative tip position L_1/L for each mode. The value of L_1/L at which k is the same for both modes is taken as the solution. Typically, $L_1/L \approx 0.96$–0.98.

Elastic properties of the sample are deduced from k by means of a second model for the contact mechanics between the tip and the sample [19]. Most commonly used is Hertzian contact mechanics, which describes the elastic interaction between a hemispherical tip with a radius of curvature R and a flat surface. The parameters involved in this model are shown in Fig. 12.1b. A normal (vertical) static load F_N is applied to the tip, causing the sample to deform by an amount δ. The contact area has radius a. The normal contact stiffness k between the tip and the sample is given by

$$k = 2aE^* = 2E^* \left[\sqrt[3]{\frac{3RF_N}{4E^*}} \right] = \sqrt[3]{6E^{*2}RF_N}. \tag{12.1}$$

Here, E^* is a reduced modulus defined by

$$\frac{1}{E^*} = \frac{1}{M_{\text{tip}}} + \frac{1}{M_s}. \tag{12.2}$$

M_s and M_{tip} correspond to the indentation (plane strain) moduli of the sample and the AFM tip, respectively. For elastically isotropic materials $M = E/(1 - v^2)$, where E is Young's modulus and v is Poisson's ratio. In anisotropic materials, M depends on direction and is calculated from the second-order elastic stiffness tensor [20].

In principle, one could measure a directly, or determine R and F_N for a given experiment and calculate a. Equation (12.1) could then be used to determine E^*, from which the modulus M_s could be found using (12.2). This approach is difficult in practice. With typical values of $R = 25$ nm, $F_N = 1$ µN, and $E^* = 80$ GPa, (12.1) yields $a \approx 6$ nm. Although the small size of a is precisely why contact-resonance AFM methods can achieve nanoscale spatial resolution, it means that direct measurements of a are very difficult. To overcome this problem, the referencing approach described below has been developed for use in experiments [15, 21].

12.2.2 Modulus Measurements

Contact-resonance AFM experiments are performed with an apparatus such as the one shown schematically in Fig. 12.2a. The apparatus consists of a commercially available AFM and a few off-the-shelf components. The specimen is bonded to an ultrasonic piezoelectric transducer mounted on the AFM translation stage. The transducer is excited with a continuous sine wave voltage by a function generator. The amplitude of the cantilever deflection is monitored by the AFM's internal position-sensitive photodiode. Lock-in techniques are used to isolate the component

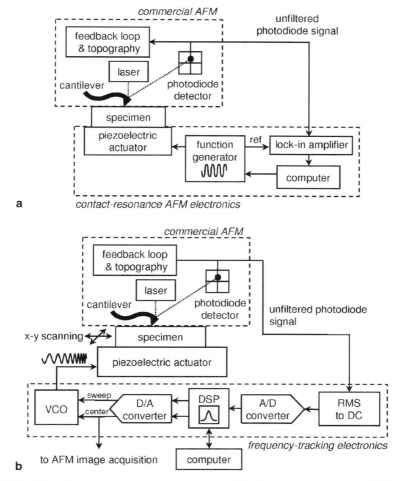

Fig. 12.2 Schematics of experimental apparatus used in the contact-resonance AFM experiments. (**a**) Apparatus for modulus measurements at a fixed sample or for qualitative imaging. (**b**) Frequency-tracking apparatus for quantitative imaging

of the photodiode signal at the excitation frequency. In this way, a spectrum of the cantilever response vs. frequency can be obtained by sweeping the transducer excitation frequency and recording the lock-in output signal. The typical frequency range involved ($\sim$0.1–3 MHz) means that it is necessary to have access to the unfiltered photodiode output signal from the AFM.

Contact-resonance spectra are acquired for transducer excitation voltages low enough to ensure a linear interaction, i.e., the tip remains in contact with the sample at all times. As described above, spectra for two different resonant modes are used to determine the correct value of k and the tip position L_1/L. Resonance frequency is measured on two samples in alternation: (1) the test or unknown sample, and (2) a reference or calibration specimen whose elastic properties have been determined by another means. The measured contact-resonance frequencies are used to calculate values of k for both the test and reference materials with the cantilever-dynamics model described above.

If measurements are performed on the test (subscript s) and reference (subscript ref) samples at the same values of F_N, it can be shown [21] that

$$E_s^* = E_{ref}^* \left(\frac{k_s}{k_{ref}} \right)^m , \tag{12.3}$$

where $m = 3/2$ for Hertzian contact and $m = 1$ for a flat punch. The indentation modulus M_s of the test sample is then determined from E_s^* by use of (12.2) and knowledge of M_{tip}. Because the true shape of the tip is usually intermediate between a hemisphere and a flat [22], the values calculated with $m = 3/2$ and $m = 1$ set upper and lower limits on M, respectively. Multiple data sets are obtained by comparing measurements on the unknown sample to those made on the reference sample immediately before and afterward. Averaging the data sets yields a single value for the indentation modulus M_s of the test sample.

The referencing approach eliminates the need for precise knowledge of R, F_N, and a (see (12.1)), which are difficult to determine accurately. Because k depends on the contact area, this approach assumes that the contact geometries for the test material and for the reference material are identical. An alternative approach that avoids this assumption by means of a tip-shape-estimation procedure has also been developed [23]. Accuracy may also be improved by use of multiple reference samples [14, 24, 25].

The accuracy of this technique has been examined experimentally by comparing contact-resonance measurements with values obtained by other methods [14, 26]. Some example results are shown in Fig. 12.3. Values for the indentation modulus M were obtained for several thin supported films of different materials by means of contact-resonance AFM (AFAM), nanoindentation (NI), and surface acoustic wave spectroscopy (SAWS). NI is destructive to the sample and has somewhat poorer spatial resolution than AFM methods, but is widely used in industry. The SAWS method is nondestructive and is used primarily in research laboratories. The values obtained by SAWS represent the average properties over a few square centimeters

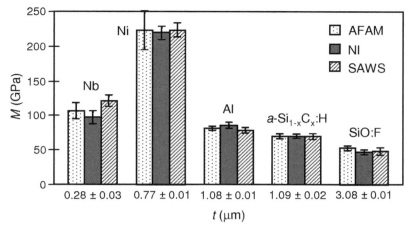

Fig. 12.3 Indentation modulus M of thin supported films obtained by contact-resonance AFM (AFAM), nanoindentation (NI), and surface acoustic wave spectroscopy (SAWS). The thickness t of each film was determined by cross-sectional SEM analysis or by stylus profilometer methods. Film materials include fluorinated silica glass (FSG), amorphous hydrogenated silicon carbide (a-Si$_{1-x}$C$_x$:H), aluminum (Al), niobium (Nb), and nickel (Ni). The *error bars* represent one standard deviation of the individual measurements

of the sample. It can be seen in Fig. 12.3 that the results obtained by all three methods are in very good agreement, with differences of less than 10% and within the measurement uncertainty for all of the samples.

The validity of contact-resonance AFM for elastic-property measurements having been established, it can be used to investigate specific systems. Work has spanned a wide range of materials, including piezoelectrics (PZT [27]), dickite clay [24], semiconductors (InP and GaAs [28]), and nanostructures (SnO$_2$ nanobelts [29] and ZnO nanowires [30]). A more complete survey of results can be found elsewhere [15]. Additional research is underway to better understand the range and limits of applicability of the method. For instance, how thin a film can be measured accurately without accounting for the properties of the underlying substrate? We have examined this question using a series of nanocrystalline nickel (Ni) films deposited on silicon (Si) substrates [31]. The results indicated that the modulus of films as thin as ∼50 nm could be accurately measured with contact-resonance AFM. The film thickness for which the substrate begins to play a role depends on the elastic properties of both the tip and sample. Other work has shown how to extend the basic approach described above in order to obtain further information about elastic properties [32]. By simultaneously measuring the contact-resonance frequencies of both flexural and torsional modes, shear elastic properties such as Poisson's ratio v or shear modulus G separately from Young's modulus E. Further work is needed to incorporate the results of these and other studies into standardized measurement procedures.

12.2.3 Imaging and Mapping

The scanning capabilities of the AFM mean that two-dimensional images of near-surface mechanical properties can be obtained with contact-resonance methods. Qualitative "amplitude images" indicative of local variations in stiffness are obtained with an apparatus like that in Fig. 12.2. The frequency of the excitation transducer is held constant while the tip is scanned across the sample. During scanning, the lock-in detector senses variations in the cantilever vibration amplitude at the excitation frequency due to changes in the local contact stiffness. The output signal of the lock-in is used as an external input to the AFM for image acquisition. Amplitude imaging has been used to investigate the nanoscale elastic properties of systems such as carbon-fiber-reinforced polymers [12], piezoelectric ceramics [33], and dislocations in graphite [34].

Examples of amplitude imaging are shown in Fig. 12.4. The sample contained a blanket film of an organosilicate glass (denoted SiOC) approximately 280 nm thick. Copper (Cu) lines were deposited into trenches created in the SiOC blanket film. The sample was etched briefly in a hydrofluoric acid solution in order to remove any protective surface layers. The topography image shown in Fig. 12.4a reveals that the sample is very flat, with features <10 nm high. Figure 12.4b, c shows amplitude images acquired at two different frequencies. Small features inside the Cu lines can be seen. These are most likely due to shifts in the contact area that arise from small topographical features (e.g., pores, polishing effects). In addition to the SiOC film and the Cu lines, bright regions can be seen at the SiOC/Cu interfaces. This feature corresponds to a thin barrier layer deposited on the sidewall of the trenches, and is not obvious in the topography image. In Fig. 12.4b, the SiOC regions of the image are brighter than the Cu regions. However, the Cu regions are brighter in the image in Fig. 12.4c, which was acquired at a higher excitation frequency. This information suggests that the contact-resonance frequency of the Cu regions is generally higher than that of the SiOC regions. Because higher contact-resonance frequencies imply greater elastic modulus, it can be inferred that the modulus of the Cu lines is higher than that of the SiOC film.

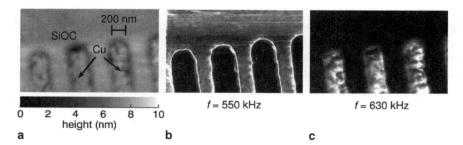

Fig. 12.4 Example of contact-resonance AFM amplitude imaging. The sample contained copper (Cu) lines in an organosilicate glass (SiOC) film. (**a**) Topography. (**b**) Amplitude of the cantilever vibration at an excitation frequency $f = 550\,\text{kHz}$. (**c**) Amplitude at $f = 630\,\text{kHz}$. Images were acquired using a cantilever with lowest free-space frequency $f_1^0 = 151.3\,\text{kHz}$

Figure 12.4 shows that the components of different elastic stiffness are easily identified by amplitude imaging. However, the figure also illustrates the difficulties involved in trying to evaluate the relative stiffness of different sample components. Instead, quantitative imaging or mapping of nansocale elastic properties is ultimately desired. Quantitative imaging involves detecting the frequency of the contact-resonance peak at each position as the tip moves across the sample. A single such contact-resonance-frequency image could provide more information than an entire series of amplitude images. However, if the sample components differ greatly in their elastic properties, the contact-resonance frequency will vary significantly across the sample making detection more difficult. Several solutions to this challenge have been demonstrated [16, 25, 26, 33, 35, 36]. Typically, there is a trade-off between the imaging speed and the amount of custom hardware and software required.

We have also developed contact-resonance-frequency imaging techniques for nanomechanical mapping [26, 37]. Our approach differs conceptually from other implementations, in that the starting frequency of the frequency sweep window is continuously adjusted to track the contact-resonance peak frequency. In this way, a high-resolution spectrum is acquired with a minimum number of data points, even if the contact-resonance frequency shifts significantly across the imaged region. Without feedback, other approaches must perform a frequency sweep at every point over the same relatively wide range that encompasses all possible peaks. Our approach also utilizes a digital signal processor (DSP) architecture. One advantage of a DSP approach is that it facilitates future upgrades, because changes are made in software instead of hardware.

A schematic of the frequency-tracking apparatus is shown in Fig. 12.2b. The circuit is described in detail elsewhere [37]. In brief, an adjustable-amplitude, swept-frequency sinusoidal voltage is applied to the piezoelectric actuator beneath the sample. As the cantilever is swept through its resonant frequency by the piezoelectric actuator, the photodiode detects the cantilever's vibration amplitude and sends this signal to the DSP circuit. Inside the circuit, the signal is converted to a voltage proportional to the root-mean-square (rms) amplitude of vibration and sent to an analog-to-digital (A/D) converter. The DSP reads the A/D converter output signal and constructs a complete resonance curve as each sweep completes. It then finds the peak in the resonance curve and uses this information in a feedback-control loop. The control loop adjusts a voltage-controlled oscillator (VCO) to tune the center frequency of vibration to maintain the cantilever response curve centered on resonance. The control voltage is also sent to an auxiliary input port of the AFM instrument for image acquisition. Each pixel in the resulting image thus contains a value proportional to the peak (resonant) frequency at that position. A frequency range can be specified in order to exclude all but the cantilever mode of interest. A total of 128 data points are acquired for each resonance curve, and at 48 kS s^{-1} the system is capable of acquiring the full cantilever resonance curve 375 times per second. The currently implemented circuit realizes approximately 17–18 bits of resolution, corresponding to a frequency resolution of about 12 Hz over the full-range span of 3 kHz–3 MHz. The AFM scan speed must be adjusted to ensure that

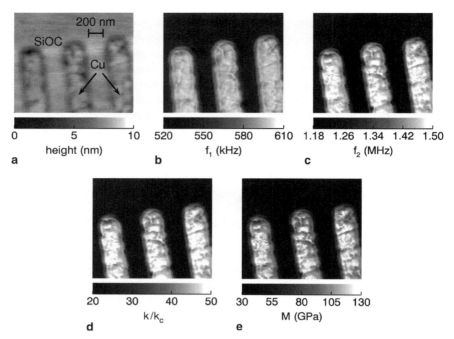

Fig. 12.5 Example of quantitative contact-resonance AFM imaging. The images correspond to approximately the same region of the SiOC/Cu sample shown in Fig. 12.4. Contact-resonance-frequency images of (**a**) first (f_1) and (**b**) second (f_2) flexural modes, respectively. (**c**) Normalized contact stiffness k/k_c calculated from (**a**) and (**b**). (**d**) Map of the indentation modulus M calculated from (**c**) assuming flat-punch contact mechanics. The free-space frequencies of the cantilever's lowest two flexural modes were $f_1^0 = 151.3\,\text{kHz}$ and $f_2^0 = 938.0\,\text{kHz}$

several spectrum sweeps are made at each image position. For scan lengths up to several micrometers, an image with 256×256 pixels is usually acquired in less than 25 min.

An example of quantitative imaging with our frequency-tracking electronics is shown in Fig. 12.5. The images correspond to the same region of the SiOC/Cu structure in Fig. 12.4. The topography image in Fig. 12.5a shows the SiOC blanket film and the slightly recessed ($<5\,\text{nm}$) Cu lines. The contact-resonance-frequency images for the two lowest flexural modes of the cantilever are shown in Fig. 12.5b, c, respectively. The frequency images reveal directly that the contact-resonance frequency in the Cu regions is higher than in the SiOC regions.

An image of the normalized contact stiffness k/k_c calculated from the images of f_1 and f_2 is shown in Fig. 12.5d. The image was calculated on a pixel-by-pixel basis with the analysis approach described above for point measurements. Depending on the sample and type of information desired, it may be sufficient to evaluate the contact-stiffness map alone. In other cases, it may be preferable to calculate a map of the indentation modulus M from the contact-stiffness image. Such calculations involve the same models and assumptions used in point measurements of M. For instance, it is necessary to choose a specific contact-mechanics model, and to assume that the model remains valid for the entire image (i.e., tip wear is not dramatic). Reference values of E^* and k/k_c are also needed. Here, we calculated a map of the indentation modulus M from the contact-stiffness image assuming that the tip was flat. We assumed that the mean value of E^* for the SiOC region corresponded to $M_{SiOC} = 44.3$ GPa. This value was obtained from point measurements made directly on the SiOC film. For the reference value of k/k_{lever}, we used the average value over the SiOC region of the image. The resulting modulus map is shown in Fig. 12.5e. Although some assumptions were made to obtain the map, it shows that quantitative values for M can be achieved.

Other mechanical properties besides elastic modulus can be imaged with contact-resonance AFM methods, if they influence the contact stiffness between the tip and the sample. One such property of technological interest is the relative bonding or adhesion between a film and a substrate. To experimentally investigate the sensitivity of contact-resonance AFM to variations in film adhesion [38], we fabricated a model system of gold (Au) and titanium (Ti) films on (001) silicon (Si). Figure 12.6a shows a cross-sectional schematic of the sample. A rectangular grid with $5\,\mu m \times 5\,\mu m$ squares (10-μm pitch) of Ti was created on Si by standard microfabrication techniques. A blanket film of gold (Au) and a 1-nm topcoat of Ti were deposited on top of the grid. The sample was intended to contain variations in the adhesion of a buried interface, but only minimal variations in topography and composition at the surface. A crude scratch test was performed by lightly dragging one end of a tweezer across the sample. Optical micrographs showed that this treatment had removed the film in the scratched regions without a Ti interlayer (squares), but left the gold intact in the scratched regions containing a Ti interlayer (grid). The result confirmed our expectation that the film adhesion was much stronger in regions containing the Ti interlayer. The Ti topcoat was included merely to prevent contamination of the AFM tip by the soft Au film.

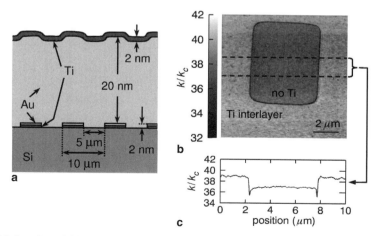

Fig. 12.6 Imaging of film/substrate adhesion. (**a**) Schematic of sample in cross section. (**b**) Map of the normalized contact stiffness k/k_c calculated from contact-resonance-frequency images. (**c**) Average stiffness vs. position across the center of (**b**)

To understand how contact-resonance methods sense variations in a buried interface, note that experiments probe the sample properties to a depth $z \approx 3a$, where a is the tip–sample contact radius [19]. For Hertzian contact mechanics, $a^3 = (3RF_N)/(4E^*)$. For $z > 3a$, the stress field beneath the tip is sufficiently small relative to its value at the surface (less than 10%) that the measurement is not sensitive to property variations. The relative depth sensitivity of contact methods is determined by the choice of experimental parameters R and F_N, as well as the sample and tip properties (E^*). We estimate that $a = 6$–8.5 nm for our experimental conditions. Therefore, the experiments should probe the film interface $(z = 22$–24 nm $\approx 3a)$.

Contact-resonance-frequency imaging experiments were performed on the sample with the methods described above. An image of the normalized contact stiffness k/k_c calculated from the experimental contact-resonance-frequency images of f_1 and f_2 is shown in Fig. 12.6b. The image reveals that the contact stiffness is lower in the square region with poor adhesion (no Ti interlayer). A line scan of the average value of k/k_c vs. position obtained from 40 lines in the center of the image is shown in Fig. 12.6c. The mean value of k/k_c is 39.1 ± 0.6 in the grid regions and 37.1 ± 0.5 in the square, a difference of 5%. Several other contact-stiffness images acquired at different sample positions consistently showed a decrease in k/k_c of 4–5% for the regions of poor adhesion that lacked a Ti interlayer.

The results are consistent with theoretical predictions for layered systems with disbonds [39]. An impedance-radiation theory was used to model the disbonded substrate/film interface by a change in boundary conditions (i.e., zero shear stress at the interface). For a disbond in a 20 nm aluminum film $(M = 78 \text{ GPa})$ on (001) Si $(M = 165 \text{ GPa})$, a reduction of approximately 4% in the contact stiffness was predicted, very similar to our results. The system modeled in Ref. [39] contained a film material different from that used in our experiments. However, the overall combination of conditions (film and substrate modulus, applied force, etc.) was sufficiently similar to ours that we believe a comparison is valid. These results represent progress toward quantitative imaging of adhesion, a goal with important implications for the development of thin-film devices in many technological applications.

12.3 Microtensile Testing

Techniques for microtensile testing of thin films have now been under development at NIST Boulder for over 10 years [40–42]. This section reviews and summarizes the rationale behind microtensile testing, and the current state of the experimental techniques available here. Only experimental techniques in use at the authors' laboratory are described in detail; many alternative techniques for obtaining information about the strength, ductility, and other mechanical properties of micro- and nanoscale materials are being explored elsewhere. An extensive list of microtensile results from the literature is given, to emphasize the usefulness of the microtensile technique.

12.3.1 Rationale for Microtensile Testing

Tensile testing is the standard means of obtaining mechanical properties of structural metals [43]. Because the stress field is uniaxial and uniform throughout the gage section until significant plastic strain occurs, unambiguous and accurate Young's modulus, yield strength, ultimate tensile strength, and ductility can be obtained from an accurate force–displacement record. For macroscale structures, these properties are often specified for a given material and used for comparison of alternative materials, development of new materials, and quality control. In addition, tensile properties are the essential input parameters for structural design and numerical modeling and simulation of mechanical behavior of structures. These modeling and simulation techniques are routinely applied to micro- and nanoscale structures used in advanced microelectronics devices and are a fundamental tool in designing applications of nanomaterials. The accuracy of the results of these analytical efforts depends on the accuracy of the material property values used as input. However, it has long been realized that micro- and nanometer-sized thin films typically have microstructures and properties, produced by their fabrication processes, that are different from the microstructures and properties of bulk materials of the same nominal chemical composition [44,45]. Usefully accurate estimates of properties of films prepared by one process often cannot be made from films prepared by another process or from bulk specimens. The properties of thin films need to be measured in the dimension and conditions that are used in actual structural applications. Therefore, accurate methods for mechanical characterization of micro- and nanoscale materials are needed, and tensile testing is being developed to meet this need. Tensile test techniques applicable to micro- and nanoscale materials are only now beginning to be used as part of the commercial design and quality control of advanced small-scale devices and structures.

12.3.2 Specialized Techniques Required

Tensile properties of macroscale materials are generally measured according to standardized procedures, such as ASTM E8 and ASTM E345 [46,47]. Specimens specified in these test methods are typically several millimeters or even centimeters in thickness for rectangular specimens (or in diameter for round specimens); these dimensions are orders of magnitude larger than thin films used in today's microelectronics industry. The testing machines, gripping devices, and specimen-preparation procedures prescribed in the standard test methods are difficult to apply in testing thin films that have thicknesses measured in micro- or nanometers. Early developments of the experimental techniques have been reviewed by Hoffman [48] and Menter and Pashley [49]. Most of the specimens used in these early approaches were still relatively large compared with those of current interest. Early attempts to pull thin films in conventional testing machines used specimens lifted from the substrate; researchers encountered problems in placing the specimen on the grips without excessive wrinkling, and depended on special separation layers beneath the specimen film, such as water soluble sodium chloride. Specimen fabrication, preparation, and especially mounting specimens to the loading devices were cited as major difficulties.

12.3.3 Bulk Micromachined Microtensile Specimens

The test methods for thin films, at NIST and elsewhere, took a big step forward with the introduction of microfabrication techniques, including lithography, deposition, and etching, to produce tensile specimens [40, 50–53]. Microfabrication of tensile specimens brought its own advantage, namely, that a large number of specimens with uniform thickness, composition, geometry, and structure are fabricated in a single process run. It became evident that since films in actual devices are always produced on substrates, the use of the substrate to support the thin film specimen is appropriate. Bulk micromachining of MEMS devices had been developed by this time, demonstrating the concept of etching away a selected portion of the original substrate to form a useful device. But the substrate is always much more massive than the film, so it must be removed at least from beneath the gage section of the specimen. The fabrication challenge here is the chemical selectivity required to etch through hundreds of micrometers of silicon without damaging the metal specimen. Aqueous hydrazine has been used, but this material is hazardous. Another disadvantage is the large width of the gage section, $100\,\mu m$ or more, compared to the line widths used in microelectronic interconnect structures and also compared to a typical film thicknesses of $1\,\mu m$. Ding et al. [50] reported the use of a silicon frame design for testing doped silicon. The first realization of this scheme for metal films was the silicon frame tensile specimen [40]. To produce the silicon frame tensile specimen, photolithographic patterning is used to form a straight and relatively narrow gage section with larger grip sections on a silicon frame. The substrate beneath the gage section is removed by a suitable etchant. The silicon frame, carrying its tensile specimen of a thin film, is mounted on a suitable test device capable of supplying force and displacement [54]. The silicon frame is cut, while leaving the specimen undamaged. This step has been accomplished manually with a dental drill, using a temporary clamp to hold the specimen in place, and by the use of a cutting wheel mounted on a moveable stage [53]. A device driven by piezoelectric blocks was used to conduct tensile and fatigue tests on aluminum and copper films using the silicon frame tensile specimen [54–56]. Similar specimens and instrumentation have been used at other laboratories.

All the microtensile (MT) techniques include measurements of force and displacement. The force is measured using a load cell, either commercial or custom-built. For the generation of bulk-micromachined specimens discussed so far, the force might amount to 0.1 N; commercial load cells with this range are available. The challenge in displacement measurement for microscale testing is the same in principle as for macroscale testing: Measurement of the stretch of the gage section, which may be much different from the displacement of the load fixture because of the variable compliance of the load train. But the adaptation of common macroscale techniques for measurement of strain in a thin film proved impossible. Adhesively bonded strain gages or displacement gages clipped onto the specimen have not been successfully scaled down to dimensions useful for testing of thin films. Optical techniques have been adapted instead. Displacement has been measured by interferometric techniques such as electron speckle pattern interferometry (ESPI), for example as in [57], or by diffraction from markers placed on the specimen surface [58].

12.3.4 Surface-Micromachined Tensile Specimens

In recent years, a new generation of smaller-scale tensile specimens produced by surface micromachining and complementary test techniques, has been developed [41, 59, 60]. The NIST-Boulder version is described here. Depending on the material to be tested and the deposition processes used, the pattern-and-etch steps may vary. The substrates used originally were (100) silicon wafers coated by suppliers with a "wet" oxide layer about 0.5 μm thick. A typical fabricated specimen ready for testing is shown in Fig. 12.7. A mask-pattern-etch procedure was used to produce a rectangular window in the oxide, exposing the underlying bare silicon where the specimen would be formed. More recently, entirely uncoated silicon wafers have

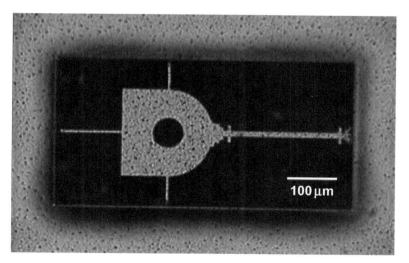

Fig. 12.7 Fabricated microtensile specimen ready for testing. The gage section, which is the long section to the right of the structure, is just under 200-μm long. The *hole*, which is used for loading, is about 50 μm in diameter. The *narrower lines* connected to the tab are the tethers, which are usually cut before testing

been used. This eliminated a processing step. The specimen material is deposited as a blanket film, aluminum in Fig. 12.7. The specimen geometry is formed, within the region of exposed silicon on oxidized wafers or at arbitrary locations on bare wafers, by subtractive photolithography using masking and wet-chemical etching. One end of the specimen's gauge section (right side in Fig. 12.7) connects to the surrounding film that remains adhered to the oxidized silicon. The other end connects to a tab, which has a hole of 50 μm in diameter for pin-loading. There are three tethers connected to the tab, as shown in Fig. 12.7. The function of the tethers is to hold the tab in place after the underlying silicon has been removed. More recent designs have eliminated some of the tethers. The gauge section, the tab, and the tethers are freed by chemically removing the underlying silicon substrate with xenon difluoride (XeF_2). This dry etchant very selectively attacks silicon, but is stopped by SiO_2, copper, aluminum, polyimide, and many other materials of interest in the electronics industry. For easy loading, the silicon needs to be removed to a depth of at least 50 μm everywhere within the rectangular etch region.

The original design [61] for the loading hole was a square because it was more convenient for the mask-design software to generate. During testing, however, failure commonly occurred at the loading hole with cracks initiated at the corners of the hole. The current specimen has a circular hole that eliminates the loading-hole failure in thick specimens, while loading-hole reinforcement through selective lithography is also used in specimens made from thinner films. At the ends of the gauge section, there are flags that facilitate the measurements of displacement and strain by the digital image correlation technique [62, 63]. The nominal dimensions for the gauge section of the rectangular specimen are 180 μm × 10 μm × 1 μm. Variants with thickness down to 0.5 μm and up to 10 μm have been tested. For stress and strain calculations, the dimensions of each specimen are individually measured. The thickness is determined by a profilometer. The length and width are measured with a scanning electron microscope (SEM). The measurement accuracy of length and width is within ±1%, while thickness is within ±3%.

12.3.5 Microtensile Apparatus

Our present test apparatus consists of the following components:

1. A three-axis micromanipulator, which is assembled from three separate commercial single-axis stages, each driven by an inchworm piezomotor. The motors are controlled by a desktop computer with a specialized interface and software. According to the manufacturer, each axis of the micromanipulator assembly has a travel range of 25 mm. The motor drive for each axis has an encoder with a resolution of 0.05 μm. We move the manipulator by specifying end points, but the displacement rates vary from the nominal values because of the variable action of the piezo-stepping of the motors. The motors have a nominal speed range from 0.004 to 1,000 μm s^{-1}. It is noted that the manufacturer's specifications of

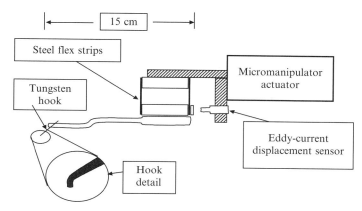

Fig. 12.8 Schematic drawing of a loading system with tungsten loading pin or hook mounted at the end of the brass rod and force sensor assembly

the micromanipulators given here are for reference only. The actual strains of a specimen during a test are determined from digital images of the specimen surface.

2. A loading system consisting of tungsten pin, brass rod, and force sensor, which is carried by the micromanipulator assembly, as shown in Fig. 12.8. The force sensor contains an eddy-current displacement sensor and two spring steel flex strips. As shown in the figure, the ends of the steel strips are screw-fastened to two ceramic blocks that serve as a heat insulator when tests are performed at elevated temperatures. The top ceramic block is attached to the micromanipulator assembly and the bottom one is supported by the flex strips. The brass loading rod with the tungsten pin at its end is attached to the bottom ceramic block. The rod is aligned to one of the micromanipulator's axes to facilitate alignment during testing. The sharp tungsten pins are commercially available as electrical contacts for wafer probes. We bend the tip 45° so it will be normal to the surface of the specimen, and blunt it to a diameter of about 40 μm to fit the hole in the specimen's tab. The steel flex strips have a fixed nominal length of 20.3 mm and a width of 6.4 mm. Two thicknesses, 0.13 mm and 0.38 mm, have been used depending on the force required. The eddy-current sensor monitors the deflection of the lower ceramic block against the stiffness of the flex strips. With the 0.13-mm plates, the sensitivity is about 8 mN for the nominal range of the eddy current sensor, which is 25 μm. For the 0.38-mm plates, the sensitivity is about 150 mN for the same deflection range.

3. An optical microscope equipped with a digital camera. High magnification is needed to view the specimen for alignment and for engagement of the loading pin to the specimen under manual control of the micromanipulator. The images of the specimen surface during the course of a test are required for calculation of specimen's extensions and strains. We use a simple 25× objective lens with a working distance of about 20 mm to view the specimen while we engage the

loading pin; then we move the microscope to view and record images of the specimen gage during the test. The 200-μm gage length of the specimen was chosen to nearly fill the field of view of the digital camera on the microscope. The long working distance is needed to leave room for the loading system between the lens and the surface of the specimen. We also have conducted tests in the SEM, which has continuous magnifications that provide sufficient flexibility for handling and testing specimens. However, the time required to pump down the SEM and its slower image acquisition has led us to prefer the optical microscope for microscale testing.

4. A second desktop computer controls a digital camera on the microscope to acquire images of the specimen surface during testing. The two computers are synchronized through communications via serial ports on the computers, so that the individual data points for force and stage movement can be associated with the proper corresponding image. Our camera acquires grayscale images of $1,280 \times 1,024$ pixels, although we save only $1,280 \times 256$ pixels.

12.3.6 Force Calibration

While force calibrations for conventional mechanical test machines are based on dead weights and gravity, dead-weight calibration with standard gram-denominated masses is not practical for the load cells used in present-day microtensile testing. For example, an aluminum specimen of the nominal $1 \times 10 \mu m^{-2}$ cross section fails at an applied load of approximately 2 mN or less, while a mass of only 1 g provides a force of about 10 mN. Along with the minuscule forces involved is the geometry issue. Microscope stages are horizontal, so in microtensile testing the force is applied in a horizontal direction; dead-weight loads would have to be redirected from vertical to horizontal, and the needed pulleys are prone to friction. We have used a force pendulum setup for calibration. With a 20 g mass and a height of a few hundred millimeters, our setup provides on the order of 1 mN per millimeter of horizontal displacement. The friction and inertia are much less than those in simple pulley systems with comparable force levels. The setup can be trivially converted to higher force levels by using a larger mass. The accuracy of our force measurement scheme, as estimated from variability among repeat calibrations, is about 4%. For low-force measurements, the system is limited by its drift rate of about 20 μN over times of tens of minutes. This drift is believed to be produced by temperature variation and air currents in the laboratory.

12.3.7 Strain Measurement

Our tensile testing practice typically records the engineering strain of the specimen, given by the change in length divided by the initial length. We obtain the needed displacements of the two ends of the gage section by use of digital image correlation [62]. Our optical references for the ends of the gage section were shown in Fig. 12.7. We track the displacement of both reference shapes through the course of the test. If the specimen stays in focus, so that the images remain sharp, the precision of each position measurement is about 0.05 pixel. The gage length is about 780 pixels, so the accuracy of the strain measurement is about 60 microstrain. The range is determined by the size in pixels of the image, and is over 20% for our gage length. Digital image correlation is a powerful technique but has its pitfalls. For our system, problems have been found from "pixellation" in the cameras we have used, meaning that the individual pixels in the camera have slightly different sensitivity, so that artificial steps in intensity are introduced into the image, and from previously uncontrolled vertical motion of the specimen because of compliance in the loading system.

12.3.8 Our Results

Stress–strain curves obtained for thin films using the NIST apparatus are shown in Fig. 12.9, for an electrodeposited copper film [64], and Fig. 12.10, for a commercial photodefinable polyimide [65].

We often measure low modulus values, especially for electrodeposited copper [64]. On the other hand, Hurley et al. [66] showed that the values of Young's modulus measured by a variety of test techniques, including microtensile testing, seem to be consistent.

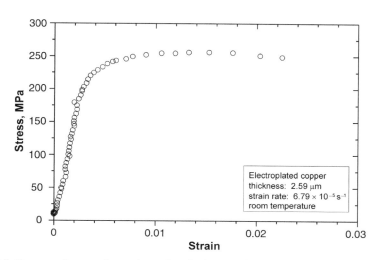

Fig. 12.9 Stress–strain curve for an electrodeposited copper film, obtained by the NIST microtensile test technique

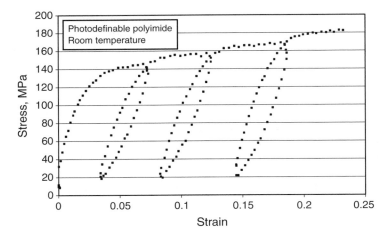

Fig. 12.10 Engineering stress–strain curve for photodefinable polyimide

12.3.8.1 Results in General

The underlying capability for and interest in standardization of microtensile testing is a relatively recent development. Until a few years ago, testing in one laboratory of specimens made elsewhere was a novelty. There is at present no standard test method for microtensile testing of thin films; individual investigators adapt the standard methods for bulk metal specimens to fit their specific specimen geometry. Standardization is hindered by the multitude of specimen sizes and designs that are in use, which has resulted from the difficulty of fabricating microtensile specimens. A recent round robin showed reasonable agreement among several laboratories in the strength of polySi, though most labs required their own unique specimen geometry. The different geometries were produced on the same MEMS chip [67]. The strength values obtained for polySi were impressively high, of the order of 1/30 of the polycrystalline Young's modulus, which is the usual estimate of the theoretical strength of a solid.

12.3.9 Other Recent Techniques

The membrane deflection tensile test was applied to a series of face-centered-cubic (FCC) metals by Espinosa et al. [68]. Another recent advance is the cofabrication of a specimen and a protective frame that includes a force sensor [69]. This specimen is suitable for use inside a transmission electron microscope (TEM).

12.4 Properties of Specific Materials

Fabrication methods and test techniques for some widely used thin film materials have become sufficiently widespread so that specific values of the properties can be usefully given. Of course, Brotzen's rule 3 still holds: the microstructure and chemical composition of the specific material at hand determine its properties. Table 12.1 shows a summary of mechanical properties for commonly used thin film materials.

Table 12.1 Mechanical properties of selected thin films as measured by microtensile testing or nanoindentation

Material	Fabrication method	Thickness (μm)	Yield strength (MPa)	Ultimate tensile strength (MPa)	Young's modulus (GPa)	Elongation to failure (%)	Ref.
Al	Sputtered	0.05	327		62[a]		[69]
Al	Sputtered	0.1	700				[69]
Al	Sputtered	0.2	330		70		[69]
Al	e-Beam evaporated	0.2	205	375[b]	65–70		[60]
Al	e-Beam evaporated	1	150		65–70		[60]
Al	e-Beam evaporated	1	94	151	24–30	22.5	[61]
Al–0.5%Cu	MOSIS[c]	1.5 and 2.4	65	74	40	1.4	[70]
Cu	e-Beam evaporated	0.2	345		125–129		[60]
Cu	e-Beam evaporated	1	160[d]		125–129		[60]
Cu	Electrodeposited	9.7	253	311	67		[71]
Au	e-Beam evaporated	0.3, 0.5	220		53–55		[60]
Au	e-Beam evaporated	1	90		53–55		[60]
Ni	Electrodeposited	4.7		1,516	102–114		[72]
Polyimide	Spun on, baked	0.6	103	181	5.5	24	[65]
PolySi	MUMPS 25, 30[c]	3.5	NA	950	157		[73]
PolySi	SUMMiT[c]	2.5	NA	3,000			[67]

The study by Espinosa [69] is quoted extensively here because it is recent; uses consistent methods on gold, aluminum, and copper films; includes specimen thickness and width effects; and includes considerable microstructural characterization and posttest observations of the specimens.
[a]The authors remark that this value shows the effect of film thickness on Young's modulus. If correct, this would be one of the first experimental demonstrations of this theoretically predicted effect
[b]Unusually high value for this material and thickness
[c]Special proprietary deposition process; sources identified in the references
[d]Unusually low value for this material and thickness

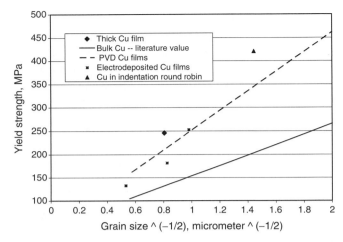

Fig. 12.11 Yield strength of copper plotted against grain size to the minus one-half power. Bulk materials are represented as the *solid line*

It is a testament to the progress of research in thin film characterization that the large uncertainty in the mechanical properties of even 1-μm thick films of common materials such as aluminum and copper, that existed as late as around the year 1990, is past. Now the mechanical properties of films with dimensions of 1 μm and larger are considered to be fully understandable based on their microstructure. As an example of this, Fig. 12.11 displays the dependence of the yield strength, obtained by microtensile testing, of some copper films [74]. The yield strength values follow the trend for bulk copper. There are no "mysterious effects" arising from the *micrometer* size scale.

12.5 Controlled Joule Heating by Alternating Current

The microelectronics industry often makes use of thin patterned films that are not easily accessed by many of the more established methods. For example, lower-level conductors may be less than 100 nm in width, deposited into trenches, fully encased within a dielectric, and located several micrometers below the surface of a device. In order to characterize these materials with such methods, either cross sections must be prepared from fabricated devices, or special test configurations must be made, approximating the materials of interest. As a result, test specimens might not accurately represent the microstructures, geometries, or mechanical constraint conditions undergone by final products. To consider such effects, approximations and extrapolations must then be incorporated, potentially leading to considerable measurement uncertainty.

Fig. 12.12 Schematic specimen suitable for electrical approach to fatigue and strength
testing. *V* represents voltage
probes and *i* represents current probes

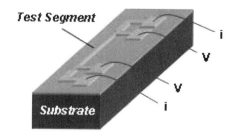

This section introduces an alternative approach to testing the fatigue and strength properties of thin-film materials in their as-fabricated condition, by application of controlled Joule heating. This approach is applicable to systems with extremely fine dimensions, meandering geometries, multiple layers, and/or passivations. The test is based on early observations by Philofsky et al. [75], who first related alternating current loading to thermal fatigue. An electrical test method for mechanical response circumvents the need for special test configurations or specimen cross sectioning. The primary experimental requirement is that the material of interest be accessible electrically, i.e., one must be able to contact it through electrical probes or wire-bonded connections, shown schematically for a nonpassivated specimen in Fig. 12.12. Careful experimental design may allow implementation in a manufacturing environment. We discuss the principle of cyclic Joule heating as a basis for thermal fatigue testing and strength testing, and conclude with descriptions of microstructural changes induced by the test.

12.5.1 Cyclic Joule Heating

Application of electric current to a thin patterned film leads to Joule heating, which refers to the increase in energy of a conductor due to interactions between the conducting particles and ions making up the conductor. Alternating current applied to a conductor under conditions of low frequency and high current density can then result in cyclic Joule heating.

Mechanical testing of a patterned film on a substrate by use of AC is based on the principle that within each half cycle of current, joule heating of the current-carrying segment causes not only that segment, but also the immediately surrounding materials, to become heated. A cyclic strain, $\Delta\varepsilon$, results from the change in temperature, ΔT, if there is a difference in coefficients of thermal expansion, $\Delta\alpha$, between the film and the substrate:

$$\Delta\varepsilon = \Delta\alpha \cdot \Delta T. \qquad (12.4)$$

For the case of a film constrained by a semi-infinite substrate, nearly all of the cyclic strain is taken up by the film. Assumption of an appropriate constitutive relation such as linear elasticity then leads to a determination of cyclic stress in the film. These cyclic variables provide a means for performing fatigue tests and, as shown later, strength tests.

Fig. 12.13 Schematic relationships between low-
frequency, high-density alternating current input into
a patterned metal segment
and resulting temperature
and mechanical stress be-
haviors as functions of time.
Values for current density,
temperature, and stress are
order-of-magnitude examples

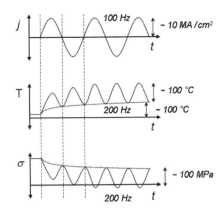

Figure 12.13 shows schematically the variations of current density, j, temperature, T, and stress, σ, in a patterned metal segment with time, upon exposure to an
AC current. A key requirement for successful application of electrical-based tests
is knowledge of specimen temperature with time. Calibration of the variation of
electrical resistance, R, with temperature can be performed by measuring the (low-
current) steady-state resistance of a test structure that is allowed to reach thermal
equilibrium in a furnace at a variety of temperatures spanning the range of interest.

This results in dR/dT for a specific specimen material and provides a means for
determining specimen temperature during the course of a test through

$$R(T) = R_0 + \Delta T \frac{dR}{dT}, \tag{12.5}$$

where $R(T)$ is the measured specimen resistance at the unknown temperature T, R_0
is the initial room-temperature resistance, and $\Delta T = T - T_0$, where T_0 is room temperature. Solving for T gives:

$$T = \frac{(R(T) - R_0)}{\left(\dfrac{dR}{dT}\right)} + T_0. \tag{12.6}$$

An example data set for the variation of temperature with time during the course
of a test, as measured by resistance, is shown in Fig. 12.14, for the case of a copper line subjected to AC stressing at an RMS current density of $17.5\,\mathrm{MA\,cm^{-2}}$,
at 100 Hz. Note that it is not difficult to attain temperature amplitudes of several
hundred degrees Celsius.

In order for this method to be useful as a test of mechanical response, it is important to identify the conditions that lead to well-controlled cyclic Joule heating without interference from phenomena that can affect the strain cycle, e.g., insufficient
temperature cycling, or cause unintended forms of damage, e.g., electromigration.

Heat flow in materials that typically compose a microelectronic device is
rapid enough to allow for nearly complete dissipation within one power cycle
for frequencies up to several kilohertz [76]. Within this range of frequencies, the

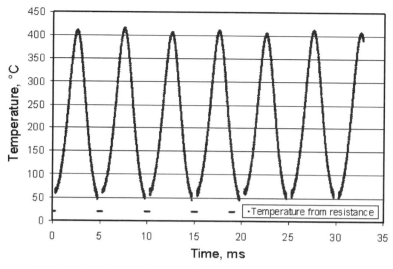

Fig. 12.14 Example of the time-resolved variation of temperature in a patterned copper segment subjected to current stressing at $17.5\,\mathrm{MA\,cm^{-2}}$, at an AC frequency of $100\,\mathrm{Hz}$

instantaneous power varies slowly enough that steady-state conditions can be maintained throughout each power cycle. For higher frequencies, this is not the case, and it can be shown for an infinitely long, narrow line on a substrate that the temperature amplitude will decrease with the logarithm of the frequency [76]:

$$\Delta T = \frac{Q}{\pi k}(-\ln(2\omega)/2 + \text{const}),\qquad(12.7)$$

where Q is the amplitude of the power per unit length, $\omega/2\pi$ is the frequency of the electrical signal, and k is the thermal conductivity of the substrate. Practically speaking, electrical signal frequencies below several kilohertz have been found to provide sufficient temperature amplitudes for mechanical testing.

A phenomenon that could potentially induce an unintended form of damage is electromigration, where atoms of the conductor can be displaced due to momentum transfer from the conducting electrons [77]. Resulting failures take the form of voids and hillocks, which can lead to open- or short-circuit events. Electromigration damage takes place by diffusive mechanisms, suggesting the importance of time during which a stress-based driving force is active. Under conditions of AC, the important factor is the time available for diffusion during one power cycle. An estimate of diffusion distance, x, during a power cycle can be made by considering $x = \sqrt{Dt}$, where D is diffusivity and t is the duration of the power cycle. For lattice diffusion, a typical value of D for metallic conductors used in microelectronics is $10^{-18}\,\mathrm{m^2\,s^{-1}}$; for grain boundary diffusion, a typical value of D is $10^{-15}\,\mathrm{m^2\,s^{-1}}$. A test running at an electrical frequency of $100\,\mathrm{Hz}$, i.e., power cycling at $200\,\mathrm{Hz}$, gives $t = 0.005\,\mathrm{s}$, resulting in diffusion distances of $0.05\,\mathrm{nm}$ for lattice diffusion and

1.5 nm for grain boundary diffusion. These are relatively short distances, producing a very small amount of atomic motion in each power cycle. With a reversal in the direction of the current, most if not all of the motion may be reversed [78]. Consideration of low enough frequencies to allow nearly complete thermal dissipation in each power cycle, combined with high enough frequencies to avoid conditions that lead to electromigration damage, suggests an effective range of testing frequencies of several tens to several thousands of hertz.

12.5.2 Fatigue Testing

Cyclic Joule heating under properly controlled conditions provides a convenient method for performing strain-control fatigue tests on patterned materials adherent to substrates. Measurement of fatigue lifetime from a specimen entails applying constant-amplitude AC to the specimen and measuring the time to open circuit failure. Conversion to current density compensates for slight variations in cross-sectional area from specimen to specimen. Generation of a lifetime curve requires testing of multiple specimens over a range of current densities, measuring their times to failure, and plotting all data on the same graph.

Raw test data may take the form of current density plotted against lifetime, as shown in Fig. 12.15 for interconnects of copper and Al-1Si [79]. Note that the form of this data is reminiscent of $S - N$ curves commonly found in fatigue testing of bulk materials. The data may be alternatively represented in terms of temperature amplitude, total strain amplitude, or stress amplitude vs. lifetime, through use of the resistance calibration described earlier. Replotting the data of this figure in terms of number of cycles to failure shows that the AC stressing method can be used for both low- and high-cycle fatigue measurements, with the conventional view that the transition between these regimes occurs for lifetimes in the range of 10^4–10^5 cycles [80].

Fig. 12.15 Fatigue lifetime plot for two types of patterned metal films on a silicon substrate. *Error bars* represent $0.3 \, \mathrm{MA \, cm^{-2}}$ uncertainty in current density. AC electrical frequency was 100 Hz

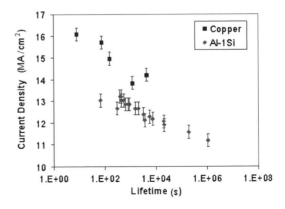

Table 12.2 Total strain amplitudes corresponding to various temperature ranges for the cases of aluminum on silicon ($\Delta\alpha$ (Al/Si) $= 2.05 \times 10^{-5}$ per $°C$) and copper on silicon ($\Delta\alpha$ (Cu/Si) $= 1.30 \times 10^{-5}$ per $°C$)

ΔT ($°C$)	$\Delta\varepsilon$ (Al/Si) (%)	$\Delta\varepsilon$ (Cu/Si) (%)
400	0.82	0.52
350	0.72	0.46
300	0.62	0.39
250	0.51	0.33
200	0.41	0.26
150	0.31	0.20
100	0.21	0.13
50	0.10	0.07

In a typical laboratory setup, current can usually be sufficiently well controlled without the need for highly specialized electronic equipment, to enable applied temperature amplitudes that are repeatable to approximately $\pm 1 °C$. This suggests control over total strain amplitude down to the range 0.001–0.002%. These ranges of temperature and strain are sufficient for characterizing the types of thermal cycling undergone by microelectronic devices during manufacture and use. Table 12.2 shows some relationships between temperature amplitude and total strain amplitude for fatigue testing of aluminum films on silicon and copper films on silicon.

With strain control down to 10^{-5} and frequencies of up to several kilohertz, AC stressing also enables the study of ultra-high-cycle fatigue behavior. For example, a test running at an AC frequency of 5 kHz can result in application of 10^{10} cycles in about 11.5 days. The facts that numerous test specimens can reside on one wafer, and wire-bonding methods can be used to control each specimen, point out another advantage of this test method – a large number of tests can be run in a relatively short period, so that statistically significant lifetime data can be generated much more quickly, compared to serial testing of individual specimens at lower frequencies. Furthermore, this approach to thermal fatigue testing is much faster than methods such as wafer curvature measurements associated with temperature cycling within a furnace.

Examples of microstructural changes observed in electrically cycled patterned metal interconnects will be covered in a later section, with emphases on grain structure and defect behavior. It will be shown that the damage induced by this test method does indeed take place by cyclic deformation mechanisms.

12.5.3 Strength Measurement

Cyclic joule heating can also be used to evaluate ultimate strength of patterned materials on substrates [81]. Predictions of fatigue lifetime from measurements of tensile properties have been made in numerous engineering materials through application of a method termed the modified universal slopes equation, which was the result of an empirical correlation study of the fatigue lifetime behaviors of 50 steels, aluminum, and titanium alloys [82]. This strain-based approach to fatigue analysis represents total strain range, $\Delta\varepsilon$, as a sum of power-law terms in the elastic and plastic strain ranges

$$\frac{\Delta\varepsilon}{2} = \left(\frac{\sigma_{\mathrm{f}}}{E}\right)(2N_{\mathrm{f}})^b + \varepsilon_{\mathrm{f}}(2N_{\mathrm{f}})^c, \tag{12.8}$$

where σ_{f} is the true fracture strength, E is Young's modulus, N_{f} is the number of cycles to failure, ε_{f} is the true fracture strain, and b and c are fitting constants. Assumption of globally elastic behavior in each cycle, reasonable for the types of strains that microelectronic devices typically undergo, leads to

$$\frac{\Delta\varepsilon}{2} = \left(\frac{\sigma_{\mathrm{f}}}{E}\right)(2N_{\mathrm{f}})^b. \tag{12.9}$$

Multiplying through by E gives a stress representation, also termed the Basquin relation

$$\Delta\sigma = \sigma_{\mathrm{f}}(2N_{\mathrm{f}})^b, \tag{12.10}$$

where $\Delta\sigma$ is the stress amplitude, σ_{f} is termed the fatigue strength coefficient in this relation, and b the fatigue strength exponent. Ultimate strength provides a reasonable estimate of σ_{f} for negligible necking [83]. Tensile data obtained from numerous materials were used to generate fatigue lifetime predictions, and those predictions compared to experimental fatigue data from the same materials [84]. Eighty percent of the fatigue data came within a factor of 3 of the predicted lifetime, which suggests that tensile data can indeed be used to reasonably predict cyclic behavior.

An inverse approach can also be taken – strength may be estimated from fatigue data such as that generated by the AC stressing method. A plot of stress amplitude vs. cycles to failure can be fit to the Basquin relation. Extrapolating such a fit to a single load reversal ($N_{\mathrm{f}} = 1/2$) leads to $\Delta\sigma = \sigma_{\mathrm{f}}$, and the stress amplitude becomes equal to the ultimate strength, providing the basis for estimating strength from electrically induced fatigue data.

The electrical method of inducing cyclic strain, in the absence of controlled substrate temperature, requires additional considerations due to the fact that it imparts a nonzero mean stress to the tested material. Mean stress must be taken into consideration when applying the Basquin relation, since a tensile mean stress is observed to reduce cyclic lifetime. Morrow [85] has suggested the following modification to the Basquin relation, to properly consider the effect of mean stress, σ_{m}:

$$\Delta\sigma = (\sigma_{\mathrm{f}} - \sigma_{\mathrm{m}})(2N_{\mathrm{f}})^b. \tag{12.11}$$

A log plot of stress amplitude vs. cycles to failure provides values for $\sigma_{\mathrm{f}} - \sigma_{\mathrm{m}}$ and b. In the absence of control of substrate temperature, fatigue testing by cyclic Joule heating inherently causes not only a nonzero mean stress, but also a nonconstant

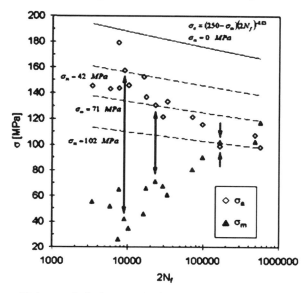

Fig. 12.16 Stress–lifetime analysis for pure aluminum interconnects subjected to electrically induced fatigue testing. *Arrows* represent correlations between stress amplitude and mean stress for each measurement. σ_a is stress amplitude and σ_m is mean stress

mean stress. In other words, the nonzero mean stress varies with applied stress amplitude. This is a result of the fact that thermal strains are added to the initial state of residual strain in the specimen. For the case of physical-vapor-deposited metals, the residual strain tends to be tensile, so that application of additional heat acts to reduce this tensile stress, resulting in a tensile mean stress. Higher current density increases stress amplitude, but decreases the mean tensile stress.

Figure 12.16 shows a plot of stress vs. lifetime for pure aluminum interconnects subjected to AC stressing, including results of a nonlinear regression used to determine the fatigue strength coefficient and exponent [81]. The solid line represents the case for zero mean stress, and the dashed lines represent cases for tensile mean stresses of 42 MPa, 71 MPa, and 102 MPa, determined by subtracting the stress corresponding to the measured temperature amplitude from the initial level of residual stress. The open symbols are applied stress amplitude, and the filled symbols are mean stress associated with each value of stress amplitude. The regression technique for this set of aluminum interconnects resulted in a fatigue strength coefficient (~ultimate strength) of 250 ± 40 MPa and a fatigue strength exponent of -0.03 ± 0.02. This estimate of ultimate strength compares nicely with the value of 239 ± 4 MPa obtained from microtensile specimens fabricated from the same films.

12.5.4 Microstructure Changes During AC Stressing

Electron-microscopic observations of the evolution of damage in aluminum and copper films subjected to cyclic Joule heating by alternating current have revealed heterogeneous behavior typical of plastic deformation in polycrystalline metals. For example, surface roughness was observed to increase with increasing cycling, but developed only in selected regions of stressed lines. The evolution of damage also included selective changes in grain size and crystallographic texture, which can be explained in terms of crystal plasticity.

Figure 12.17 shows a sequence of SEM images taken in plan-view from one area of one specimen after it was subjected to 0 s, 10 s, 20 s, and 40 s of thermal cycling. Note the progressive development of damage, observable after even just several seconds of cycling, taking the form of surface undulations in selected areas.

Continued cycling led to the formation of severe surface topography that eventually caused open-circuit failure. Figure 12.18 shows a sequence of SEM images that

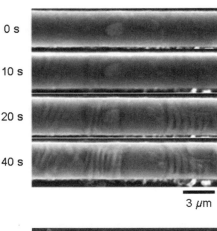

Fig. 12.17 SEM sequence showing development of surface topography in selected regions of an Al–1Si interconnect subjected to the indicated levels of accumulated cycling. Current density of 12.2 MA cm^{-2}, cyclic straining frequency of 200 Hz

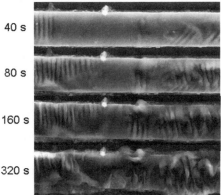

Fig. 12.18 SEM sequence showing advanced stages of surface damage in a different region of same specimen as that shown in Fig. 12.17

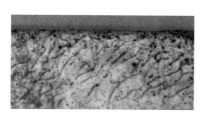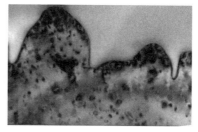

300 nm

Fig. 12.19 TEM images showing variations in surface topography from one region to another, after 1.39×10^5 thermal cycles, or 697 s of cycling, within the same specimen of Al–1Si shown in Figs. 12.17 and 12.18. The images also show differences in dislocation arrangements, with the image on the *left* containing more isolated dislocation segments and fewer dislocation loops than the image on the *right*. The *bottom edge* of both images corresponds to the substrate surface

demonstrate more advanced degrees of damage in the same specimen; the particular specimen under observation failed after 697 s of thermal cycling. Cross-sectional TEM observations showed that the surface topography that developed as a result of thermal cycling varied considerably from place to place within one specimen, as shown in Fig. 12.19, corresponding to natural variations in localized plasticity from grain to grain [86]. Both grains shown in the figure were part of a specimen cycled for 1.39×10^5 thermal cycles. The grain in the left image of the figure shows a flat surface, virtually unchanged from its original thickness of 0.5 μm. Surface upsets corresponding to a local film thickness of approximately 0.7 μm and depressions corresponding to a local film thickness of approximately 0.3 μm are visible in the grain shown in the right image of the figure. Figure 12.19 also shows the presence of dislocations in the film – the grain with a flatter surface contains residual dislocation segments, while the grain with surface undulations is free of isolated segments, but contains debris in the form of prismatic loops. In the latter case, dislocations could easily glide and leave the surface, resulting in upsets and depressions akin to intrusions and extrusions seen in purely mechanical fatigue; prismatic loop debris is typically left behind in face-centered cubic metals as a result of extensive dislocation glide [87]. In the case of the flatter surface, residual segments remained in the interior of the grain due to localized variations in grain size, dislocation sources, or resolved shear stress, any of which could inhibit dislocation motion.

Mapping of the grain structure by automated electron backscatter diffraction (EBSD) showed that with increased cycling, some grains grew considerably, while others disappeared [86]. An example is shown in Fig. 12.20, where after just 2,000 thermal cycles, grain "A" increased in area by 114%, representing an equivalent circular diameter increase from 3.0 to 4.4 μm. Grain "B" shows an area increase of 60%, corresponding to an equivalent circular diameter increase from 4.0 to 5.2 μm.

Measurements made over a 100 μm × 3 μm region of the tested interconnect showed that the mean grain diameter increased from 1.4 ± 0.4 μm at the start of the

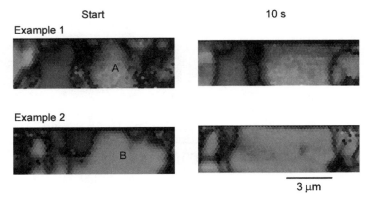

Fig. 12.20 EBSD maps showing two examples of grain growth in an Al–1Si interconnect subjected to 2,000 thermal cycles at 200 Hz. *Left images* represent grains at start of test; *right images* represent grains after cycling

test to 2.4 ± 0.4 μm after 6.4×10^4 thermal cycles. Correlations of residual dislocation content with EBSD pattern sharpness [86] showed that grains that grew tended to be those that had lower residual dislocation content. Grains that shrunk tended to be those that had higher residual dislocation content. This set of observations suggested that during rapid thermal cycling as induced by alternating current, grain growth takes place by strain-induced boundary migration [88, 89], wherein stored plastic strain energy differences from grain to grain drive growth.

12.6 Summary and Outlook

In a discussion of metrologies developed at NIST for determination of mechanical response of micro- and nanoscale systems, we have addressed three emerging approaches. First, we discussed new tools to measure and image mechanical properties that exploit the nanoscale spatial resolution of the atomic force microscope. Contact-resonance AFM methods examine the resonant modes of a vibrating AFM cantilever when its tip is in contact with a specimen. The frequency information is used to deduce the tip–sample contact stiffness, from which near-surface properties such as local elastic modulus can be extracted. In addition to the basic physical principles involved, the experimental apparatus and measurement methods were discussed. Examples were also given of the different information that can be obtained with contact-resonance AFM methods. This includes not only measurements of the indentation modulus M at a single sample location, but also elastic-property imaging and mapping, and evaluating thin-film adhesion at a buried interface. The results discussed here, as well as those from groups around the world, show significant progress in advancing the state of the art. It is anticipated that contact-resonance AFM techniques will continue to develop, and will play a significant role in future nanotechnology efforts by providing quantitative nanomechanical information.

In regard to microtensile testing for direct determination of film properties, the practical issue facing experimentalists now is the need for imaging at magnifications higher than optical microscopy can provide. The logical path forward would be to test routinely within the SEM or TEM. Haque et al. [59] have shown one possible path, but it requires a big advance in specimen fabrication over the one-mask, micrometer-scale lithography that suffices for specimens for testing under the optical microscope.

Scientific impetus for additional advances in mechanical test techniques may come from a new frontier of materials research – mechanical behavior of small volumes of material. The effects of "length scale" on mechanical behavior have received intense study recently [90]. A complementary viewpoint is that fine-scale microstructural variation may become significant when an experiment samples the behavior of materials of very small volumes, for example, the material under the sharp tip of a nanoindenter [91]. Tensile testing of *films* has not contributed any of the surprising results in this new field, but eventually the question of whether and when small volume effects appear in the behavior of tensile specimens will have to be answered.

The ultimate length scale for mechanical property measurements is the few-atom scale, which may be a source of insight into the behavior of practical materials, and may also be important in establishing the accuracy of atomistic modeling approaches such as molecular dynamics. In this regard, the results reported by Rubio-Bollinger et al. [92], which seem to show that a chain of gold atoms can have a strength value (calculated by the authors from the results in [92]) surpassing that of high-strength steel, stand as a challenge to material testers and material developers alike. Another provocative recent result by Uchic et al. [93] seems to show that small volume effects can begin to appear in stressed cross sections with an area of hundreds of square micrometers. At the present time this result seems to be at odds with the typical results of microtensile testing of specimens with the typical $1 \times 10\,\mu m^2$ cross section, because the tensile results are generally considered to be consistent with the behavior of the bulk material when the Hall–Petch rule is used to account for the grain size.

Extending our test methodologies to approaches that do not require special specimen configurations, we have presented a method for inducing thermal fatigue in patterned thin films on substrates, by application of alternating current under conditions of low frequency and high current density. The sole requirement is that the specimen be accessible electrically. The AC stressing method makes use of differences in coefficients of thermal expansion between the test film of interest and its substrate. Measurements of fatigue lifetime are straightforward with this approach, when coupling the raw data with calibration of specimen resistance with temperature. Ultimate strength can also be measured, by extrapolation of fatigue data back to a single load reversal. At this point, the electrical approach is still under development, but holds promise for accessing the mechanical response of extremely narrow structures, unusual geometries, and films buried beneath other layers, all of which pose significant difficulties for the more established methods for measuring mechanical response.

Acknowledgments The contributions of current and former NIST coworkers (R. Geiss, M. Kopycinska-Müller, and E. Langlois) are gratefully acknowledged. We also value interactions with the research groups of J. Turner (Univ. Nebraska–Lincoln) and W. Arnold and U. Rabe (Fraunhofer Institut für Zerstörungsfreie Prüfverfahren, Saarbrücken, Germany). Nanoindentation measurements for comparison to contact-resonance AFM results were provided by N. Jennett (National Physical Laboratory, UK), A. Rar (then at Univ. Tennessee–Knoxville), and D. Smith (NIST). We thank E. Arzt (Max-Planck-Institute for Metals Research, Stuttgart, Germany), C. Volkert (Univ. Göttingen, Germany), and R. Mönig (Forschungszentrum Karlsruhe, Germany) for fruitful collaborations on development of electrical test methods for thermal fatigue. We are grateful for support from the NIST Office of Microelectronics Programs.

References

1. Binning G, Quate CF, Gerber Ch (1986) Phys. Rev. Lett. 56:930
2. Maivald P, Butt HJ, Gould SAC, Prater CB, Drake B, Gurley JA, Elings VB, Hansma PK (1991) Nanotechnology 2:103
3. Burnham NA, Kulik AJ, Gremaud G, Gallo PJ, Oulevey F (1996) J. Vac. Sci. Technol. B 14:794
4. Troyon M, Wang Z, Pastre D, Lei HN, Hazotte A (1997) Nanotechnology 8:163
5. Rosa-Zeiser A, Weilandt E, Hild S, Marti O (1997) Meas. Sci. Technol. 8:1333
6. Butt HJ, Cappella B, Kappl M (2005) Surf. Sci. Rep. 59:1
7. Zhong Q, Inniss D, Kjoller K, Elings VB (1993) Surf. Sci. 290:L688
8. Yamanaka K, Ogiso H, Kolosov OV (1994) Appl. Phys. Lett. 64:178
9. Cuberes MT, Assender HE, Briggs GAD, Kolosov OV (2000) J. Phys. D: Appl. Phys. 33:2347
10. Yamanaka K, Nakano S (1996) Jpn. J. Appl. Phys. 35:3787
11. Rabe U, Arnold W (1994) Appl. Phys. Lett. 64:149
12. Yamanaka K, Nakano S (1998) Appl. Phys. A 66:S31
13. Crozier KB, Yaralioglu GG, Degertekin FL, Adams JD, Minne SC, Quate CF (2000) Appl. Phys. Lett. 76:195
14. Hurley DC, Shen K, Jennett NM, Turner JA (2003) J. Appl. Phys. 94:2347
15. Rabe U (2006) In: Bushan B, Fuchs H (eds) Applied Scanning Probe Methods, vol. II. Springer, New York, chap. 2
16. Huey BD (2007) Annu. Rev. Mater. Res. 37:351
17. Rabe U, Amelio S, Kester E, Scherer V, Hirsekorn S, Arnold W (2000) Ultrasonics 38:430
18. Arinero R, Lévêque G (2003) Rev. Sci. Instrum. 74:104
19. Johnson KL (1985) Contact Mechanics. Cambridge University Press, Cambridge, UK
20. Vlassak JJ, Nix WD (1993) Philos. Mag. A 67:1045
21. Rabe U, Amelio S, Kopycinska M, Hirsekorn S, Kempf M, Göken M, Arnold W (2002) Surf. Interface Anal. 33:65
22. Kopycinska-Müller M, Geiss RH, Hurley DC (2006) Ultramicroscopy 106:466
23. Yamanaka K, Tsuji T, Noguchi A, Koike T, Mihara T (2000) Rev. Sci. Instrum. 71:2403
24. Prasad M, Kopycinska M, Rabe U, Arnold W (2002) Geophys. Res. Lett. 29:13
25. Stan G, Price W (2006) Rev. Sci. Instrum. 77:103707
26. Hurley DC, Kopycinska-Müller M, Kos AB, Geiss RH (2005) Meas. Sci. Technol. 16:2167
27. Rabe U, Amelio S, Kopycinska M, Hirsekorn S, Kempf M, Göken M, Arnold W (2002) Surf. Interface Anal. 33:65
28. Passeri D, Bettucci A, Germano M, Rossi M, Alippi A, Orlanducci S, Terranova ML, Ciavarella M (2005) Rev. Sci. Instrum. 76:093904
29. Zheng Y, Geer RE, Dovidenko K, Kopycinska-Müller M, Hurley DC (2006) J. Appl. Phys. 100:124308

30. Stan G, Ciobanu CV, Parthangal PM, Cook RF (2007) Nano Lett. 7(12):3691
31. Kopycinska-Müller M, Geiss RH, Müller J, Hurley DC (2005) Nanotechnology 16:703
32. Hurley DC, Turner JA (2007) J. Appl. Phys. 102:033509
33. Rabe U, Kopycinska M, Hirsekorn S, Muñoz Saldaña J, Schneider GA, Arnold W (2002) J. Phys. D: Appl. Phys. 35:2621
34. Tsuji T, Yamanaka K (2001) Nanotechnology 12:301
35. Yamanaka K, Maruyama Y, Tsuji T, Nakamoto K (2001) Appl. Phys. Lett. 78:1939
36. Passeri D, Bettucci A, Germano M, Rossi M, Alippi A, Sessa V, Fiori A, Tamburri E, Terranova ML (2006) Appl. Phys. Lett. 88:121910
37. Kos AB, Hurley DC (2008) Meas. Sci. Technol. 19:015504
38. Hurley DC, Kopycinska-Müller M, Langlois ED, Kos AB, Barbosa N (2006) Appl. Phys. Lett. 89:021911
39. Sarioglu AF, Atalar A, Degertekin FL (2004) Appl. Phys. Lett. 84:5368
40. Read DT, Dally JW (1993) J. Mater. Res. 8:1542
41. Cheng YW, Read DT, McColskey JD, Wright JE (2005) Thin Solid Films 484:426
42. Read DT, Volinsky AA (2007) In: Suhir E, Lee YC, Wong CP (eds) Materials and Structures: Physics, Mechanics, Design, Reliability, Packaging: Volume 1. Materials Physics / Materials Mechanics. Springer, New York, chap. 4, pp. 135–180
43. Dieter G (1986) Mechanical Metallurgy. McGraw-Hill, New York
44. Hoffman RW (1989) In: Bravman JC, Nix WD, Barnett DM, Smith DA (eds) Mater. Res. Soc. Symp. Proc., vol. 130. Materials Research Society, Warrendale, pp. 295–306
45. Brotzen FR (1994) Int. Mater. Rev. 39:24
46. ASTM (2004) E8-04 Standard Test Methods for Tension Testing of Metallic Materials. American Society for Testing and Materials, West Conshohocken, PA
47. ASTM (2004) E345-93(2002) Standard Test Methods of Tension Testing of Metallic Foil. American Society for Testing and Materials, West Conshohocken, PA
48. Hoffman RW (1966) In: Hass G, Thun RE (eds) Physics of Thin Films. Academic, New York, pp. 211–273
49. Menter JW, Pashley DW (2007) In: Neugebauer CA, Newkirk JB, Vermilyea DA (eds) Structure and Properties of Thin Films. Wiley, New York, pp. 111–148
50. Ding XY, Ko WH, Mansour JM (1990) Sens. Actuators A Phys. 23:866
51. Ruud JA, Josell D, Spaepen F, Greer AL (1993) J. Mater. Res. 8:112
52. Steinwall JE (1994) Ph.D. thesis, Cornell University, Ithaca, NY
53. Sharpe WN, Yuan B, Edwards RL (1997) J. Microelectromech. Syst. 6:193
54. Read DT (1998) J. Test. Eval. 26:255
55. Read DT, Dally JW (1995) J. Electron. Packaging 117:1
56. Read DT (1998) Int. J. Fatigue 20:203
57. Read DT (1998) Meas. Sci. Technol. 9:676
58. Fox JC, Edwards RL, Sharpe WN (1999) Exp. Tech. 23:28
59. Haque MA, Saif TA (2004) Proc. Natl. Acad. Sci. USA 101:6335
60. Espinosa HD, Prorok BD, Peng B (2004) J. Mech. Phys. Solids 52:667
61. Read DT, Cheng YW, Keller RR, McColskey JD (2001) Scr. Mater. 45:583
62. Bruck HA, McNeill SR, Sutton MA, Peters WH (1989) Exp. Mech. 29:261
63. Read DT, Cheng YW, Sutton MA, McNeill SR, Schreier H (2001) In: Shukla A, O'Brien EW, French RM, Ramsay KM (eds) Proceedings of the SEM Annual Conference and Exposition on Experimental and Applied Mechanics. Society for Experimental Mechanics, Bethel, CT, pp. 365–368
64. Read DT, Cheng YW, Geiss R (2004) Microelectron. Eng. 75:63
65. Read DT, Cheng YW, McColskey JD (2002) In: Proceedings of the SEM Annual Conference and Exposition on Experimental and Applied Mechanics. Society for Experimental Mechanics, Bethel, CT, pp. 64–67
66. Hurley DC, Geiss RH, Kopycinska-Muller M, Muller J, Read DT, Wright JE, Jennett NM, Maxwell AS (2005) J. Mater. Res. 20:1186

67. LaVan DA, Tsuchiya T, Coles G, Knauss WG, Chasiotis I, Read DT (2001) In: Muhlstein C, Brown SB (eds) ASTM STP 1413: Mechanical Properties of Structural Films. American Society for Testing and Materials, West Conshohoken, PA, pp. 16–27
68. Espinosa HD, Prorok BC, Peng B (2004) J. Mech. Phys. Solids 52:667
69. Haque MA, Saif TA (2002) In: Shukla A, French RM, Andonian A, Ramsey K (eds) SEM Annual Conference and Exposition on Experimental and Applied Mechanics. Society of Experimental Mechanics, Bethel, CT, pp. 134–138
70. Read DT, Cheng YW, McColskey JD, Keller RR (2002) In: Ozkan CS, Freund LB, Cammarata RC, Gao H (eds) Mater. Res. Soc. Symp. Proc., vol. 695. Materials Research Society, Warrendale, PA, pp. 263–268
71. Read DT, Geiss R, Ramsey J, Scherban T, Xu G, Blaine J, Miner B, Emery RD (2003) In: Bahr DF (ed) Mater. Res. Soc. Symp. Proc., vol. 778. Materials Research Society, Warrendale, PA, pp. 93–98
72. Yeung B, Lytle W, Sarihan V, Read DT, Guo Y (2002) Solid State Technol. 45(6):125
73. Sharpe WN, Jackson KM, Coles G, Eby MA, Edwards RL (2001) In: Muhlstein C, Brown SB (eds) ASTM STP 1413: Mechanical Properties of Structural Films. American Society for Testing and Materials, West Conshohoken, PA, pp. 229–247
74. Read DT, Keller RR, Barbosa N, Geiss R (2007) Metall. Mater. Trans. A 38A:2242
75. Philofsky E, Ravi K, Hall E, Black J (1971) In: Proc. 9th Annual Reliability Physics Symposium. IEEE, New York, pp. 120–128
76. Mönig R, Keller RR, Volkert CA (2004) Rev. Sci. Instrum. 75:4997
77. Ho PS, Kwok T (1989) Rep. Prog. Phys. 52:301
78. Ting LM, May JS, Hunter WR, McPherson JW (1993) In: Proc. 31st Annual Reliability Physics Symposium. IEEE, New York, pp. 311–316
79. Keller RR, Barbosa III N, Geiss RH, Read DT (2007) Key Eng. Mater. 345:1115
80. Manson SS, Halford GR (2006) Fatigue and Durability of Structural Materials. ASM International, Materials Park, OH, p. 64
81. Barbosa III N, Keller RR, Read DT, Geiss RH, Vinci RP (2007) Metall. Mater. Trans. 38A:2160
82. Muralidharan U, Manson SS (1988) J. Eng. Mater. Technol. Trans. ASME 110:55
83. Suresh S (1998) Fatigue of Materials, 2nd edn. Cambridge University Press, Cambridge, UK, p. 223
84. Nachtigall AJ (1975) Properties of Materials for Liquefied Natural Gas Tankage, ASTM STP 579. American Society for Testing and Materials, West Conshohocken, PA, pp. 378–396
85. Morrow JD (1968) Fatigue Design Handbook – Advances in Engineering. Society of Automotive Engineers, Warrendale, PA, pp. 21–29
86. Keller RR, Geiss RH, Barbosa III N, Slifka AJ, Read DT (2007) Metall. Mater. Trans. 38A:2263
87. Kuhlmann-Wilsdorf D, Wilsdorf HGF (1963) Electron Microscopy and Strength of Crystals. Interscience, New York, pp. 575–604
88. Beck PA, Sperry PR (1950) J. Appl. Phys. 21:150
89. Battaile CC, Buchheit TE, Holm EA, Wellman GW, Neilsen MK (1999) In: Mater. Res. Soc. Symp. Proc., vol. 538. Materials Research Society, Warrendale, PA, pp. 267–273
90. Haque MA, Saif MTA (2003) Acta Mater. 51:3053
91. Greer JR, Oliver WC, Nix WD (2005) Acta Mater. 53:1821
92. Rubio-Bollinger G, Bahn SR, Agrait N, Jacobsen KW, Vieira S (2001) Phys. Rev. Lett. 87:026101
93. Uchic MD, Dimiduk DM, Florando JM, Nix WD (2004) Science 305:986

Chapter 13
Mechanical Characterization of Low-Dimensional Structures Through On-Chip Tests

Alberto Corigliano, Fabrizio Cacchione, and Sarah Zerbini

13.1 Introduction

The field of microelectromechanical systems (MEMS) [27,41,42,50,51] is moving from its pioneering period to a growing diffusion phase. Many large-scale applications can nowadays be found in various fields of engineering: automotive, aerospace, consumer.

The large-scale industrial production obliges producers to more carefully focus on reliability issues related to various causes of failures and in particular on mechanical failures such as fatigue and fracture induced by accidental drop. It is therefore of paramount importance to measure and control the mechanical properties of materials used in MEMS and nanoelectromechanical systems (NEMS), in primis of polysilicon, which is by far the most diffused material in the production of MEMS. The successful fabrication and the reliable use of structures with feature sizes in the range of 1 µm to 1 mm is strongly contingent on a sufficiently rigorous understanding of their *length-scale*-dependent and *process*-dependent mechanical properties. In turn, such understanding requires the ability to measure the mechanical properties of micro-scale structures.

There exist today many different micro-scale mechanical test techniques for polysilicon at the micro-scale, researchers have explored a variety of physical principles and experimental setup which allow to determine mechanical properties of almost invisible structures [1, 3–5, 14–17, 19–21, 23, 28, 32–34, 38, 43, 44, 47, 49, 53, 55, 61, 62, 68, 69]. A major distinction can be done between so called *off-chip* [1,3,14–16,38,53,55,62] and *on-chip* [6,19,21,28,32,33,44] methodologies. In both cases, the micro-device is generally produced by the deposition and etching procedures. An off-chip tensile test generally resorts to some sort of external

A. Corigliano and F. Cacchione
Department of Structural Engineering, Politecnico di Milano, piazza Leonardo da Vinci 32, 20133 Milano, Italy
alberto.corigliano@polimi.it, cacchione@stru.polimi.it

S. Zerbini
MEMS Product Division, STMicroelectronics, via Tolomeo 1. 20010 Cornaredo, (Milano), Italy
sarah.zerbini@st.com

F. Yang and J.C.M. Li (eds.), *Micro and Nano Mechanical Testing of Materials and Devices*,   349
doi: 10.1007/978-0-387-78701-5, © Springer Science+Business Media, LLC, 2008

gripping mechanism actuating the force together with an external sensor which measures the response of the specimen. On the contrary, on-chip test devices are real MEMS in which actuation and sensing are performed with the same working principles of MEMS.

The purpose of the present chapter is first to give a brief review of state of the art related to the mechanical characterization of polysilicon at the micro-scale and second to discuss some recent results related to a completely on-chip approach for the mechanical characterization of thin and thick polysilicon layers pursued recently in [8–11, 21, 22, 24, 65]. Three different MEMS devices for on-chip testing are here discussed, which load up to rupture under bending thin $(0.7\,\mu m)$ or thick $(15\,\mu m)$ polysilicon specimens. The first one is based on a rotational electrostatic actuator which contains a series of interdigitated comb-fingers and loads a couple of thin polysilicon specimens in bending in the plane parallel to the substrate. The second one loads the couple of thin polysilicon specimens in the plane orthogonal to the substrate by a parallel plate actuator which moves in the direction orthogonal to the substrate. The third device is conceived to cause initiation and propagation of a crack in a thick polysilicon specimen. It is based on a large number of comb-finger electrostatic actuators which deform a notched specimen through a lever system. In all three cases discussed, the data reduction procedure is based on the measurement of the capacitance variation of a displacement sensor, which makes it possible to determine the Young's modulus of various specimens and the maximum stress at rupture.

An outline of the chapter is as follows. Section 13.2 is dedicated to a brief discussion of state of the art concerning off-chip and on-chip methodologies recently proposed for MEMS mechanical characterization. Section 13.3 contains a brief description of the fabrication process and of the data reduction procedure adopted in the experimental tests carried out by the authors. Section 13.4 is devoted to the description of the Weibull approach for the determination of the failure probability of polysilicon MEMS. Sections 13.5 and 13.6 are, respectively, dedicated to the description of the rotational and parallel plate electrostatic actuators used for mechanical characterization of thin polysilicon films as well as relevant experimental results. Section 13.7 concerns the description of devices and the discussion on experimental results relevant to the test structures used for mechanical characterization of thick polysilicon films. Finally, Section 13.8 contains some closing remarks.

13.2 Mechanical Characterization of Polysilicon as a Structural Material for MEMS

13.2.1 Polysilicon as a Structural Material in MEMS

Polysilicon is by far the most common structural material in MEMS applications. It is used for a huge variety of applications, mainly because of two factors (1) the existence of well-established deposition technologies, in which polycrystalline silicon has a very important role since the beginning of the microelectronics era and (2) the excellent physical properties of this material. Its Young's modulus is higher than that of titanium and comparable with that of steel. The rupture resistance of silicon at the micro-scale is within the resistance range of the best construction steels, while its density is less than aluminium's. Silicon has high thermal conductivity and small thermal expansion coefficient, and its melting point is just $100\,^{\circ}C$ less than iron's. The excellent thermal properties make it a very good material for high temperature applications.

Since polysilicon is an aggregate of mono-crystalline silicon grains, its properties depend on the behaviour of the grains, on their shapes, on their crystal orientation, and on the physical characteristics of grain boundaries. This means that the overall properties of polysilicon are strongly influenced by the process used for the deposition. Process data such as the deposition temperature and the deposition pressure affect the final properties of the material.

Besides the dependence of physical properties on the processes, it must be emphasised that the measure of these properties is a very difficult and challenging task at the micro-scale. Earlier work conducted by several researchers revealed significant differences in the elastic properties and the nominal strength of polysilicon without providing in-depth explanation. A main question was arisen, therefore, as to whether the newly evolving test methodologies are adequately precise. In pursuing this question, a round robin study [54] demonstrated the inconsistency of the measured modulus and strength values, even when specimens from the same source were examined. The material used in that work was fabricated in close physical proximity from the same wafer of the same run and in the same deposition reactor at the Microelectronics Center of North Carolina (MCNC, now Cronos-JDS Uniphase). In the round robin study, values of the elastic modulus differed considerably, namely from 132 to 174 GPa, and the strength also had a rather wide scattering, ranging from 1.0 GPa, for specimens tested in tension, to 2.7 GPa for specimens tested in bending. A second round robin examination, conducted on the material fabricated at the Sandia National Laboratories [39] also showed signs of inconsistent rupture strengths due to the dependence on the specimen size and the measurement technique. A relatively new round robin test was carried out by Tsuchiya et al. [64]. The specimens, produced by the same process, in the same wafer, were distributed to five different research groups. In this case the measured modulus varied from 134 to 173 GPa, while the fracture strength varied from 1.44 to 2.51 GPa. As a result of these findings, it appears advisable for any micro-fabrication facility not to use the properties cited in the literature for final design and verifications but to identify the most feasible measurement technique and to conduct measurements for every fabrication run.

13.2.2 Testing Methodologies

According to some definitions provided by ASTM (American Society for Testing and Materials) standards for testing at the macro-scale, among material properties of interest in the context of the present discussion are (1) the *Young's modulus*, defined as the slope of the linear part of the stress–strain curve; (2) the *Poisson's ratio*, associated with the lateral expansion or contraction when the material is subjected to uniaxial test in the linear range; (3) the *fracture strength*, defined as the normal stress at the beginning of fracture and (4) the response to cyclic loading in terms of the *S–N curve*, related to the number of cycles at rupture. In order to measure material properties, one should be able to construct a specimen according to a given design, apply an external input in terms of forces or displacements and measure the specimen response using direct procedures, in the sense that the variable of interest should be (almost) directly measured. All these steps are fully standardised at the macro-scale and are currently applied for testing construction materials like steel and concrete. Unfortunately, these practices cannot be easily fulfilled at the scale of MEMS. In particular, one has to resort to fully or partially indirect approaches, e.g. to measure the Young's modulus, the bending of cantilever beams is often utilised; from which deflection is measured and the property of interest is calculated on the basis of an analytical or numerical model. Even during on-chip tension tests some sort of inverse analysis has to be performed since, in general, only capacitance variations are measured directly while deformations are obtained on the basis of a numerical model. Many testing methodologies have been proposed in scientific literature for the extraction of static mechanical properties of polysilicon [56]. Limiting the attention to silicon MEMS, a general classification of test procedures can be made between *off-chip* and *on-chip* devices. In both cases the micro-device is generally produced by deposition and etching procedures.

On-chip test devices [21] are real MEMS in which actuation and sensing are performed with the same working principles of MEMS. On-chip devices rely on the fact that all the mechanical parts needed to load the specimen and the majority of the ones (electrical or optical) needed for the measure of displacements and strains are built together with the specimen during the micromachining fabrication process. Usually, in these structures two main parts (the actuator and the sensing devices) can be found. In many cases these consist of a large number of capacitors that can be used for the creation of an electric field, which in turn causes a force to act onto the specimen, or for the measure of a capacitance which is directly related to the displacement of the specimen itself. The advantage of on-chip testing methods is linked to the ease of fabrication and use (usually without costly equipments) and to the fact that complex handling and alignment of the specimen are avoided. The major drawback is that the force developed by on-chip actuators can be insufficient to break specimens in quasi-static conditions and that the maximum displacement is also limited in the order of micrometre. On-chip testing of MEMS devices is especially advocated since the thin-film microstructure and the state of residual stress is a *strong* function of the micro-fabrication processes. Nevertheless it requires accurate modelling and numerical/analytical analyses of the whole device.

An off-chip test [56] generally resorts to some sort of external gripping mechanism actuating the force and an external sensor measuring the response of the specimen. All the experimental setups that use an external apparatus (load cells, micro-regulation screws, etc.) in order to create a stress state into the specimen are usually included in the category of off-chip testing. In this case a lot of attention has to be paid during the handling of tiny MEMS specimens during the system–specimen alignment and to the specimen gripping systems. The challenge of picking a specimen only a few micron thick, place it into a test machine and perform the test is a formidable task. The main advantage of off-chip methodologies is that the forces and the displacements can be relatively high to break even several micrometres thick specimens in pure tension and that many different configurations can be set up to create an a priori desired multi-axial stress state into the specimen.

In principle, material parameters for MEMS, and in primis the Young's modulus, E, can be determined using several test devices. Among others: tension tests, bending of cantilever beams, resonant devices, bulge tests, buckling tests. Clearly, the most direct approach is the tension test, but unfortunately this is not always applicable since it requires the deployment of considerable forces at the micro-scale in order to produce sensible deformation in the specimen. Hence a wealth of alternative solutions has appeared in the literature.

In Sects. 13.2.3 and 13.2.4, the mechanical characterization of polysilicon as a structural material for MEMS is discussed. It is decided to group the most important experiments in two principal families: quasi-static testing, used for the characterization of Young's modulus, Poisson's ratio and fracture properties (Sect. 13.2.3) and high-frequency testing, used for the characterization of fatigue properties (Sect. 13.2.4).

13.2.3 Quasi-static Testing

The very first test carried out for the quasi-static, mechanical characterization of silicon can be traced back to the 1980s [18]. The volume of the specimens used was approximately $1 \, \mathrm{cm}^3$ huge if compared with typical MEMS dimensions. From the first half of 1990s the increasing interest for mechanical characterization of polysilicon arose and the consequence was that a large number of test typologies were designed.

In the following, a selection of test devices and setups considered as the most common and interesting is presented. The classification is based upon the actuation mechanism and on the way the system response is sensed.

13.2.3.1 Off-Chip Tension Test

This is the most common and important technique for the measurement of mechanical properties for MEMS applications [3,12–16,48,52,55,56,60,63,64]. A MEMS specimen is produced and then placed on a testing system. Usually the specimen is gripped to the system with the aid of UV curing adhesives or via electrostatic gripping. This is the way to re-create common macro-scale testing techniques at the micro-scale. The displacement is imposed onto the specimen by the use of piezo-transducers with a resolution in the order of nanometre. The load cells measure the applied load with an accuracy of some μN. Specimen displacements can be measured either with the use of optical systems [48] or using laser interferometry [52] or via digital image correlation [12].

The results obtained from off-chip tension tests cover the most important quantities for mechanical design using polysilicon. The measure of Young's modulus and in some cases of Poisson's ratio [53] together with the rupture strength are the most common for all the research groups that worked with the off-chip tension testing. Besides, it is important to underline that with this kind of setup it is possible to study the scale effects due to specimen's dimension and the stress gradient acting in the specimen.

Recently Chasiotis [12] also demonstrated the determination of the fracture toughness K_{IC} of polysilicon. This result was achieved by means of a nano-indentation nearby the specimen that caused a crack to propagate through the substrate and partially involve the specimen. At the end of the process it was thus possible to have a pre-cracked specimen, necessary for a complete fracture mechanics characterization.

Other possible configurations for off-chip tension tests [56] can be obtained, e.g. by first patterning a tensile specimen onto the surface of a wafer and then exposing the gauge section by the etching of the wafer. Alternatively, a specimen can be fixed to a die at one end and actuated by means of an electrostatic probe at the other end.

13.2.3.2 Off-Chip and On-Chip Bending Test

Out-of-plane bending of test specimens is generally performed via an off-chip apparatus, e.g. by a diamond stylus which deflects a cantilever beam [29]. The deflection of the free end is measured and the Young's modulus is obtained through inverse modelling of the cantilever beam. However, if the beam is long, forces are small and difficult to calibrate; if the beam is short, forces are higher but inverse analysis of the beam is more involved.

Doubly supported silicon-micro-machined beams [37] can also be used to study the out-of-plane bending of materials. In this case a voltage is applied between the conductive polysilicon or micro-machined beam and the substrate to pull the beam down. The voltage that causes the beam to make contact can be related to the beam stiffness. Residual stresses in the beams and support compliance cause significant vertical deflections, which affect the performance of these micro-machined devices. Tests need to be supported by models of the devices that take into account the compliance of the supports and the geometrical nonlinear dependence of the vertical deflections on the stress in the beam.

In-plane bending is a classical test for MEMS since several structural parts of accelerometers are subjected to this kind of deformation. A typical on-chip layout is presented in Fig. 13.1, where a cantilever polysilicon beam attached to a moving mass (on the left) is subjected to bending induced by the fixed rectangular block (on

Fig. 13.1 On-chip in-plane bending test (courtesy of MEMS Production Division STMicroelectronics)

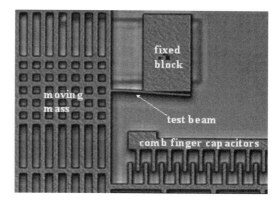

the right). Actuation is performed by means of interdigitated comb-finger capacitors. This test is often conducted to establish the flexural strength of the cantilever beam as a structural component, but it requires considerable care when employed for the evaluation of the Young's modulus due to the uncertainties in geometrical parameters and the model of the beam.

13.2.3.3 Test on Membranes (Bulge Test)

The *bulge test* is one of the earliest techniques used to measure the Young's modulus, Poisson's ratio and/or residual stress of non-integrated, free-standing thin structures. This method relies on the use of thin polysilicon membranes (circular, square, or rectangular in shape [70]), relatively easy to design and realize, bonded along their periphery to a supporting frame [31, 59, 67, 70]. Micro-fabrication techniques are particularly well suited for the creation of such test structures with reproducible and well-defined boundary conditions. During the test the membrane is loaded with a pressure difference acting on the top and bottom surfaces. The membrane deforms and its profile is measured by a profilometer. Usually the deflection at the centre is recorded as a function of the applied pressure. There are several analytical or semi-empirical formulae correlating the deflection to elastic properties. From this test it is possible to measure (1) the biaxial elasticity modulus; (2) the Poisson's ratio; (3) the nominal rupture strength and (4) the internal stresses (measuring the buckled configuration of the membrane). One of the shortcomings of this methodology is that sometimes the membrane separates from the substrate before the end of the test. Moreover, since the mechanical response of the membrane varies with the third power of its thickness and the fourth power of its lateral dimension, it is necessary to have a good fabrication technology and a very accurate measure of the thickness of the layer.

13.2.3.4 Nano-Indenter Driven Test

Nano-indenters are often used for the mechanical characterization of thin films. Basically there are two ways of using nano-indenters: film nano-indentation and the use of a nano-indenter as an actuator to load MEMS structures. Hardness (indentation) tests are routinely used to characterise large-scale structures. In direct analogy, considerable efforts have been made to develop nano-indentation technique to characterise micro-scale structures, and commercial instruments have been developed [40]. Nano-indentation experiments [25, 35, 40, 48] are performed allowing the tip to penetrate the film under study. During the penetration in the layer, an elastic–plastic stress state is generated. This is the main reason why the elastic characterization is done during the unloading phase, when the tip starts going backward to reach the rest position. The results of this experimental test are a force vs. penetration depth plot and the projected area of contact under the indenter. The Young's modulus and fracture properties are computed using some semi-empirical formulae. The main advantage of this method is that there is no need for an ad hoc designed specimen; it is sufficient to have a portion of material large enough to apply the nano-indenter. However, the application of this technique to thin films is complicated by several factors including substrate effects and pile-up of material around the indenter.

Another possibility is to use the nano-indenter tip as an actuator to perform a sort of off-chip test [26, 58]. In these tests the tip moves an extremity of the specimen, creating a stress state in it. The applied load is measured with a piezo-scanner, while there are different ways to measure the displacement of the specimen, like interferometry [26] or the measure of the displacement of the tip with aid of a laser beam and photodiodes [58]. This methodology could be very accurate, but it needs a very expensive instrumentation.

13.2.4 High-Frequency Testing

Polysilicon in the MEMS technology is also used for the fabrication of resonators, gyroscopes and other devices that oscillate at high frequencies during their whole life. One of the most important failure mechanisms for such systems is fatigue. Fatigue is usually interpreted as a phenomenon caused by the motion of the dislocations present in the material that can coalesce during the stress cycles to form micro-cracks. Micro-cracks then can lead to the formation of one or more macro-cracks which cause the failure of the structure. Polysilicon is a brittle material and there is no dislocation motion below temperature of about $900\,^{\circ}$C, therefore it is not expected to be prone to fatigue in usual operating conditions. Nevertheless, in the second half of 1990s some groups started working on this subject and found that also polysilicon can undergo fatigue after a large number of cycles, typically more than 10^9, combined with high stress levels. In order to reproduce experimentally fatigue failure, it is very important to work with experimental setups that can allow a relatively high-frequency testing (at least 1 kHz) which in turn allow to reach a large number of cycles in reasonable time. On-chip tests are usually the best choice for this kind of study because they can work at high frequencies and it is possible to perform multiple fatigue tests at the same time with some electrical control system.

On-chip test systems make in general use of electrostatic actuation between a fixed (stator) and a movable part (rotor) to load the specimen with a desired level of stress. The force developed by the actuator is proportional to the actuation area and inversely proportional to the gap between the rotor and the stator. It turns out that to have a force sufficiently large to induce fatigue into the specimens one should have a highly scaled lithography, a thick polysilicon layer and a design area big enough for the thousands of capacitors needed for the actuation. These requirements are the main reason why on-chip testing is not common for quasi-static characterization. Since the MEMS is a dynamic system, the force needed to *move* the seismic mass decreases if one loads the structure with a time-varying force at a frequency close to the resonant frequency of the system. This is what is usually done in fatigue tests for MEMS, in which a reasonable low voltage is used to cause the specimen to fatigue rupture. Very interesting results were obtained e.g. in [2].

13.2.4.1 Fatigue Mechanisms

As discussed in Sect. 13.2.4, it has been experimentally shown that fatigue in polysilicon is a possible failure mechanism. Nevertheless, the reasons why fatigue rupture occurs in polysilicon are not yet completely clear and understood. Among the most active groups in the study of fatigue in polysilicon are those of Pennsylvania State University and of Case Western Reserve University. These two groups proposed the most known and accepted interpretations for fatigue mechanisms in polysilicon.

The first group [44–47] proposed a mechanism called *reaction layer fatigue* which can be summarized as follows: when the polysilicon is first exposed to air the native oxide is formed, as one of the final steps of the process. The oxide thickens in the highly stressed regions and becomes the site for environmentally assisted cracks which grow in a stable way in the layer. When the critical size is reached, the silicon itself fractures catastrophically by trans-granular cleavage.

The second group [32–34] showed that even an high stress state cannot cause an appreciable growth of the native oxide, thus excluding the pure environmentally assisted fatigue. On the other hand, they noticed that fatigue life decreases with increasing the level of humidity. They did not propose a specific fatigue mechanism for polysilicon.

Fatigue testing remains one of the most open research area in the field of mechanical characterization of polysilicon. The mentioned researches are only examples of a wider discussion now active in the scientific literature [1].

13.3 On-Chip Testing with Electrostatic Actuation: Description of the General Methodology

13.3.1 Fabrication Process

The on-chip test devices discussed in this chapter can be produced following the surface micromachining process ThELMA^{TM} (Thick Epipoly Layer for Microactuators and Accelerometers) which has been developed by STMicroelectronics to realize in-silicon inertial sensors and actuators (see also [21, 24, 42] for further details). The Thelma process permits the realization of suspended structures anchored to substrate through very compliant parts (springs) which are capable of moving in a direction orthogonal to the plane of the wafer, such as the structure described in Sect. 13.6, or in a plane parallel to the underlying silicon substrate, such as those described in Sects. 13.5 and 13.7. The process flow exploits several state-of-the-art integrated circuit technology steps, together with dedicated MEMS operations, like high aspect ratio (trench), dry etch and sacrificial layer removal for structure release.

This technology is more complex than the *Surface Micromachining* but allows obtaining silicon structures with a relatively large thickness. This, in turn, increases the area of side surfaces and the global capacitance in electrostatic actuators which move parallel to the substrate.

The process consists of the phases described concisely hereafter and illustrated schematically in Fig. 13.2:

1. *Thermal oxidation of substrate.* The silicon substrate is covered by a layer of permanent oxide of 2.5 μm in thickness which is formed by a thermal treatment at 1,100 °C.
2. *Deposition and patterning of horizontal interconnections.* The first polysilicon layer is deposited above the thermal oxide. This layer (*poly1*) is used to define the *buried runners* which are used to bring potential and capacitance signals outside the device and can be used as structural layer in *thin polysilicon* devices, as done in Sects. 13.5 and 13.6.
3. *Deposition and patterning of a sacrificial layer.* An oxide layer of 1.6 μm in thickness is deposited by means of a *Plasma Enhanced Chemical Vapour Deposition* (PECVD) process. This layer, together with the thermal oxide layer, forms a 4.1-μm thick layer which separates the moving part from the substrate and is analogous with the sacrificial layer in a *Surface Micromachining* process.
4. *Epitaxial growth of the structural layer (thick polysilicon).* The polysilicon grows to a thickness of 15 μm in the reactors.
5. *Structural layer patterning by trench etching.* The parts of the mobile structure are fabricated by deep trench etch to reach the oxide layer.
6. *Sacrificial oxide removal and contact metallization deposition.* The sacrificial oxide layer is removed using a chemical reaction. In order to avoid *stiction* due to the capillary interaction, this is done in dry conditions. The contact metallization is deposited which is used to make the wire-bonding between the device and the metallic frame.

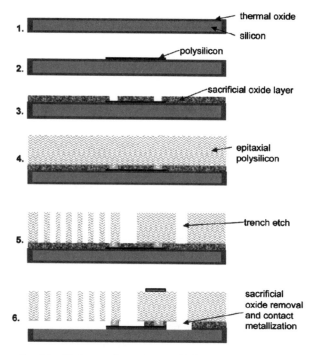

Fig. 13.2 Schematic of the Thelma surface micromachining process: (1) Substrate thermal oxidation; (2) Deposition and patterning of horizontal interconnections; (3) Deposition and patterning of a sacrificial layer; (4) Epitaxial growth of the structural layer (thick polysilicon); (5) Structural layer patterning by trench etching and (6) Sacrificial oxide removal and contact metallization deposition

Here, the focus is on the mechanical characterization of the thin polysilicon film named *poly1* and of the thick structural layer named *epipoly* as described previously in the Thelma process (see phases 2 and 4). The devices for on-chip testing described in Sects. 13.5 and 13.6 contains 0.7-μm thick *poly1* specimens (see Fig. 13.3) suspended over the substrate, which is obtained by removing part of the initial 2.5-μm thick oxide layer in the etching phase (phase 1). The device discussed in Sect. 13.7 concerns the mechanical characterization of thick epipoly specimens. Figure 13.4 shows a vertical section of the material, in which a clear columnar pattern can be recognized.

13.3.2 Test Structures and Data Reduction Procedures

In the test devices to be described in Sects. 13.5–13.7, the specimens are cofabricated with the actuator in order to obtain precise alignment and gripping of the specimens and also to reduce the setup size. The devices have an integrated system

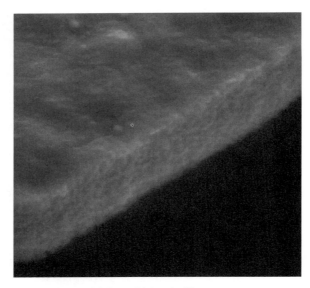

Fig. 13.3 SEM image of the tested 0.7-μm thick polysilicon

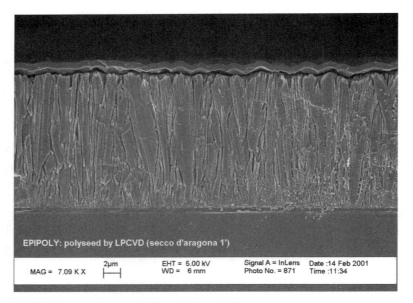

Fig. 13.4 SEM image of the tested 15-μm thick polysilicon

of electrostatic actuation: inter-digitized *comb-finger* actuators in the bending tests, and parallel plate actuator in the out-of-plane bending test.

During the test an input voltage V is applied to the actuator and a capacitance variation C is measured. The capacitance variation can be related to some significant displacement (or rotation) of the specimen through simplified analytical formulae or through finite element simulations of the complete device. The corresponding electrostatic force can then be determined as a function of the displacement from the derivative of the electrostatic energy, which, in turn, is proportional to the derivative of the capacitance. This general scheme for the data reduction will be detailed in Sects. 13.5–13.7 for each on-chip test structure. In Fig. 13.5, the meaningful

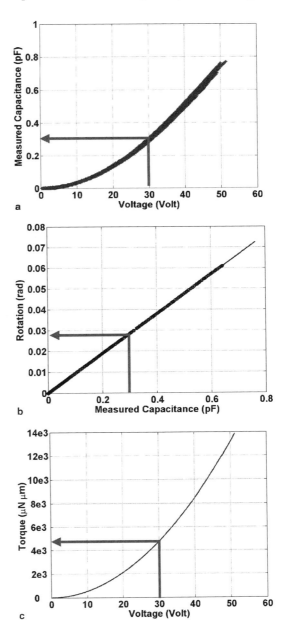

Fig. 13.5 Data reduction procedure applied to the rotational electrostatic actuator. (**a**) Experimental capacitance vs. voltage plots; (**b**) Rotation vs. measured capacitance plot; (**c**) Torque vs. applied voltage plot

Fig. 13.6 Rotational electrostatic actuator. Experimental
torque vs. rotation plots

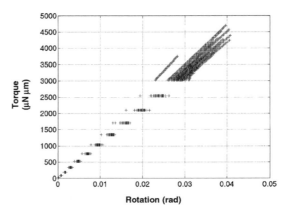

experimental plots used in the data reduction procedure are shown for the *rotational*
device in Sect. 13.5, while Fig. 13.6 shows the final torque vs. rotation plots.

Tests are carried out at room temperature and at atmospheric humidity, with a
probe station mounted on an optical microscope. A slowly increasing voltage is
applied in order to create quasi-static loading conditions in the specimen. Details on
the experimental setup can be found in the papers of the authors [21].

13.4 Weibull Approach

The Weibull approach [7,57,66] is widely used in the study of brittle materials, such
as ceramics. This is the reason why it has been recently applied also to the study
of rupture phenomena in polysilicon MEMS [15, 16, 21, 28, 30]. The Weibull theory
essentially provides a way to estimate the failure probability of a mechanical system,
starting from the computation of the probability of failure of its weakest part. The
theory is therefore also known as the *weakest link approach*. The Weibull cumulative
distribution function is found, applying the theorem of joint probability, by first
computing the probability of survival of a system composed by a large number of
elementary parts. This basic idea is extended to a general case, after introduction
of limiting hypotheses. The choice of the function giving the survival probability
of a single part was not originally based on a precise mechanical interpretation of
the rupture process. A recent and interesting discussion on the applicability of the
Weibull approach can be found in [7].

By means of the Weibull approach, it is possible to take into account the experimental scatter of the strength values of brittle materials, the statistical size effect and
the dependence of the probability of failure on the stress distribution. The application of the Weibull approach to a uniformly stressed uniaxial bar gives the following
equation for the probability of failure P_f:

$$P_f = 1 - \exp\left[-\frac{\Omega}{\Omega_r}\left\langle\frac{\sigma - \sigma_u}{\sigma_0}\right\rangle_+^m\right],\tag{13.1}$$

where Ω is the volume of the bar, Ω_r is a statistically uniform representative volume, σ_u, σ_0 and m are material parameters, $\langle \bullet \rangle_+$ denotes the positive part of $\bullet$ ($< \bullet >_+= \bullet$, if $\bullet > 0$; $< \bullet >_+= 0$, if $\bullet \leqslant 0$). It is important to recall that (13.1) is based on the assumption of statistically uniformity of every element Ω_r. From (13.1), the parameter σ_0 can be interpreted as the increase of the stress level with respect to the value σ_u to which corresponds a probability of failure of 63.2% in a tensile specimen, uniformly stressed, with the volume Ω_r.

In a multi-axial, non-uniform stress state, it is usually assumed that cracks form in the planes normal to the principal stresses $\sigma_1(\mathbf{x}), \sigma_2(\mathbf{x}), \sigma_3(\mathbf{x})$; the probability of failure is given by:

$$P_f = 1 - \exp\left[-\frac{1}{\Omega_r} \int_\Omega \sum_{i=1}^{3} \left\langle \frac{\sigma_i(\mathbf{x}) - \sigma_u}{\sigma_0} \right\rangle_+^m d\Omega \right]. \tag{13.2}$$

Equation (13.2) is obtained from (13.1) assuming statistical uniformity of every volume and computing the joint probability of survival for every infinitesimal volume.

The general expression (13.2) is here applied under the assumption that $\sigma_u = 0$, which means that all level of stresses have an influence on the probability of failure. Equation (13.2) is then re-written in a more compact way as:

$$P_f = 1 - \exp\left[-\frac{1}{\Omega_r} \int_\Omega \left(\frac{\tilde{\sigma}(\mathbf{x})}{\sigma_0} \right)^m d\Omega \right]. \tag{13.3}$$

The equivalent stress $\tilde{\sigma}(\mathbf{x})$ is defined by:

$$\tilde{\sigma}(\mathbf{x}) \equiv \left(\sum_{i=1}^{3} \langle \sigma_i(\mathbf{x}) \rangle_+^m \right)^{\frac{1}{m}}. \tag{13.4}$$

The above relations can be used to estimate the probability of failure P_f of a given structure or solid once the *Weibull parameters* m and σ_0 are known and the elastic distribution of stresses has been computed via analytical formulae or numerical solutions, e.g., the finite element (FE) method.

The parameters m and σ_0 are usually experimentally determined starting from a series of uniaxially tensile tests on cylindrical specimens of volume Ω and surface area A. In this simple case (13.3) reduces to:

$$P_f = 1 - \exp\left[-\frac{\Omega}{\Omega_r} \left(\frac{\sigma}{\sigma_0} \right)^m \right]. \tag{13.5}$$

The Weibull parameters can be identified also from a specimen or structure loaded in a multi-axial situation with a non-uniform stress distribution. Let us re-write (13.3) in a form similar to (13.5):

$$P_{\mathrm{f}} = 1 - \exp\left[-\frac{1}{\Omega_{\mathrm{r}}}\int_{\Omega}\left(\frac{\tilde{\sigma}(\mathbf{x})}{\sigma_0}\right)^m \mathrm{d}\Omega\right] \equiv 1 - \exp\left[-\frac{\Omega}{\Omega_{\mathrm{r}}}\left(\frac{\sigma_{\mathrm{nom}}}{\sigma_0}\right)^m \beta^m\right], \quad (13.6)$$

where β is defined by

$$\beta^m \equiv \frac{1}{\sigma_{\mathrm{nom}}^m \Omega}\int_{\Omega}\sum_{i=1}^{3}\langle\sigma_i(\mathbf{x})\rangle_+^m \mathrm{d}\Omega \equiv \frac{1}{\Omega}\int_{\Omega}(h(x))^m \mathrm{d}\Omega \qquad (13.7)$$

and σ_{nom} represents a nominal stress in the non-uniformly stressed specimen or structure, which acts as a scaling parameter for the elastic response. Note that function $h(\mathbf{x})$, defined by the relations (13.7), depends only on the normalized stress distribution in the linear elastic response and is therefore independent of the load level.

To compare the behaviour of different structures, it is possible to define a critical stress level as the nominal stress level σ_{nom0} evaluated in the structure when the probability of failure is equal to 63.2%, in equivalence to σ_0 for a uniaxially, uniformly loaded specimen. From the relations (13.6), there is

$$\sigma_{\mathrm{nom0}} = \frac{\sigma_0}{\beta}\left(\frac{\Omega_{\mathrm{r}}}{\Omega}\right)^{1/m}. \qquad (13.8)$$

Given two structures (13.1) and (13.2), it is therefore possible to write

$$(\sigma_{\mathrm{nom0}})_1 = \frac{\sigma_0}{\beta_1}\left(\frac{\Omega_{\mathrm{r}}}{\Omega_1}\right)^{\frac{1}{m}}; \quad (\sigma_{\mathrm{nom0}})_2 = \frac{\sigma_0}{\beta_2}\left(\frac{\Omega_{\mathrm{r}}}{\Omega_2}\right)^{\frac{1}{m}}; \quad \frac{(\sigma_{\mathrm{nom0}})_1}{(\sigma_{\mathrm{nom0}})_2} = \frac{\beta_2}{\beta_1}\left(\frac{\Omega_2}{\Omega_1}\right)^{\frac{1}{m}}. \qquad (13.9)$$

The relation (13.9) allows a direct comparison of the behaviour of structures with different volumes and stress distributions. The variation of parameter σ_{nom0} with the volume determines the size effect related to the statistical uniform distribution of defects as described by the Weibull approach. σ_{nom0} is inversely proportional to the volume, and this dependence increases with decreasing m. At the limit, by letting m be infinite, the statistical size effect disappears. Noteworthy is also the dependence of σ_{nom0} on the parameter β, which in turn depends on the stress non-uniformity.

13.5 On-Chip Bending Test Through a Comb-Finger Rotational Electrostatic Actuator

13.5.1 General Description

The first on-chip device discussed in this chapter is shown in Fig. 13.7 [8–10]. It is made of a central ring connected to the substrate by means of two tapered 0.7-μm thick specimens (Fig. 13.7b) which also act as the suspension springs of the whole

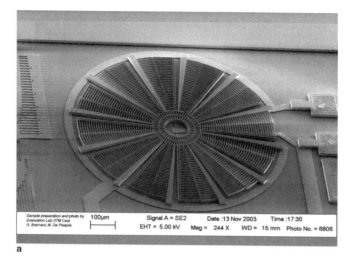

Fig. 13.7 Rotational electrostatic actuator for on-chip bending tests, (**a**) general view; (**b**) the detail of the bending specimen

device. Rigidly connected to the central ring are 12 arms with a total of 384 comb-fingers capacitors which move, due to the electrostatic attraction, towards the stators connected rigidly to the substrate. When a voltage is applied to the device, the comb-fingers develop a force distributed along the 12 arms, equivalent to a torque applied to the central ring. This in turn loads the two specimens in bending in the plane parallel to the substrate. The force developed by the system of the comb-fingers is sufficient to load the specimens up to rupture. The specimens are a pair of doubly clamped slender beams, with a length of 34 μm and a trapezoidal cross section.

Their width decreases linearly from 5.3 to 1.8 μm. This shape is ad hoc designed to localize the fracture of the specimen in a specified area through stress concentration.

13.5.2 Data Reduction Procedure

The rotational on-chip device is tested in order to experimentally determine the Young's modulus and the rupture strength of a 0.7-μm thick thin polysilicon. The general data reduction procedure described in Sect. 13.3.2 is applied. In the present case, simplified analytical formulae are used in order to transform the applied voltage to the torque applied by the electrostatic actuator on the rotational device and the measured capacitance variation (with respect to a reference level) to the angle of rotation of the central ring of the rotational device.

The electrostatic attraction of the 32 comb-fingers placed along each arm is assumed to be uniformly distributed along the arm, with a resultant force F_{arm}; the total torque is therefore evaluated as:

$$M = n\left(l_0 + \frac{l_{\mathrm{arm}}}{2}\right)F_{\mathrm{arm}} = n_{\mathrm{a}}\left(l_0 + \frac{l_{\mathrm{arm}}}{2}\right)n_{\mathrm{cf}}\varepsilon_0 \frac{tV^2}{g}, \qquad (13.10)$$

where n_{a} is the number of arms, l_0 the distance between the external part of the central ring and the centre of the whole device, l_{arm} the length of each arm, n_{cf} the number of comb-fingers for each arm, ε_0 the dielectric constant of vacuum, t the thickness of the arms in the direction orthogonal to substrate, g the gap between the rotor and the stator in each comb-finger, and V the applied voltage. Equation (13.10) allows us to compute the value of the global torque M for a given voltage V.

Let us consider the 32 comb-fingers distributed along each arm defined by index $i = 0, 1, \ldots, 31$, being $i = 0$ the finger nearest to the centre. The contribution to the total capacitance of the ith comb-finger is given by

$$C_i = C_{0i} + 2\frac{\varepsilon_0 t}{g}\Delta x_i = 2\frac{\varepsilon_0 t}{g}\theta(R_0 + i\Delta R), \qquad (13.11)$$

where C_{0i} is the capacitance corresponding to the initial configuration, Δx_i the value of the displacement in the direction orthogonal to the arm due to electrostatic attraction, θ the angle of rotation of the device, R_0 the distance between the rotor finger $i = 0$ and the centre, ΔR the distance between two stator fingers. The total capacitance of the rotational device can be computed by summing the contribution of each finger and of each arm:

$$\begin{aligned}
C &= n_{\mathrm{a}}\sum_{i=0}^{31}C_i = n_{\mathrm{a}}\sum_{i=0}^{31}C_{0i} + 2n_{\mathrm{a}}\frac{\varepsilon_0 t}{g}\theta\sum_{i=0}^{31}(R_0 + i\Delta R) \qquad (13.12) \\
&= C_0 + \left[2n_{\mathrm{a}}\frac{\varepsilon_0 t}{g}(32R_0 + 496\Delta R)\right]\theta.
\end{aligned}$$

Table 13.1 Parameters used for the rotational electrostatic actuator

$n_a = 12$	$n_{cf} = 32$
$l_0 = 100\,\mu m$	$l_{arm} = 307.65\,\mu m$
$t = 15\,\mu m$	$g = 2.5\,\mu m$
$R_0 = 108\,\mu m$	$\Delta R = 9.5\,\mu m$
$\varepsilon_0 = 8.854 \times 10^{-6}\,pF\,\mu m^{-1}$	

From the above equation, it is possible to compute the rotation of the device from the measured total capacitance C; since only the capacitance variations are relevant and the value of C_0 is assumed to be zero.

The total torque and the total capacitance are computed starting from experimental values of the voltage and capacitance variation by introducing the data of Table 13.1 in (13.10) and (13.12).

The resulting values are

$$M = 5.18V^2\ \mu N\,\mu m; \quad C = C_0 + 10.41\theta\ pF. \tag{13.13}$$

An example of the experimental capacitance vs. applied voltage plots and the corresponding torque vs. rotation plots obtained after using the relations (13.13) is also shown in Figs. 13.5 and 13.6.

Starting from the torque vs. rotation plot, it is possible to calculate the Young's modulus and the rupture strength of the material using linear elastic FEM analysis performed on a 3D FEM mesh. This result is obtained after assuming (1) the deformation of the device is only due to the beam specimens at the centre, i.e. the external part built with a 15-μm thick polysilicon is assumed to be rigid; (2) the specimen is linear elastic and homogeneous up to rupture; (3) displacements and strains are small and (4) the global behaviour is linear and the global stiffness is proportional to the Young's modulus of the specimen.

It is important to remark that, besides the above-mentioned hypotheses, the geometry of the specimen must be carefully reproduced in the FEM model. The final FEM mesh is therefore obtained from SEM images in order to reproduce the real geometry.

13.5.3 Experimental Results

Figure 13.8 shows the experimental distributions of the Young's modulus obtained using the rotational devices described in Sect. 13.5.2. The mean value is 178 GPa. An image of a specimen loaded up to rupture in the plane parallel to the substrate by means of the rotational actuator is shown in Fig. 13.9.

The Weibull approach briefly described in Sect. 13.4 is applied to 50 experimental results, computing the volume integral in (13.7) starting from a linear elastic FEM analysis. The volume integrals are computed by using a numerical Gaussian integration on each FEM.

Fig. 13.8 Rotational actuator
for the in-plane bending test:
The distribution of the value
of the Young's modulus

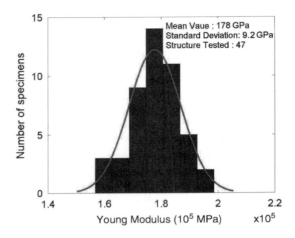

Fig. 13.9 Rotational actuator for the in-plane bending tests: Example of rupture

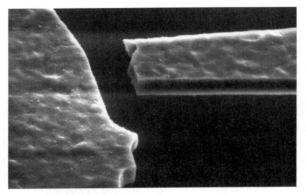

The Weibull parameters $\sigma_0 = 1.84\,\text{GPa}$; $m = 6.2$ were obtained and the nominal stress value in (13.8) was $\sigma_{nom0} = 2.89\,\text{GPa}$. These results are shown in Fig. 13.10.

13.6 On-Chip Bending Test Through a Parallel Plate Electrostatic Actuator

13.6.1 General Description

The second on-chip device discussed in this chapter is shown in Fig. 13.11 [8,9,11]. Figure 13.11a shows the whole device, and Fig. 13.11b is a zoom view of the central part where the 0.7-μm thick beam specimens are placed. A holed plate of 15-μm thick polysilicon is suspended over the substrate by means of four elastic springs

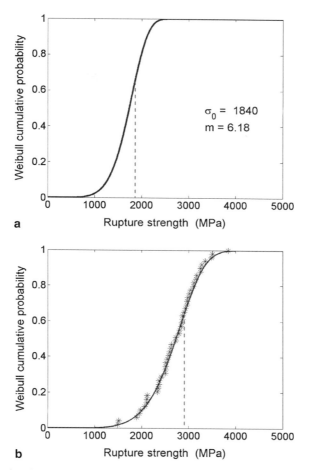

Fig. 13.10 Rotational actuator for the in-plane bending tests: The Weibull cumulative probability densities. (a) Equivalent Weibull plot for a uniaxial, unit volume specimen; (b) Experimental data obtained from 50 tests on the rotational structure, $\sigma_{nom0} = \sigma_0(\Omega_r/\Omega)^{1/m}/\beta = 2.89\,\text{GPa}$

placed at the four corners. The holed plate is also connected to the thin polysilicon film specimens placed at the centre as shown in Fig. 13.11. The two symmetric specimens are in turn connected on one side to the holed plate, while on the other are rigidly connected to the substrate.

The two specimens are therefore equivalent to a couple of doubly clamped beams. The holes in the plate are due to the etching process for the elimination of the sacrificial layer, thus allowing for movement of the holed plate with respect to the substrate. The movement in the direction orthogonal to the substrate is obtained by electrostatic attraction of the holed plate towards the substrate. The whole plate and the substrate thus act as a parallel plate electrostatic actuator. When the plate moves towards the substrate, the couple of specimens bend.

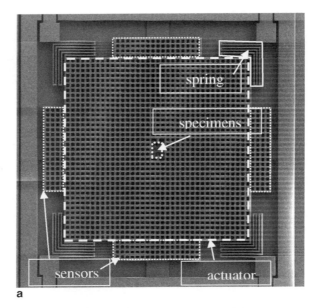

a

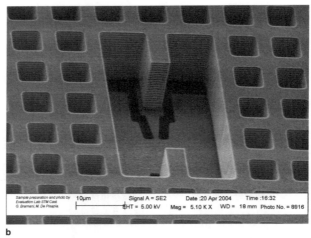

b

Fig. 13.11 Parallel plate actuator for the out-of-plane bending tests: (**a**) General view; (**b**) detail of one of the specimen

It is important to remark that only the squared part of the holed plate acts as an actuator (see Fig. 13.11a), while the holed rectangular parts added to each side of the plate act as sensors. These in turn allow for the experimental determination of the capacitance variation and vertical movement, as discussed in Sect. 13.6.2. The length of each specimen is 7 μm. In order to force the rupture in a section, their cross section changes linearly with the decrease in the width from 3 to 1 μm (see Fig. 13.11b).

13.6.2 Data Reduction Procedure

Similar to the device described in Sect. 13.5, the parallel plate actuator is used to experimentally determine the Young's modulus and the rupture strength of the 0.7-μm thick thin polysilicon.

The general data reduction procedure, as described in Sect. 13.3.2, is again applied. The experimentally determined capacitance vs. voltage plots are transformed into the force vs. displacement plots by using the relationships between capacitance and displacement and between voltage and electrostatic force, respectively. These relations are determined by the electrostatic boundary element (BEM) and FEM simulations. In particular, a series of electrostatic BEM simulations on one of the lateral sensors allows for the determination of the capacitance variation vs. vertical gap plot, which is directly used to convert experimental capacitance variation data to the vertical displacements.

The FEM electrostatic simulations are used in order to obtain the vertical force of attraction on the square holed plate acting as a rotor. From the results of the FEM simulations it is deduced that the attractive vertical force can be computed by making use, with negligible error, of the analytical relation for a parallel plate actuator with the same surface and with the correction due to edge effects.

Starting from the force–displacement plot, the force acting on the specimens is obtained by subtracting the part equilibrated by the elastic suspension springs in the four corners of the holed plate (see Fig. 13.11a). An elastic 3D FEM solution of the specimen under bending in the vertical plane is then used to relate the global stiffness of the specimen to the Young's modulus and the force at rupture to the maximum tensile stress in the specimen. Experimental values of the Young's modulus and rupture stress are therefore finally obtained.

A key point in the data reduction procedure, which deserves further study and careful examination, is the sensitivity of the results to the value of the vertical gap between the holed plate and the substrate. The vertical gap cannot be easily measured on the real device and it can strongly depend on the quality of the etching process. In the results presented here a mean value of 1.65 μm is used.

13.6.3 Experimental Results

Figure 13.12 shows the distributions of the Young's modulus obtained using the device described in Sect. 13.6.1. The mean value is 174 GPa. It can be observed that the mean values of the Young's modulus (178 GPa and 174 GPa) obtained from the two devices as described in Sects. 13.5.1 and 13.6.1, differ less than the standard deviation of the two distributions. It can therefore be concluded that the two sets of experimental results give in practice the same value for the Young's modulus. The conclusion implies that the possible non-uniformity of the thin polysilicon film along its thickness does not influence sensibly the value of the average Young's modulus obtained.

Fig. 13.12 Parallel plate actuator for the out-of-plane bending tests – experimental distribution of the Young's modulus

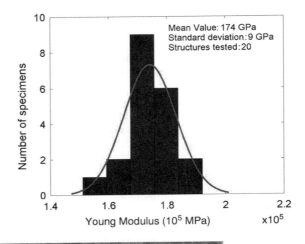

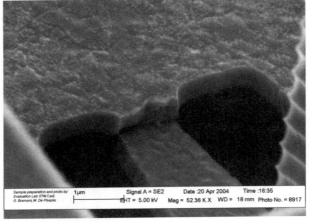

Fig. 13.13 Parallel plate actuator for the out-of-plane bending tests: Example of rupture

For the present device, the Weibull approach described in Sect. 13.4 is applied. Figures 13.13 and 13.14 have the same meaning of Figs. 13.9 and 13.10 as obtained in the case of the rotational actuator. The obtained Weibull parameters are $\sigma_0 = 2.24\,\mathrm{GPa}$, $m = 5.1$ in this case, while the nominal stress value in (13.8) is $\sigma_{nom0} = 3.03\,\mathrm{GPa}$. The remarkable difference between σ_0 and σ_{nom0} obtained from both test structures demonstrates the importance of the effects of the stress distribution. In a highly non-uniformly stressed structure like the ones here tested, the apparent value of rupture is higher than in a uniformly stressed specimen.

The difference in the Weibull parameters obtained from the two sets of experimental results may be explained in various ways. A first explanation could be the influence of different loading conditions (in plane and out of plane) which could locally initiate different rupture mechanisms, related to the anisotropy and

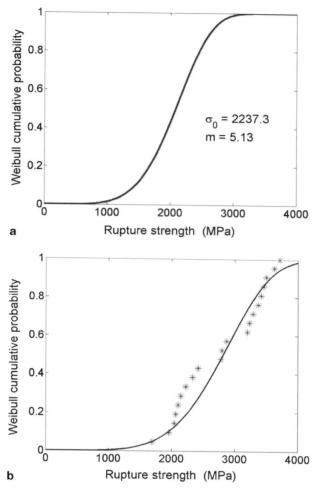

Fig. 13.14 Parallel plate actuator for the out-of-plane bending tests: The Weibull cumulative probability densities. (**a**) Equivalent Weibull plot for a uniaxial, unit volume specimen; (**b**) Experimental data obtained from 21 tests on the parallel plate structure, $\sigma_{nom0} = \sigma_0(\Omega_r/\Omega)^{1/m}/\beta = 3.03\,\text{GPa}$

non-uniformity of distribution of the polysilicon film along its thickness. This conclusion is in partial contrast with the one derived with reference to the elastic behaviour.

A second explanation could be associated with the assumption of $\sigma_u = 0$ as introduced in Sect. 13.4 in order to simplify the Weibull approach. It is in fact known that a non-zero σ_u value has an influence on the final value of the Weibull parameters σ_0, m. In addition, the possible discrepancies in the results obtained from the two on-chip tests can be due to the influence of geometrical parameters like the gap between stator and holed plate in the device described in Sect. 13.6.1.

13.7 Test Structure for the Mechanical Characterization of Thick Polysilicon

13.7.1 General Description

The third device is shown in Fig. 13.15. Compared to the one presented in the previous section, this structure presents many unique features. It is designed to test the

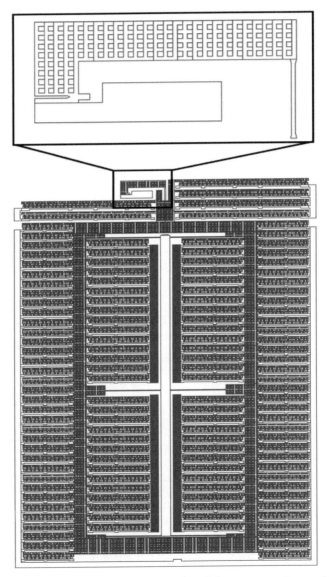

Fig. 13.15 Test structure for thick polysilicon – top view of the structure and zoom of the specimen

mechanical properties of the epitaxial polysilicon, *epipoly*, the *thick* layer deposited with the ThELMA process. This film is about 20 times thicker than the *poly1* film; therefore it requires a larger force to break the specimen. This explains the reason for the relatively big dimensions of the structure that takes up a $1{,}600 \times 2{,}250\,\mu m^2$ rectangular area.

The actuation mechanism is controlled by a number of comb-finger capacitors in a plane parallel to the one of the wafer, a very common choice for MEMS structures. In the next subsections the structure will be divided into four parts, which are separately discussed in detail.

13.7.1.1 The Actuator

The electrostatic actuation is realized by over 4,000 comb-finger actuators, grouped on specific structures called *arms* (see Fig. 13.16). Every arm contains 31 comb-fingers actuator. Its capacitance is a function of the seismic mass displacement x, as:

$$C_{\text{arm}} = C_0 + 31 C_{\text{comb}} = C_0 + 31 \frac{2\varepsilon_0 t}{g} x, \tag{13.14}$$

where C_0 is the capacitance of the system at rest, C_{comb} is the capacitance of a single comb-finger actuator. t is equal to $15\,\mu m$, representing the thickness of the layer, and g is equal to $2.2\,\mu m$, representing the gap between stator finger and rotor finger. The force from every arm is:

$$F_{\text{arm}} = \frac{1}{2} \frac{\partial C_{\text{arm}}}{\partial x} V^2 = 31 \frac{\varepsilon_0 t}{g} V^2, \tag{13.15}$$

where V is the applied voltage. The total number of arms n_a is 130. Hence the total force developed by the actuator can be expressed as:

$$F_{\text{arm}} = n_a \frac{1}{2} \frac{\partial C_{\text{arm}}}{\partial x} V^2 = 4{,}030 \frac{\varepsilon_0 t}{g} V^2. \tag{13.16}$$

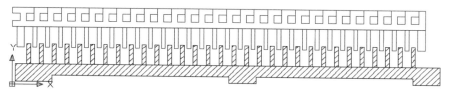

Fig. 13.16 Actuation arm of the test structure for thick polysilicon

13.7.1.2 Frame

The frame is a suspended structure that supports the actuator arms (see Fig. 13.17). Six suspension springs are placed in the central part of the frame. These springs avoid the collapse of the structure onto the substrate during actuation. They are rectangular cross-sectioned slender beams of 291 μm in length. The in-plane width is 3.2 μm, and the out-of plane thickness is 15 μm. Using a Young's modulus of $E = 145$ GPa, as determined in [21], the linear stiffness of the six springs for a movement parallel to the substrate as shown in Fig. 13.17 can be easily computed:

$$k_{spring} = 6 \left(12 \frac{EJ}{l^3} \right) = 17.35 \ \mu N \ \mu m^{-1}. \tag{13.17}$$

With reference to Fig. 13.15, the upper part of the frame is clamped to the specimen, which is thus loaded with the force created by the actuator that is not absorbed by the spring system. It is worth noting that the suspension system is designed in order to be as compliant as possible to exert almost all the force (the 90% of the force developed) onto the specimen.

13.7.1.3 Sensing

In the upper part of the structure shown in Fig. 13.15, there is the sensing system. It is made up with six arms with 80 comb-finger capacitors on each (Fig. 13.18). The total sensing capacitance is a function of the displacement of the specimen as:

$$C_{sens} = C_0^{sens} + 480 \frac{2\varepsilon_0 t}{g} x. \tag{13.18}$$

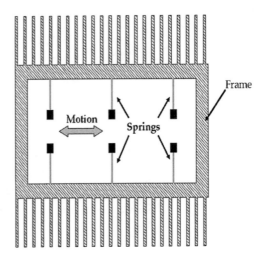

Fig. 13.17 Schematic of the frame (filled with *dashed line*) for the test structure of thick polysilicon – in *black* the anchor points, and in *gray* the six suspension springs

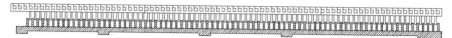

Fig. 13.18 Test structure for thick polysilicon. Sensing arm (*dashed* the fixed part)

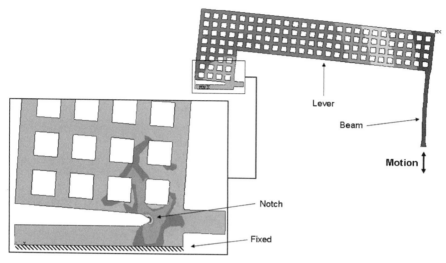

Fig. 13.19 Test structure for thick polysilicon. Deformed shape of the specimen and contour plot of principal tensile stress of the notched zone

Using the FEM analysis, a simulation of the sensing system is performed. The results of the simulations show that the analytical formula holds for a displacement up to 10 μm, which has never been reached during the experimental test.

13.7.1.4 Specimen

The specimen is designed to perform both quasi-static and fatigue tests. It consists in a lever system that causes a stress concentration in a localized region (Fig. 13.19). The specimen can be divided into four parts: a beam that is the physical link between the frame and the specimen; the lever, that transforms the axial action coming from the beam into a bending moment acting in the notched zone; a notch, that is the most stressed part, where the crack nucleates; and a part fixed to the substrate.

13.7.2 Data Reduction Procedure

The measured capacitance change (Fig. 13.20) is used to compute the displacement imposed on the extremity of the loaded beam in (13.18). This causes the rotation

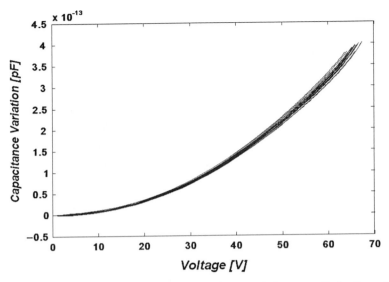

Fig. 13.20 Test structure for thick polysilicon – capacitance variation vs. applied voltage plot

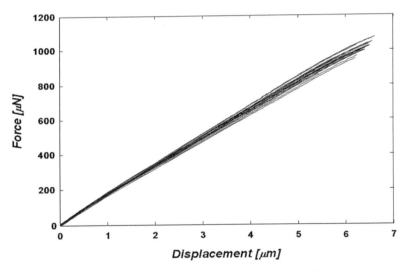

Fig. 13.21 Test structure for thick polysilicon – force vs. displacement plots

of the lever arm and the creation of the desired state of stress in the notch. The force produced by the actuator for every imposed voltage is computed using (13.16). From this information it is then possible to plot the force vs. displacement curves (Fig. 13.21). Thus one can combine the experimental results and the FEM simulations to determine the Young's modulus and the maximum tensile stress at which rupture occurs in every test.

13.7.3 Experimental Results

A number of 31 structures, deposited on the same wafer, are tested. As seen from Fig. 13.20, the measurements are very repeatable. The force vs. displacement plots shown in Fig. 13.21 appear to be linear, implying that the electrostatic behaviour of the rotor and sensor parts could be described by the analytical formulae used in the data reduction procedure.

From the slope of the force vs. displacement plots it is possible, as done with the out-of-plane structure, to compute the Young's modulus. The values deduced are in agreement with the ones obtained by Corigliano et al. [21], confirming the overall quality of the data reduction procedure. The mean value measured is 143 GPa with a standard deviation of ± 3 GPa.

Even in this case, the data concerning the rupture of the specimens are interpreted in the framework of the Weibull statistics as discussed in Sect. 13.4. From Fig. 13.22, the experimental results are clearly well interpolated by the Weibull cumulative distribution function. This allows for the computation of the Weibull modulus, $m = 25.76$ and the Weibull stress σ_0. The Weibull stress, representing the level of stress, which gives 63.2% of failure probability for a pure tension specimen with the same size as the reference volume, is 3.62 GPa.

After the testing, the specimens are investigated using an optical microscope. The images show that the fracture starts from the notch as predicted from the FEM simulations carried out during the design phase. From the pictures shown in Fig. 13.23, there are two main aspects.

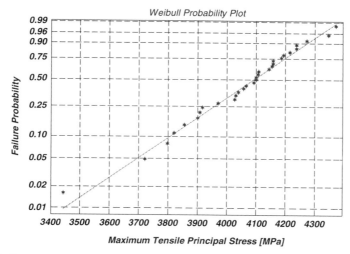

Fig. 13.22 Test structure for thick polysilicon – the Weibull plot of the experimental data (*asterisks*) and interpolating Weibull cumulative distribution (*dashed line*)

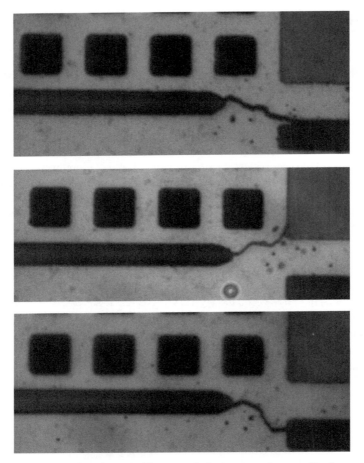

Fig. 13.23 Test structure for thick polysilicon – optical microscope images of broken specimens

1. The crack path is often irregular and can be quite different from one structure to another. This can be due to the crystalline structure and grains orientation in the notched area. The grain morphology and orientation is different from one structure to another and has a very important impact on the direction of the crack propagation.
2. The crack starting point is not always the same. This is due to the non-uniform flaw distribution on the notch surface, caused by the fabrication process. The distribution of flaws is supposed to be responsible of the scatter of the experimental results.

13.8 Conclusions

The focus of this chapter is on the mechanical characterization of a variety of small polysilicon structures using the on-chip testing methodology. After reviewing recent results in literature, the discussion has addressed the general on-chip testing procedure, as developed recently by the authors, with reference to three designed and fabricated on-chip structures. As a useful approach to interpret the rupture values, the Weibull theory is used and issues related to the size effect and stress gradient effect are discussed.

The results and list of references presented in this chapter are by no means complete and are highly based on the author's experience. It is nevertheless the author's opinion that the general overview and technical results here presented can be a useful introduction to the use of the on-chip testing devices, not only for the mechanical characterization of polysilicon but also for the mechanical testing of other materials at the micro- and nano-scale.

Acknowledgments The contributions of EU NoE Design for Micro & Nano Manufacture (PATENT-DfMM), contract No.: 507255 and of Cariplo foundation contract no. 2005–06–49 "Innovative models for the study of the behaviour of solids and fluids in micro/nano-electromechanical systems" are gratefully acknowledged.

References

1. Ando T, Shikida M, Sato K (2001) Sens. Actuators A 93:70
2. Bagdahn J, Sharpe Jr WN (2003) Sens. Actuators A Phys. 103:9
3. Bagdahn J, Sharpe Jr WN, Jadaan O (2003) J. Microelectromech. Syst. 12:302
4. Ballarini R, Mullen RL, Yin Y, Kahn H, Stemmer S, Heuer AH (1997) J. Mater. Res. 12(4):915
5. Ballarini R, Mullen RL, Kahn H, Heuer AH (1998) In: Proceedings MRS Symposium, vol. 518, 13–17 April, San Francisco, CA, USA, p. 33
6. Ballarini R, Kahn H, Heuer AH, De Boer MP, Dugger MT (2003) In: Milne I, Ritchie RO, Karihaloo B (eds) Comprehensive Structural Integrity, vol. 8. Elsevier, Amsterdam, 2003, chap. 9, pp. 325–360
7. Bažant Z, Xi Y, Reid S (1991) J. Eng. Mech. ASCE 117:2609
8. Cacchione F, Corigliano A, De Masi B, Riva C (2005) Microelectron. Reliab. 45:1758
9. Cacchione F, De Masi B, Corigliano A, Ferrera M, Vinay A (2005) In: Proceedings NSTI Nanotech, 8–12 May, Anaheim, CA, USA
10. Cacchione F, De Masi B, Corigliano A, Ferrera M (2006) Sensor Lett. 4(1):38
11. Cacchione F, Corigliano A, De Masi B, Ferrera M (2006) Sensor Lett. 4(2):184
12. Chasiotis I (2006) J. Appl. Mech. 73:714
13. Chasiotis I, Cho SW, Jonnalagadda K (2006) J. Appl. Mech. 73:714
14. Chasiotis I, Knauss WG (2002) Exp. Mech. 42(1):51
15. Chasiotis I, Knauss WG (2003) J. Mech. Phys. Solids 51:1533
16. Chasiotis I, Knauss WG (2003) J. Mech. Phys. Solids 51:1551
17. Chen GS, Ju MS, Fang YK (2000) Sens. Actuators A 86:108
18. Chen MH, Leipold (1980) Ceram. Bull. 59:469
19. Chi SP, Wensyang H (1999) J. Microelectromech. Syst. 8:200
20. Cho SW, Chasiotis I (2007) Exp. Mech. 47:37

21. Corigliano A, De Masi B, Frangi A, Comi C, Villa A, Marchi M (2004) J. Microelectromech. Syst. 13(2):200
22. Corigliano A, Cacchione F, De Masi B, Riva C (2005) Meccanica 40:485
23. Corigliano A, Domenella L, Espinosa HD, Zhu Y (2007) Sensor Lett. 5:1
24. De Masi B, Villa A, Corigliano A, Frangi A, Comi C, Marchi M (2004) In: Proceedings MEMS04, 25–29 January, Maastricht
25. Ding JN, Meng YG, Wen SZ (2001) Mater. Sci. Eng. B 83:42
26. Espinosa HD, Prorok BC, Fischer M (2003) J. Mech. Phys. Solids 51:47
27. Gardner JW, Varadan VK, Awadelkarim OO (2001) Microsensors MEMS and Smart Devices. Wiley, Chichester
28. Greek S, Ericson F, Johansson S, Schweitz JA (1997) Thin Solid Films 292:247
29. Hollman P, Alahelisten A, Olsson M, Hogmark S (1995) Thin Solid Films 270:137
30. Jadaan O, Nemeth N, Bagdahn J, Sharpe WN (2003) J. Mater. Sci. 38:4087
31. Jayaraman S, Edwards RL, Hemker KJ (1998) Mater. Res. Soc. Symp. Proc. 505:623
32. Kahn H, Tayebi N, Ballarini R, Mullen RL, Heuer AH (2000) Sens. Actuators A 82:274
33. Kahn H, Ballarini R, Bellante JJ, Heuer AH (2002) Science 298:1215
34. Kahn H, Ballarini R, Heuer AH (2004) Science 8:71
35. Kim JH, Yeon SC, Jeon YK, Kim JG, Kim YH (2002) Sens. Actuators A Phys. 101:338
36. Knauss WG, Chasiotis I, Huang Y (2003) Mech. Mater. 35:217
37. Kobrinsky M, Deutsch E, Senturia SD (1999) MEMS Microelectromech. Syst. 1:3
38. Kramer T, Paul O (2000) Sens. Actuators A 92:292
39. La Van DA, Tsuchiya T, Coles G, Knauss WG, Chasiotis I, Read D (2001) In: Muhlstein C, Brown SB (eds) Mechanical Properties of Structural Films, ASTM STP 1413. American Society for Testing and Materials, West Conshohocken, PA, pp. 1–12
40. Li X, Bhushan B (1999) Thin Solid Films 144:210
41. Lyshevski SE (2002) MEMS and NEMS. Systems, Devices and Structures. CRC, New York
42. Madou MJ (2002) Fundamentals of Microfabrication. CRC, New York
43. McCarty A, Chasiotis I (2007) Thin Solid Films 515:3267
44. Muhlstein CL, Brown SB, Ritchie RO (2001) Sens. Actuators A 94:177
45. Muhlstein CL, Stach EA, Ritchie RO (2002) Appl. Phys. Lett. 80(9):1532
46. Muhlstein CL, Stach EA, Ritchie RO (2002) Acta Mater. 50:3579
47. Muhlstein CL, Howe RT, Ritchie RO (2004) Mech. Mater. 36(1–2):13
48. Oha CS, Lee HJ, Ko SG, Kim SW, Ahn HG (2005) Sens. Actuators A 117:151
49. Oostemberg PM, Senturia SD (1997) J. Microelectromech. Syst. 6:107
50. Prorok BC, Zhu Y, Espinosa HD, Guo Z, Bažant Z, Zhao Y, Yakobson BI (2004) In: Nalwa HS (ed) Encyclopedia of Nanoscience and Nanotechnology, vol. 5. American Scientific, Stevenson Ranch, CA, pp. 555–600
51. Senturia SD (2001) Microsystem Design. Kluwer, Dordrecht
52. Sharpe Jr WN, Yuan B, Edwards RL (1997) J. Microelectromech. Syst. 6(3):193
53. Sharpe Jr WN, Vaidyanathan, Bin Yuan, Edwards RL (1997) In: Proceedings IEEE. The Tenth Annual International Workshop on Micro Electro Mechanical Systems, 26–30 January, Nagoya, Japan, pp. 424–429
54. Sharpe Jr WN, Brown JS, Johnson GC, Knauss WG (1998) In: Materials Research Society Proceedings, vol. 518, San Francisco, CA, pp. 57–65
55. Sharpe WN, Turner KT, Edwards RL (1999) J. Exp. Mech. 39:162
56. Sharpe WN (2002) In: The MEMS Handbook. CRC, New York
57. Stanley P, Inanc EY (1984) In: Probabilistic methods in the mechanics of solids and structures, Proc. Symposium to the memory of W. Weibull, 19–21 June, Stockholm, Springer, Berlin
58. Sundarajan S, Bhushan B (2002) Sens. Actuators A Phys. 101:338
59. Tabata O, Kawahata K, Sugiyama S, Igarashi I (1989) Sens. Actuators A 20:135
60. Tai Y, Muller RS (1990) In: IEEE Micro Electro Mechanical Systems, 11–14 February, Napa Valley, CA, pp. 147–152
61. Tsuchiya T, Sakata J, Taga Y (1997) In: Proceedings MRS Symposium, vol. 505, 1–5 December, Boston, MA, USA, pp. 285–290

62. Tsuchiya T, Tabata O, Sakata J, Taga Y (1998) J. Microelectromech. Syst. 7(1):106
63. Tsuchiya T, Shikida M, Sato K (2002) Sens. Actuators A Phys. 97–98:492
64. Tsuchiya T, Hirata M, Chiba N, Udo R, Yoshitomi Y, Ando T, Sato K, Takashima K, Higo Y, Saotome Y, Ogawa H, Ozaki K (2005) J. Microelectromech. Syst. 14(5):1178
65. Villa A, De Masi B, Corigliano A, Frangi A, Comi C (2003) In: Bathe KJ (ed) Proceedings Second MIT Conference on Computational Fluid and Solid Mechanics, vol. 1, June 2003, Boston, Elsevier, pp. 722–726
66. Weibull W (1951) J. Appl. Mech. 18:293
67. Yang J, Paul O (2002) Sens. Actuators A 89:1
68. Yi T, Li L, Kim CJ (2000) Sens. Actuators A 83:172
69. Zhu Y, Corigliano A, Espinosa HD (2006) J. Micromech. Microeng. 16:242
70. Ziebat V (1999) PhD Dissertation, ETH Zurich

Index